T0406298

Lecture Notes in Mobility

Series Editor

Gereon Meyer, VDI/VDE Innovation + Technik GmbH, Germany
e-mail: gereon.meyer@vdivde-it.de

For further volumes:
http://www.springer.com/series/11573

Jan Fischer-Wolfarth · Gereon Meyer
Editors

Advanced Microsystems for Automotive Applications 2014

Smart Systems for Safe, Clean and Automated Vehicles

Editors
Jan Fischer-Wolfarth
VDI/VDE Innovation + Technik GmbH
Berlin
Germany

Gereon Meyer
VDI/VDE Innovation + Technik GmbH
Berlin
Germany

ISSN 2196-5544 ISSN 2196-5552 (electronic)
ISBN 978-3-319-08086-4 ISBN 978-3-319-08087-1 (eBook)
DOI 10.1007/978-3-319-08087-1
Springer Cham Heidelberg New York Dordrecht London

Library of Congress Control Number: 2014941880

Printed on acid-free paper

Springer is part of Springer Science+Business Media (www.springer.com)

Preface

There are two major trends that coincide in the current discussion about the automobile of the future: The electrification of the powertrain on one hand and highly automated driving on the other. Both technology paths seem to fit very well with the sustainability objectives of transportation policy as they aim to increase the efficiency of mobility and energy consumption while improving the quality of life in urban areas. Besides just coinciding, however, there is a great potential for these trends to interact: Automation potentially increases driving ranges, usability and affordability of electric vehicles, for example, by optimising traffic flow, valet parking, and inductive charging or anticipative use of the battery capacity. At the same time, electrification enables automated driving, for example, through easy integration of drive-by-wire controls, appropriate electric and electronic architecture, or accurate actuation and speed adjustments without gear shifts. There are also similarities in the requirements of data communication and maps. One might argue that highly automated functionalities can be easily implemented into a conventional car and similarly into an electric one. It is, however, questionable if the public would accept a robot car based on an internal combustion engine as easily as one based on a clean and silent electric powertrain.

The two future trends, electrification and automation, are essentially built on innovations in the domain of smart systems such as sensors and actuators, adaptive components and MEMS devices, novel electric and electronic architectures, intelligent interfaces, and their successful and synergetic integration. Therefore, the future technology development in both of these trends is a main subject of the Strategic Research Agenda of the European Technology Platform on Smart Systems Integration (EPoSS) as well as the specific roadmaps derived from it, for example, the one on road vehicle electrification jointly developed with the European Green Vehicles Initiative PPP or the one on road vehicle automation which is currently being established by a task force of EPoSS. The role of electronic components and systems for smart mobility is also emphasised in the Multi-Annual Strategic Research and Innovation Agenda of the new Joint Technology Initiative ECSEL in the framework of Horizon 2020.

Novel trends in key enabling technologies for the automobile and the synergies between them have always been discussed at the International Forum on Advanced Microsystems for Automotive Applications (AMAA) at an early stage. Thus, the topic of the AMAA 2014, held in Berlin on June 23–24, 2014, is "Smart Systems for Safe, Clean and Automated Vehicles". The AMAA organisers, VDI/VDE Innovation + Technik GmbH together with the European Technology Platform on Smart Systems Integration (EPoSS), greatly acknowledge the support given for this conference from the European Union through the Coordination Actions CAPIRE, Smart EV-VC, and GO4SEM.

The papers in this book, a volume of the Lecture Notes in Mobility series, were written by leading engineers and researchers who have attended the AMAA 2014 conference to report their recent progress in research and innovation. The papers were peer-reviewed by the members of the AMAA Steering Committee and are made accessible worldwide.

As the co-chairman and chairman of the AMAA 2014, we would like to express our deepest gratitude to all the authors for their high quality contributions to the conference and also to this book. We would also like to gratefully acknowledge the tremendous support we have received from our colleagues at VDI/VDE-IT, especially Beate Müller, Christian Martin, and Sebastian Stagl.

Berlin, June 2014

Jan Fischer-Wolfarth
Gereon Meyer

Advanced Microsystems for Automotive Applications 2014

Funding Authority

European Commission

Supporting Organisations

European Council for Automotive R&D (EUCAR)
European Association of Automotive Suppliers (CLEPA)
Strategy Board on Electric Mobility (eNOVA)
Advanced Driver Assistance Systems in Europe (ADASE)
Zentralverband Elektrotechnik- und Elektronikindustrie e.V. (ZVEI)
Mikrosystemtechnik Baden-Württemberg e.V.

Organisers

European Technology Platform on Smart Systems Integration (EPoSS)
VDI|VDE Innovation + Technik GmbH

Steering Committee

Mike Babala	TRW Automotive, Livonia MI, USA
Serge Boverie	Continental AG, Toulouse, France
Geoff Callow	Technical & Engineering Consulting, London, UK
Kay Fürstenberg	Sick AG, Hamburg, Germany
Wolfgang Gessner	VDI|VDE-IT, Berlin, Germany
Roger Grace	Roger Grace Associates, Naples, FL, USA
Klaus Gresser	BMW Forschung und Technik GmbH, Munich, Germany
Riccardo Groppo	Ideas & Motion, Cavallermaggiore, Italy
Hannu Laatikainen	Murata Electronics Oy, Vantaa, Finland
Jochen Langheim	ST Microelectronics, Paris, France
Günter Lugert	Siemens AG, Munich, Germany
Steffen Müller	NXP Semiconductors, Hamburg, Germany
Roland Müller-Fiedler	Robert Bosch GmbH, Stuttgart, Germany
Andy Noble	Ricardo Consulting Engineers Ltd. Shoreham-by-Sea, UK
Pietro Perlo	IFEVS, Sommariva del Bosco, Italy
Christian Rousseau	Renault SA, Guyancourt, France
Jürgen Valldorf	VDI|VDE-IT, Berlin, Germany
David Ward	MIRA Ltd., Nuneaton, UK

Conference Chairs

Gereon Meyer	VDI|VDE-IT, Berlin, Germany
Jan Fischer-Wolfarth	VDI|VDE-IT, Berlin, Germany

Contents

Part I Driver Assistance and Vehicle Automation

Part II Networked Vehicles, ITS and Road Safety

Contributors

Mohammed Ahmed DFKI GmbH - Robotics Innovation Center, Robert-Hooke-Str. 5, 28359 Bremen, Germany, email: mohammed.ahmed@dfki.de

Sagar Akhegaonkar INTEDIS GmbH & Co. KG, Max-Mengeringhausen-Str. 5, 97084 Wuerzburg, Germany, email: Sagar.Akhegaonkar@intedis.com

Estrela Álvarez Centro Tecnológico de Automoción de Galicia, Pol. Industrial A Granxa, Calle A, Parcela 249-250, 36400, Porriño (Pontevedra), Spain, email: estrela.alvarez@ctag.com

Dag Andersson Swerea IVF AB, PO Box 104, 43122 Mölndal, Sweden, email: dag.andersson@swerea.se

Eric Armengaud AVL List GmbH, Hans-List-Platz 1, 8020 Graz, Austria, email: eric.armengaud@avl.com

Stefano Arrigoni Politecnico di Milano, Department of Mechanics, via La Masa 1, Milan, Italy, email: stefano.arrigoni@polimi.it

Jan Bellin Volkswagen AG, *elenia* TU-Braunschweig, Rosenweg 11, 30627 Hannover, Germany, email: jan.bellin@volkswagen.de

Tjerk Bijlsma TNO, Steenovenweg 1, 5708 HN Helmond, The Netherlands, email: tjerk.bijlsma@tno.nl

Timo Birnschein DFKI GmbH - Robotics Innovation Center, Robert-Hooke-Str. 5, 28359 Bremen, Germany, email: timo.birnschein@dfki.de

Hannes Böhm AVL List GmbH, Hans-List-Platz 1, 8020 Graz, Austria, email: hannes.boehm@avl.com

Francesco Braghin Politecnico di Milano, Department of Mechanics, via La Masa 1, Milan, Italy, email: francesco.braghin@polimi.it

Manfred Brandl ams AG, Tobelbaderstr. 30, 8141 Unterpremstaetten, Austria, email: manfred.brandl@ams.com

Matthias Breuel Volkswagen AG, Volkswagen AG, EEF, 38440 Wolfsburg, Germany, email: matthias.breuel@volkswagen.de

Klas Brinkfelt Swerea IVF AB, PO Box 104, 43122 Mölndal, Sweden, email: klas.brinkfeldt@swerea.se

Lutz Bürkle Robert Bosch GmbH, Corporate Research (CR/AEV2) 12, 70442 Stuttgart, Germany, email: lutz.buerkle@de.bosch.com

Federico Cheli Politecnico di Milano, Department of Mechanics, via La Masa 1, Milan, Italy, email: federico.cheli@polimi.it

Caizhen Cheng AVL List GmbH, Hans-List-Platz 1, 8020 Graz, Austria, email: caizhen.cheng@avl.com

Po-Cheich Chiu National Tsing Hua University, Department of Electrical Engineering, Guangfu Rd. Sec. 101, Hsinchu, Taiwan 30013, R.O.C., email: tennis4558581@gmail.com

Ching-Te Chiu National Tsing Hua University, Department of Computer Science, Institute of Communications Engineering, Guangfu Rd. Sec. 101, Hsinchu, Taiwan 30013, R.O.C.

Markus Conrath Daimler AG, Architecture & Body Controller, HPC G007-BB, 71059 Sindelfingen, Germany, email: markus.m.conrath@daimler.com

Jürgen Fabian Graz University of Technology, Institute of Automotive Engineering, Inffeldgasse 11/2, 8010 Graz, Austria, email: juergen.fabian@tugraz.at

Markus Friebe Continental AG, BU ADAS, Segment Surround View, Johann-Knoch-Gasse 9, 96317 Kronach, Germany, email: markus.friebe@continental-corporation.com

Jens Gächter Graz University of Technology, Institute of Automotive Engineering, Inffeldgasse 11/2, 8010 Graz, Austria, email: jens.gaechter@tugraz.at

Harald Gall ams AG, Tobelbaderstr. 30, 8141 Unterpremstaetten, Austria, email: harald.gall@ams.com

Benjamin Girault DFKI GmbH - Robotics Innovation Center, Robert-Hooke-Str. 5, 28359 Bremen, Germany, email: benjamin.girault@dfki.de

Sebastien Glaser LIVIC, 77 rue de Chantiers, 78000 Versailles, France, email: sebastien.glaser@ifsttar.fr

Gerhard Griessnig AVL List GmbH, Hans-List-Platz 1, 8020 Graz, Austria, email: gerhard.griessnig@avl.com

David Grünwald DFKI GmbH - Robotics Innovation Center, Robert-Hooke-Str. 5, 28359 Bremen, Germany, email: david.gruenwald@dfki.de

Oleg Gusikhin Ford Motor Company, P.O. Box 6248, Dearborn, MI 48126, USA, email: ogusikhi@ford.com

Thomas Gußner Robert Bosch GmbH, Corporate Research (CR/AEV2) 12, 70442 Stuttgart, Germany, email: thomas.gussner@de.bosch.com

Tomas Gustafsson Volvo Group Trucks Technology, Advanced Technology and Research, Götaverksgatan 10, 40508 Göteborg, Sweden, email: tomas.gustafsson.3@volvo.com

Feyza Haskaraman Massachusetts Institute of Technology, Department of Mechanical Engineering, Cambridge, MA, USA, email: feyzahas@mit.edu

Bernd Hedenetz Daimler AG, Architecture & Body Controller, HPC G007-BB, 71059 Sindelfingen, Germany, email: bernd.hedenetz@daimler.com

Peter Hemmerle Daimler AG, HPC: 059-X832, 71063 Sindelfingen, Germany, email: peter.hemmerledaimler.com

Jernej Herman Elaphe Propulsion Technologies Ltd., Babno 20a, 3000 Celje, Slovenia, email: jernej@elaphe-ev.com

Gerhard Hermanns Universität Duisburg-Essen, Physik von Transport und Verkehr, Lotharstr. 1, 47057 Duisburg, Germany, email: gerhard.hermanns@uni-due.de

Michael Heuer Otto-von-Guericke-University of Magdeburg, Universitätsplatz 2, 39106 Magdeburg, Germany, email: michael.heuer@ovgu.de

Florian Hilpert Fraunhofer Institute for Integrated Systems and Device Technology IISB, Drives and Mechatronics, Schottkystr. 10, 91058 Erlangen, Germany, email: florian.hilpert@iisb.fraunhofer.de

Mario Hirz Graz University of Technology, Institute of Automotive Engineering, Inffeldgasse 11/2, 8010 Graz, Austria, email: mario.hirz@tugraz.at

Frederic Holzmann INTEDIS GmbH & Co. KG, Max-Mengeringhausen-Str. 5, 97084 Wuerzburg, Germany, email: Frederic.Holzmann@intedis.com

Robert Hoyer University of Kassel, Department of Traffic Engineering and Transport Logistics, Mönchebergstr. 7, 34125 Kassel, Germany, email: robert.hoyer@uni-kassel.de

Yar-Sun Hsu National Tsing Hua University, Department of Electrical Engineering, Guangfu Rd. Sec. 101, Hsinchu, Taiwan 30013, R.O.C.

Arno Huss AVL List GmbH, Hans-List-Platz 1, 8020 Graz, Austria, email: arno.huss@avl.com

Martin Jaiser ams Germany GmbH, Erdinger Str. 14, 85609 Aschheim b. München, Germany, email: martin.jaiser@ams.com

Stephen Jones AVL List GmbH, Hans-List-Platz 1, 8020 Graz, Austria, email: stephen.jones@avl.com

Eberhard Kaulfersch Berliner Nanotest und Design GmbH, Volmerstr. 9B, 12489 Berlin, Germany, email: eberhard.kaulfersch@nanotest.org

Jurij Kern Elaphe Propulsion Technologies Ltd., Luznarjeva 3, 4000 Kranj, Slovenia, email: jurij.kern@elaphe-ev.com

Boris S. Kerner Universität Duisburg-Essen, Physik von Transport und Verkehr, Lotharstr. 1, 47057 Duisburg, Germany, email: boris.kerner@uni-due.de

Ata M. Khan Carleton University, 1125 Colonel By Drive, Ottawa, Ontario, K1S 5B6, CANADA, ata_khan@carleton.ca

Frank Kirchner, DFKI GmbH - Robotics Innovation Center, Department of Mathematics and Computer Science, University of Bremen, Robert-Hooke-Str. 1, 28359 Bremen, Germany, email: frank.kirchner@dfki.de

André Körner Hella KGaA Hueck & Co., Beckumer Str. 130, 59555 Lippstadt, Germany, email: andre.koerner@hella.de

Micha Koller IT-Designers GmbH, Entennest 2, 73730 Esslingen, Germany, email: micha.koller@it-designers.de

Sven Kroffke DFKI GmbH - Robotics Innovation Center, Robert-Hooke-Str. 5, 28359 Bremen, Germany, email: sven.kroffke@uni-bremen.de

Emre Kural AVL List GmbH, Hans-List-Platz 1, 8020 Graz, Austria, email: emre.kural@avl.com

Michael Kurrat *elenia* TU Braunschweig, Schleinitzstr. 23, 38106 Braunschweig, Germany, email: m.kurrat@tu-bs.de

Maurice Kwakkernaat TNO, Steenovenweg 1, 5708 HN Helmond, The Netherlands, email: maurice.kwakkernaat@tno.nl

Heinz Lanzenberger MAGNA Powertrain AG & Co KG, Project House Europe, Frank Stronach-Straße 3, 8200 Albersdorf, Austria, email: heinz.lanzenberger@magnapowertrain.com

Aziz Makklya Ford Motor Company, P.O. Box 6248, Dearborn, MI 48126, USA, email: amakkiya@ford.com

Stefano Mangosio, Centro Ricerche Fiat, Strada Torino 50, 10043 Orbassano, Italy, email: stefano.mangosio@crf.it

Ferdinando Luigi Mapelli Politecnico di Milano, Department of Mechanics, via La Masa 1, Milan, Italy, email: ferdinando.mapelli@polimi.it

Laura Mazzola Politecnico di Milano, Department of Mechanics, via La Masa 1, Milan, Italy, email: laura.mazzola@polimi.it

Marc-Michael Meinecke Volkswagen AG, Brieffach 1777, Berliner Ring 2, 38440 Wolfsburg, Germany, email: marc-michael.meinecke@volkswagen.de

Thomas Paul Michalke Robert Bosch GmbH, Corporate Research (CR/AEV2) 12, 70442 Stuttgart, Germany, email: thomas.michalke@de.bosch.com

Steffen Müller NXP Semiconductors Germany GmbH, Stresemannallee 101, Hamburg, Germany, st.mueller@nxp.com

Klaus Neumaier Fairchild Semiconductor GmbH, Technology Development Center, Einsteinring 28, 85609 Aschheim, Germany, email: klaus.neumaier@fairchildsemi.com

Mihai Nica AVL List GmbH, Hans-List-Platz 1, 8020 Graz, Austria, email: nica.mihai@avl.com

Frank Niewels Robert Bosch GmbH, Corporate Research (CR/AEV2) 12, 70442 Stuttgart, Germany, email: frank.niewels@de.bosch.com

Stefan Nord Volvo Group Trucks Technology, Advanced Technology and Research, Götaverksgatan 10, 40508 Göteborg, Sweden, email: stefan.nord@volvo.com

Lydie Nouveliere LIVIC, 77 rue de Chantiers, 78000 Versailles, France, email: lydie.nouveliere@ifsttar.fr

Christian Oekermann DFKI GmbH - Robotics Innovation Center, Robert-Hooke-Str. 5, 28359 Bremen, Germany, email: christian.oekermann@dfki.de

Frank Ophelders TNO, Steenovenweg 1, 5708 HN Helmond, The Netherlands, email: frank.ophelders@tno.nl

Alexander Otto Fraunhofer Institute for Electronic Nano Systems ENAS, Department Micro Materials Center, Technologie-Campus 3, 09126 Chemnitz, Germany, email: alexander.otto@enas.fraunhofer.de

Federico Perelli Politecnico di Milano, Department of Mechanics, via La Masa 1, Milan, Italy, email: federico.perelli@polimi.it

Johannes Petzold Continental AG, BU ADAS, Segment Surround View, Johann-Knoch-Gasse 9, 96317 Kronach, Germany, email: markus.friebe@continental-corporation.com

Stefan Raue Daimler AG, Architecture & Body Controller, HPC G007-BB, 71059 Sindelfingen, Germany, email: stefan.raue@daimler.com

Hubert Rehborn Daimler AG, HPC: 059-X832, 71063 Sindelfingen, Germany, email: hubert.rehborn@daimler.com

Wolfgang Reinprecht ams AG, Tobelbaderstr. 30, 8141 Unterpremstaetten, Austria, email: wolfgang.reinprecht@ams.com

Werner Rößler Infineon Technologies AG, Automotive System Engineering, Am Campeon 1-12, 85579 Neubiberg, Germany, email: werner.roessler@infineon.com

Wolfgang Rosenstiel University of Tübingen, Computer and Engineering Department, Auf der Morgenstelle 8, 72076 Tübingen, Germany,
email: rosenstiel@informatik.uni-tuebingen.de

Sven Rzepka Fraunhofer Institute for Electronic Nano Systems ENAS, Department Micro Materials Center, Technologie-Campus 3, 09126 Chemnitz, Germany,
email: sven.rzepka@enas.fraunhofer.de

Marga Sáez Centro Tecnológico de Automoción de Galicia, Pol. Industrial A Granxa, Calle A, Parcela 249-250, 36400, Porriño (Pontevedra), Spain,
email: marga.saez@ctag.com

Francisco Sánchez Centro Tecnológico de Automoción de Galicia, Pol. Industrial A Granxa, Calle A, Parcela 249-250, 36400, Porriño (Pontevedra), Spain,
email: francisco.sanchez@ctag.com

Andreas Schmidhofer MAGNA Powertrain AG & Co KG, Project House Europe, Frank Stronach-Straße 3, 8200 Albersdorf, Austria,
email: andreas.schmidhofer@magnapowertrain.com

Michael Schreckenberg Universität Duisburg-Essen, Physik von Transport und Verkehr, Lotharstr. 1, 47057 Duisburg, Germany,
email: michael.schreckenberg@uni-due.de

Nikolay Shilov St. Petersburg Institute for Informatics and Automation of the Russian Academy of Sciences (SPIIRAS), 39, 14 Line, St. Petersburg, 199178, Russia, email: nick@iias.spb.su

Alexander Smirnov St. Petersburg Institute for Informatics and Automation of the Russian Academy of Sciences (SPIIRAS), 39, 14 Line, St. Petersburg, 199178, Russia, email: smir@iias.spb.su

Christoph Stamprath *elenia* TU-Braunschweig, Am Flughafen 13, 38110 Braunschweig, Germany, email: c.stamprath@tu-bs.de

Thomas Suermann NXP Semiconductors Germany GmbH, Stresemannallee 101, Hamburg, Germany, email: thomas.suermann@nxp.com

Roman Szczuka DFKI GmbH - Robotics Innovation Center, Robert-Hooke-Str. 5, 28359 Bremen, Germany, email: roman.szczuka@dfki.de

Davide Tarsitano Politecnico di Milano, Department of Mechanics, via La Masa 1, Milan, Italy, email: davide.tarsitano@polimi.it

Carsten Thun Hella Fahrzeugkomponenten GmbH, Dortmunderstr. 5, 28199 Bremen, Germany, email: carsten.thun@hella.com

Faical Turki Paul Vahle GmbH & Co. KG, Westicker Strasse 52, 59174 Kamen, Germany, email: faical.turki@vahle.de

Toni Weisheit University of Kassel, Department of Traffic Engineering and Transport Logistics, Mönchebergstr. 7, 34125 Kassel, Germany, email: toni.weisheit@uni-kassel.de

Norbert Weiss Volkswagen AG, Volkswagen AG, EAEF/2, 38440 Wolfsburg, Germany, email: norbert.weiss@volkswagen.de

Johann Winter ams Germany GmbH, Erdinger Str. 14, 85609 Aschheim b. München, Germany, email: johann.winter@ams.com

Yong-Ho Yoo DFKI GmbH - Robotics Innovation Center, Robert-Hooke-Str. 1, 28359 Bremen, Germany, email: Yong-Ho.Yoo@dfki.de

Mehmed Yüksel DFKI GmbH - Robotics Innovation Center, Robert-Hooke-Str. 5, 28359 Bremen, Germany, email: mehmed.yueksel@dfki.de

Josef Zehetner Virtual Vehicle Research Center, Inffeldgasse 21/A, 8010 Graz, Austria, email: josef.zehetner@v2c2.at

Zhi Zhang Politecnico di Milano, Department of Mechanics, via La Masa 1, Milan, Italy, email: zhangzhiroom@163.com

Olaf Zschieschang Fairchild Semiconductor GmbH, Technology Development Center, Einsteinring 28, 85609 Aschheim, Germany, email: olaf.zschieschang@fairchildsemi.com

[illegible] Department of Mechanics, via La Masa [illegible]
[illegible]

Carsten [illegible] Fahrzeugkomponenten GmbH, [illegible]
[illegible]

[illegible] GmbH & Co. KG, [illegible] Strasse 52, [illegible]
[illegible]

[illegible], University of Kassel, Department of Traffic Engineering and Transport Logistics, [illegible], Kassel, Germany
[illegible]

[illegible] Volkswagen AG, [illegible] Wolfsburg, [illegible]
[illegible]

[illegible]

[illegible] GmbH [illegible] Center [illegible]

[illegible]

Part I
Driver Assistance and Vehicle Automation

Evolution in Advanced Driver Assistance: From Steering Support in Highway Construction Zones to Assistance in Urban Narrow Road Scenarios

Thomas Paul Michalke, Thomas Gußner, Lutz Bürkle and Frank Niewels

Abstract. With the advances in environment sensor technology, advanced driver assistance systems (ADAS) that target increasingly complex scenarios such as inner-city traffic get into focus. Such novel ADAS will offer assistance in a wide range of urban traffic scenarios and, thus, will further decrease the number and severity of accidents. In this contribution, the evolution of an ADAS for lateral guidance in highway construction zones (i.e. the "construction zone assistant") towards assistance in narrow urban road scenarios (i.e. the "urban narrow road assistant") is presented. The focus of the contribution will be on the challenges of these two scenario types and their respective requirements on the system concept and design. While steering support in highway construction zones will be available on the market soon, its functional extension to inner-city traffic is still characterized by numerous technological challenges. Due to that, the emphasis in terms of algorithmic details will be on the "urban narrow road assistant".

Keywords: Construction zones, UR:BAN, automated steering support, driver assistance in inner-city, automated lateral control.

1 Introduction

In recent years, numerous prototypical ADAS have been developed that target the automated longitudinal and lateral control of vehicles (refer e.g. to [1-3]). Although highly demanding in terms of the complexity and number of use-cases to be covered, such systems can benefit from the restricted interaction between driver and system. Hence, on an operational level a correctly operating automated vehicle does not have to interact and cooperate with the driver's intentions. Therefore,

T.P. Michalke(✉) · T. Gußner · L. Bürkle · F. Niewels
Robert Bosch GmbH, Corporate Research (CR/AEV2), 70442 Stuttgart, Germany
e-mail: {thomas.michalke,thomas.gussner,lutz.buerkle,frank.niewels}@de.bosch.com

J. Fischer-Wolfarth and G. Meyer (eds.), *Advanced Microsystems for Automotive Applications 2014*, Lecture Notes in Mobility,
DOI: 10.1007/978-3-319-08087-1_1, © Springer International Publishing Switzerland 2014

in such systems no driver intention recognition is required. In addition to still unresolved technical issues, also well-known legal restrictions apply [4] which might considerably delay the realization of highly automated vehicles. Due to these restrictions, the realization of driver assistance as well as partly automated systems is a necessary intermediate development step on the path to full automation. The evolution of close-to-market systems for lateral guidance assistance is in the focus of this contribution. Since the driver remains involved and responsible, especially in complex scenarios driver intention recognition becomes mandatory.

Driver assistance systems for supporting the driver to stay in the lane have been available on the market since several years. Such lane keeping systems intervene by an acoustic warning or a haptic feedback on the steering wheel (refer to [5]). Furthermore, an active steering intervention based on superimposing steering torques (refer to [6, 7]) or by asymmetric braking interventions are on the market. These and other earlier lane keeping systems depend on the existence and precise detection of lane markings. More recent systems support the driver in construction sites on highways (refer to [8, 9]) and therefore rely on the detection of static and dynamic obstacles. Also the here referenced "construction zone assistant" (see Section 2) belongs to that category.

Novel systems (as e.g. the "urban narrow road assistant" presented in Section 3) currently under development explicitly focus on actively supporting the driver in inner-city traffic scenarios. As a common characteristic, such systems rely on the detection and measurement of obstacles. Based on that, the optimal vehicle trajectory is inferred. Dependent on the trajectory, lateral steering support is offered in order to prevent collisions with objects positioned laterally. Lateral guidance systems are typically classified as comfort systems while still having inherent characteristics of safety systems. When focusing on close-to-market systems of the latter type, only few related contributions exist. For example, in the context of the publicly-funded project "V-Charge" (refer to [10, 11]) an ADAS was developed that supports the driver in stationary traffic and parking maneuvers in inner-city. Different from the here presented two ADAS, the system is restricted to an ego-velocity in the range of walking speed and to scenarios without dynamic obstacles.

In the following, an overview on the two functions "construction zone assistant" and "urban narrow road assistant" is given. The focus is on the major use-cases as well as the resulting respective challenges in system design.

2 Construction Zone Assistant

This section focuses on the "construction zone assistant" (CZA). First, we briefly describe the function and then discuss necessary sensors as well as actuators and give an overview on the system architecture. Then, with the environment modeling and the driving corridor estimation two key algorithms of the system are described in more detail. Finally, the typical use-cases the system is capable to deal with as well the main limitations of the system are briefly discussed. For further information on the CZA refer to [8,9].

2.1 Function Description

The objective of the CZA is to support the driver in highway construction zones in order to keep a safe lateral distance to static infrastructure objects such as walls or traffic cones as well as to dynamic objects such as other cars or trucks travelling in neighboring lanes. In a typical use-case the driver is supported while overtaking a slower truck in a construction site resulting in a narrow driving corridor (see Fig.2b). In contrast to usual highway scenarios with marked lanes that define the driving corridor, lane markings in construction sites are often ambiguous or even not present at all. The system is supporting the driver in maintaining a collision free path by applying an appropriate steering torque whenever the driver steers in the direction of an obstacle. If appropriate, the system additionally informs the driver about the available lateral distance to objects and the width of the driving corridor ahead of the vehicle.

The function assists the driver at velocities ranging from 60 kph to 100 kph, which are typical for construction zones on highways. The availability of the system is limited to highway construction zones, which allows for simplifying assumptions on the infrastructure and the dynamics of other vehicles.

2.2 System Overview

The environment sensor setup of the CZA consists of a sensor capable of observing the space ahead of the vehicle as well as sensors observing the area next to the vehicle on both sides. A front facing sensor shall provide information on dynamic as well as static obstacles in order to be able to determine the driving corridor through the construction zone and to assist the driver in maintaining a collision free path. In addition, it is necessary to be able to detect and measure the distance to dynamic objects (i.e. other cars, trucks, etc.) driving outside of the field of view of the front sensor. Based on these requirements, we decided to equip a test-vehicle with a stereo-vision-camera (SV-camera) as a front sensor as well as four ultrasonic sensors (US-sensors) covering the space on both sides next to the vehicle. Furthermore, onboard sensors measuring yaw-rate, velocity, etc. are necessary to estimate and predict the motion of the ego-vehicle. Since the purpose of the system is to provide assisting steering torques in critical scenarios in construction zones, we use electrical power steering (EPS) as actuator. It can provide an additional torque on the steering column that is superimposed to the torque provided by the driver on the steering wheel. To make sure that the driver can always overrule the system, the superimposed steering torque is limited in terms of the absolute value as well as the rate of change.

Fig. 1 shows the system architecture that is employed to determine supporting steering torques from the sensor input signals.

Based on the two simultaneously recorded images of the SV-camera a disparity map (see e.g. [12]) is computed in sub-system 1. Afterwards, the amount of data to be processed in subsequent sub-systems is reduced by clustering the disparity map. Additionally, from pairs of images of subsequent frames, an optical flow field (see e.g. [13]) is estimated in order to be able to determine motion relative to the SV-camera. In parallel to the stereo-signal processing, the raw data of the US-sensors are processed resulting in measured radial distances of objects for each of the four US-sensors.

The purpose of the subsequent sub-system 2 is to provide a temporally stabilized representation of the free-space in front of and next to the vehicle based on the preprocessed sensor signals. Additionally, an object-list keeps track of the dynamic objects surrounding the vehicle. Using this representation, sub-system 3 computes a driving corridor based on several assumptions derived from the knowledge of being in a highway construction zone. Next, sub-system 4 predicts the motion of the ego-vehicle and checks the predicted trajectory for collisions with the borders of the driving corridor. If such a collision is predicted to occur, a collision-free trajectory is planned in sub-system 5, which finally results in a correcting steering torque superimposed by the EPS.

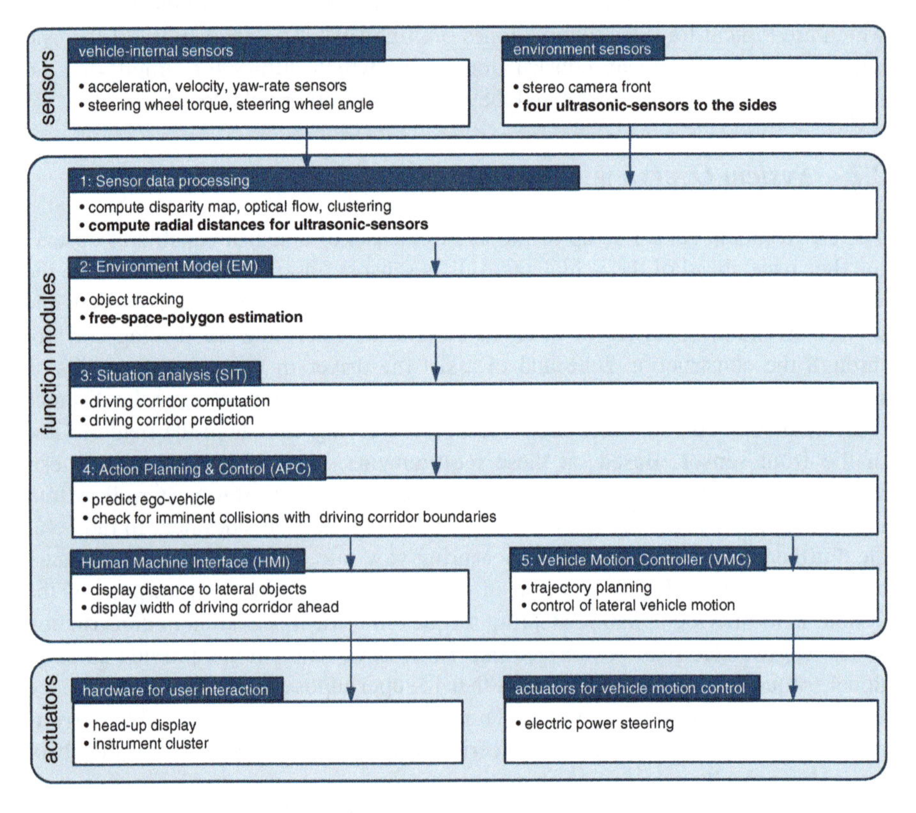

Fig. 1 System architecture of the "construction zone assistant" (bold elements differ from architecture of the "urban narrow road assistant")

2.3 Environment Model and Driving Corridor Estimation

This section describes the environment model in sub-system 2 and the estimation of the driving corridor, which represents the space, the vehicle can safely drive on, in sub-system 3. As soon as the vehicle is predicted to leave the driving corridor, a correcting steering torque is applied. Thus, the driving corridor plays a similar role as lanes do in lane assistance systems. However, the boundaries of the driving corridor are rather defined by static and dynamic obstacles than by lane markings.

The environment modeling algorithm uses the preprocessed sensor data from the SV-camera and the US-sensors to infer a compact representation of the environment employing certain assumptions especially on the temporal behavior of obstacles surrounding the vehicle. The first step in determining this representation is an object-tracking based on the assumption that all objects approximately maintain a constant velocity. We use a classical Kalman-Filter approach to estimate position, dimensions, orientation and velocity of moving objects. Secondly, all stationary obstacles are represented by means of a polygon. This polygon describes either the boundary between the free-space in front of and next to the vehicle and any type of obstacle or the boundary of the field of view of the environment sensors, respectively. This approach differs from standard approaches using a Cartesian obstacle grid map representation e.g. described in [14]. Similar to the well known grid map estimation, we assume all obstacles to be static when temporally stabilizing the polygon representation applying recursive Bayes filtering techniques. The advantage of our approach is a lower demand of computation and memory resources, since only the corners describing the boundary polygon need to be stored and computed in contrast to storing and updating the state of each cell of the grid map. The downside of this approach is that the complexity of the environment that can be represented by the polygon model is limited. For instance, a region of free-space that is not connected to the free-space in front of the vehicle cannot be described. The environment is now represented by a list of dynamic objects and a polygon describing the static world surrounding the vehicle.

Based on the stationary free-space-polygon and the list of dynamic objects, we determine the driving corridor, which consists of two polygons describing the corridors left hand and right hand side boundary, respectively. The driving corridor contains only those parts of the environment that the vehicle can reach from its current position. To this end, first, all paths that are too narrow for the vehicle to drive through are excluded. Furthermore, we exclude all areas that cannot be reached from the current position of the ego-vehicle without exceeding a certain velocity dependent yaw rate. This threshold does not necessarily coincide with physical limits derived from vehicle dynamics, but is rather derived from yaw-rates which are usually not exceeded in driving through highway construction zones. The driving corridor can also be used to inform the driver about narrow spots ahead, e.g. between a wall and a truck the driver intents to overtake.

This representation of the driving-corridor still is time-dependent, since it may contain corners resulting from dynamic objects with non-zero velocity. Depending on the trajectory-planning-algorithm employed to find a collision free trajectory in case of an imminent collision with the driving-corridor boundaries, it might be necessary to eliminate this time-dependency by predicting all non-stationary polygon corners assuming constant velocities and afterwards rearranging the polygon corners appropriately.

2.4 *Use-Cases and Limitations*

The described algorithms are capable of dealing with the typical use cases occurring in highway construction zones. The main use-cases, which were also used to evaluate the system on a test track, are depicted in Fig 2.

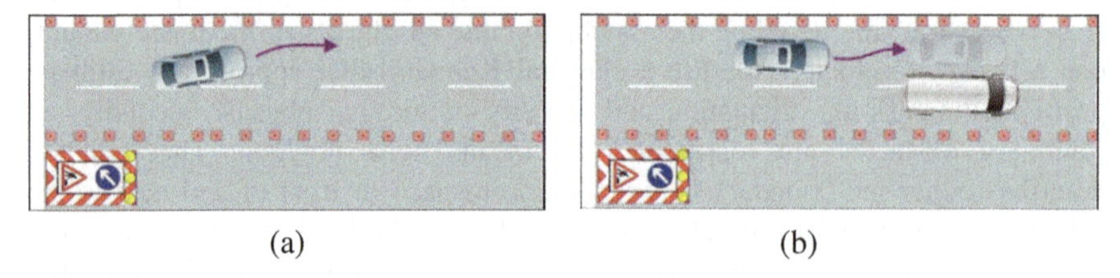

(a) (b)

Fig. 2 Main use-cases of the CZA: (a) Lateral guidance based on static infrastructure. (b) Lateral guidance based on static and dynamic objects while overtaking.

However, the system is limited to driving through highway construction zones and cannot deal with scenarios typically occurring in inner-city traffic. For instance, oncoming traffic is ignored in the driving corridor estimation for robustness reasons. Also, the assumption of all objects approximately maintaining their velocity, which is well suited for highway scenarios, is often violated in urban traffic. Finally, as discussed in section 2.3, the polygon representation of the driving corridor is not capable to precisely model many of the complex scenarios which can occur in inner-city traffic.

Clearly, the assumptions made for developing a highway construction zone assistant under the constraint of rather low computation costs are not compatible to urban scenarios. However, a similar type of driver assistance is desirable in inner-city scenarios as well. With the "urban narrow road assistant" such a system is described in the next section.

3 Urban Narrow Road Assistant

In this section, an overview of the "urban narrow road assistant" (uNRA) is given. After a brief function description, an overview of the system architecture is provided. As a major system module, the situation analysis is described in more detail. Finally, the typical use-cases and main limitations of the system are discussed. For further details on the uNRA refer to [15].

3.1 Function Description

Similar to the CZA, the uNRA provides information about the lateral distance to static and dynamic objects. Typical use-cases are depicted in Fig.5. In case a safe lateral distance to obstacles is violated the system applies a steering torque. However, in contrast to the CZA, the uNRA supports the driver in inner-city traffic. Consequently, the function is active at lower velocities ranging from 0 kph to 60 kph. The uNRA is required to handle unstructured surroundings typically found in cities (parking cars, poles, traffic islands, curb stones, etc.). As a consequence, the uNRA requires more sophisticated means to represent and interpret the environment. These will be a major focus of this contribution.

3.2 System Overview

Following the system architecture in Fig. 3, the uNRA relies on a multi-camera system consisting of a front-facing stereo video camera and distributed body cameras as environment sensors. In addition, various vehicle-internal sensors are used (e.g. velocity, acceleration, yaw-rate, and steering wheel angle sensors). The sensor data is post-processed in the data preparation and validation module (DPV), after which it is transformed to standard units and validated with respect to signal range and consistency.

The environment sensors provide a 3D representation of the vehicle surrounding which is used as input for an environment model (EM). Here the data of individual sensors is fused and represented as an occupancy grid, a free-space representation, and object models, respectively. The occupancy grid represents the static collision-relevant environment. The free-space represents the road surface that was measured to be drivable. Dynamic traffic participants are represented and tracked as objects. The application of a grid representation stems from the fact that the uNRA is required to operate in unstructured inner-city surroundings that cannot be represented with classical object-model-driven approaches (refer to [16] for more details).

In the situation analysis module (SIT) the occupancy grid and free-space are post-processed and combined resulting in the driving corridor. Furthermore, within the SIT a use-case classification and driver intention recognition are realized. Finally, different criticality measures are computed that determine the system state in the following action planning and control (APC) module. The APC furthermore checks if system boundaries are exceeded (e.g. thresholds for lateral and longitudinal ego-acceleration and road curvature). Dependent on the system state the ADAS will remain inactive (e.g. no critical situation), offer information (e.g. narrow road section ahead), a collision warning or steering support, respectively. Information and warnings are displayed through the human machine interface (HMI) module. Steering support is realized by the vehicle motion control (VMC) module. The latter two modules control the required hardware (e.g. displays or electric power steering).

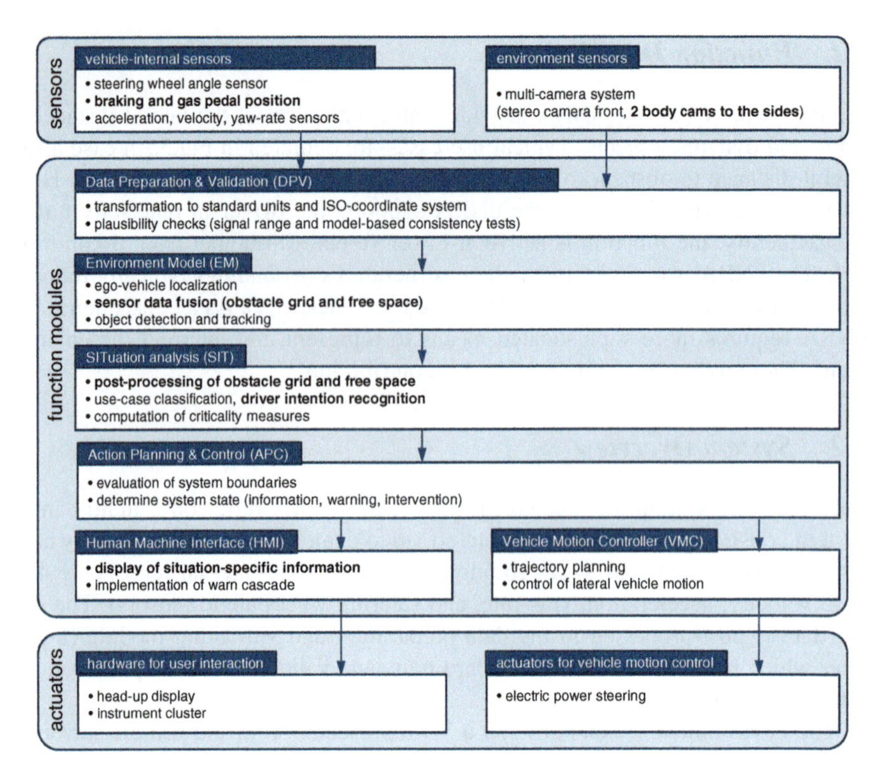

Fig. 3 System overview "urban narrow road assistant" (bold elements differ from CZA architecture)

3.3 Situation Analysis

The SIT module has the task to interpret the scenario by determining the relations between objects. Referring to the SIT sub-system overview in Fig. 4, in a first step the stationary occupancy grid (representing the static collision-relevant environment) is fused to the grid representing drivable free-space. These representations are not necessarily complementary. For example, free-space is not necessarily limited by collision-relevant objects, since the latter can be measured in larger distances than the free-space. As a preparation step, on the occupancy grid a ray tracing can be realized. Hence, it is assumed that all space in the line-of-sight to an object is free. For a combination, a simple super-position of the ray-traced grid and the free-space can be realized, which results in the driving corridor. On the driving corridor a search for possible future ego-vehicle paths is realized. Only those paths are used for further processing that match the measurement-driven, predicted ego-trajectory. As a result, a collision-free path with or without branches is now available. For example, a branch can result when the driver has the two options: (a) Stop his vehicle on the right side of the road, (b) Overtake a parking vehicle. Depending on the recognized use-case, certain branches might be rejected (e.g. if the road is too narrow and parking is no option). At path positions with a branch, a decision point is defined. Here recognition of the driver intention is required to resolve the ambiguity and hence prevent a possible false system intervention or futile warning / information. The driver intention recognition can be realized relying on a Hidden Markov Model, modeling important aspects of the driver's decision processes (refer to [15] for more details). After resolving the decision points, the resulting path is combined with the measurement-driven predicted ego-trajectory. Based on this trajectory, relevant critical borders (enclosing the collision-relevant environment) are derived, which will be the input to the VMC module that controls the steering actuator.

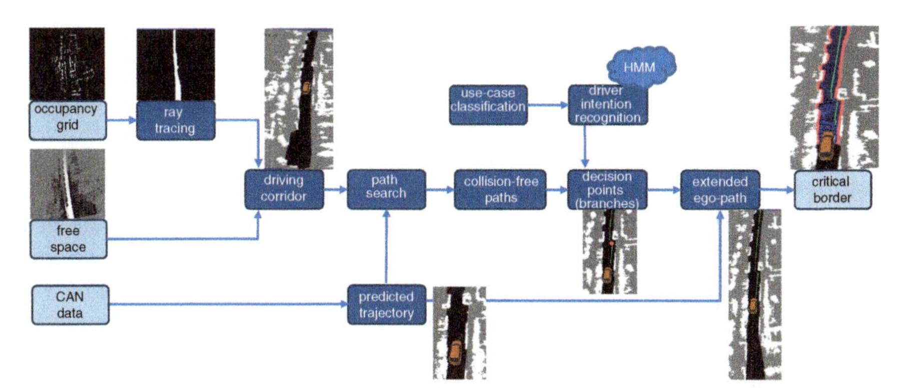

Fig. 4 Sub-system overview SIT

3.4 Use-Cases and Limitations

The “urban narrow road assistant” supports the driver in unmarked narrow road sections in inner-city traffic, informing about upcoming collision risks and offering steering support in a way that matches the driver’s intention (see Fig. 5 for typical use-cases in a static environment as well as in an environment with dynamic objects).

(a) (b)

Fig. 5 Use-cases for the uNRA (ego vehicle in red): (a) Narrow passage caused by static surroundings, (b) Narrow passage caused by dynamic objects

However, it is important to note that the uNRA provides steering support in case of imminent collision risks only. The driver always stays in the loop. In other words, the uNRA does not offer continuous, automated lateral guidance on inner-city roads.

4 Comparison and Conclusion

All major differences in system design between the CZA and uNRA stem from differences in the addressed use-cases (highways as compared to inner-city scenarios) and, as a consequence, different functional requirements.

When focusing on demanding use-cases, the uNRA is required to keep a safe distance to lateral obstacles of small height (as e.g. curb-stones) and laterally positioned dynamic obstacles of small length (as e.g. bicycles). Due to that, side cameras (instead of ultrasonic sensors for the CZA) are required.

As the uNRA offers steering support in inner-city scenarios, the system must be capable of representing the typically unstructured inner-city environment. Obviously, a classical object-driven representation would be highly inefficient and te-dious. Although the CZA already is representing the borders of the driving corridor in a grid-like fashion (polygons of discrete resolution that already offer a restricted flexibility), the uNRA still requires more flexible means of representing the multitude of objects classes typically present in inner-city scenarios. After analyzing the uNRA's functional requirements, a grid-maps-based representation turned out to be most suitable.

As compared to the CZA, for the uNRA more complex approaches for driver intention recognition are required (e.g. the CZA does not consider branching trajectories that are determined by the driver's course of action). This is due to the fact that in inner-city scenarios a higher number of relevant driver intentions exist, which increases the challenge of assuring that the ADAS reaction is in accordance to the current driver intention.

The development of the uNRA was conducted within the scope of the German research initiative UR:BAN (refer to [17]). It was supported under the code 19 S 12008 I by the German Federal Ministry of Economics and Energy on the basis of a decision by the German Bundestag. The authors are responsible for the content of this publication.

References

[1] Ardelt, M., Coester, C., Kaempchen, N.: Highly automated driving on freeways in real traffic using a probabilistic framework. IEEE Transactions on Intelligent Transportation Systems 13, 1576–1585 (2012)

[2] Dolgov, D., Thrun, S.: Autonomous driving in semi-structured environments: Mapping and planning. In: IEEE International Conference on Robotics and Automation, ICRA 2009, pp. 3407–3414 (2009)

[3] Dang, T., Kammel, S., Duchow, C., Hummel, B., Stiller, C.: Path planning for autonomous driving based on stereoscopic and monoscopic vision cues. In: 2006 IEEE International Conference on Multisensor Fusion and Integration for Intelligent Systems, pp. 191–196 (2006)
[4] Gasser, M.: Rechtsfolgen zunehmender Fahrzeugautomatisierung. Bundesanstalt für Straßenwesen (2012)
[5] Gayko, J.: Lane Departure Warning. In: Winner, H., Hakuli, S., Wolf, G. (eds.) Handbuch Fahrerassistenzsysteme, pp. 543–553 (2009)
[6] Rohlfs, M., et al.: Gemeinschaftliche Entwicklung des Volkswagen Lane Assist. In: VDI-Berichte, vol. 2048, pp. 15–33 (2008)
[7] Gayko, J.: Lane Keeping Support. In: Winner, H., Hakuli, S., Wolf, G. (eds.) Handbuch Fahrerassistenzsysteme, pp. 554–561 (2009)
[8] AKTIV: Adaptive und kooperative Technologien für den intelligenten Verkehr (2010), http://www.aktiv-online.org
[9] Bosch Construction Zone Assist. (2012) http://www.springerprofessional.de/der-assistent-faehrt-kuenftig-immer-mit/3580528.html, http://www.springerprofessional.de
[10] V-Charge: Autonomous Valet Parking and Charging for e-Mobility (2011), http://www.v-charge.eu/?page_id=28, http://www.v-charge.eu
[11] Furgale, P.: Toward autonomous driving in cities using closeto-market surround sensors an overview of the v-charge project. In: IEEE Intell. Veh. Symp., Gold Coast, Australia (2013)
[12] Hirschmüller, H.: Accurate and Efficient Stereo Processing by Semi-Global Matching and Mutual Information. In: Proceedings of the 2005 IEEE Computer Society Conference on Computer Vision and Pattern Recognition, vol. 2, pp. 807–814 (2005)
[13] Mallot, H.: Computational vision: Information processing in perception and visual behavior. MIT Press Robotica (2002)
[14] Moravec, H., Elfes, A.: High resolution maps from wide angle sonar. In: Proc. IEEE Int. Conf. on Robotics and Automation, vol. 2, pp. 116–121 (1985)
[15] Michalke, T., Gläser, C., Bürkle, L., Niewels, F.: The Narrow Road Assistant - Next Generation Advanced Driver Assistance in Inner-City. In: Proceedings of the IEEE International Conference on Intelligent Transport Systems (2013)
[16] Gläser, C., Michalke, T., Bürkle, L., Niewels, F.: Environment Perception for Inner-City Driver Assistance and Highly-Automated Driving. In: Proceedings of the IEEE Intelligent Vehicles Symposium (2014)
[17] Urbaner Raum: Benutzergerechte Assistenzsysteme und Netzmanagement (UR:BAN) (2012), http://urban-online.org/de/urban.html, http://www.urban-online.org

Smart and Green ACC: As Applied to a Through the Road Hybrid Electric Vehicle

Sagar Akhegaonkar, Sebastien Glaser, Lydie Nouveliere and Frederic Holzmann

Abstract. The Smart and Green ACC (SAGA) or simply Green ACC (GACC) may be defined as a system which autonomously generates longitudinal control commands for a vehicle while balancing the safety and efficiency factors. In previous studies, the SAGA function is investigated as applied to a battery electric vehicle. As a continuation of the SAGA function development, this paper investigates the behavior of the autonomous longitudinal controller as applied to a "Through the Road" (TtR) hybrid electric vehicle. Given the presence of two power sources, the implementation of a SAGA system in HEV/PHEV has a higher level of complexity as compared to pure EV. As an autonomous longitudinal driver command generating system, SAGA acts as a surficial controller which is then combined with a core powertrain management system. The Equivalent Consumption Minimization Strategy (ECMS) is used to determine the optimum power split between the IC engine and electric motor.

Keywords: Driving assistance, longitudinal control, energy efficiency.

1 Introduction

Advanced driver assistance systems are primarily aimed at increasing road safety (e.g. lane keeping assistance) or reducing the driver fatigue by partially or completely taking over the driving task (e.g. cruise control). The Eureka PROMETHEUS project (PROgraMme for a European Traffic of Highest Efficiency

---

S. Akhegaonkar(✉) · F. Holzmann
INTEDIS GmbH & Co. KG, Max-Mengeringhausen-Strasse 5,
97084 Wuerzburg, Germany
e-mail: {Sagar.Akhegaonkar,Frederic.Holzmann}@intedis.com

S. Glaser · L. Nouveliere
LIVIC, 77 rue de Chantiers, 78000 Versailles, France
e-mail: {Sebastien.glaser,lydie.nouveliere}@ifsttar.fr

J. Fischer-Wolfarth and G. Meyer (eds.), *Advanced Microsystems for Automotive Applications 2014*, Lecture Notes in Mobility,
DOI: 10.1007/978-3-319-08087-1_2, © Springer International Publishing Switzerland 2014

and Unprecedented Safety, 1987-1995) [1] was one of the early demonstrators of the fully automated road vehicle. The significant advancement in on board computing capacity has enabled real time image processing such as object detection, recognition, classification, and tracking using vehicle mounted cameras. Combined with long and short range sensor data, a fairly satisfactory three dimensional virtual model of the real surrounding environment can be generated and updated in real time. This presents a possibility to apply automated lateral and longitudinal control for a road vehicle with significantly higher level of safety. Most of the time, this safety level exceeds that achieved during conventional human vehicle control.

While driver and pedestrian safety form an important part of human mobility, it is also imperative to balance it with environmental safety. Thus the newly coined term "Eco-driving" comes into existence. Research projects like eCoMove [2] and EcoDriver [3] have focused their efforts on coaching the driver to drive as efficiently as possible. However, for many individuals, their driving style closely resembles their personality and constant advice on driving style may often become irritating. The modern car is being loaded with so much technology that information, more than what a human brain can process at a time, is being presented to it every instant through the human machine interface. This might, on the contrary lead to reduced safety. Hence, in order to balance safety and efficiency, a more logical solution may be to eliminate the driver from the equation and load him with an easy supervisory task while the autonomous car controls itself laterally and longitudinally.

The SAGA function concentrates on autonomously generating the longitudinal control commands while balancing the safety and efficiency. The concept of SAGA and GACC is defined and elaborated in [4] and [5]. In [4] the authors describe the basic concept with a detailed strategy definition and system behavior in specific cases. In [5], dedicated system approaches towards city and highway scenarios are elaborated. In both these papers the SAGA system is investigated with respect to a battery electric vehicle (BEV) powertrain. This paper attempts to apply the SAGA concept to a hybrid electric vehicle (HEV). Because the SAGA system is basically a cruise control, it acts on a surficial level and only generates the acceleration and braking commands. But unlike the BEV, the HEV has two sources of power and must be appropriately controlled to maximize the efficiency. The Equivalent Consumption Minimization Strategy (ECMS) [6] is used to determine the optimum power split between the IC engine and electric motor.

The paper is structured as follows. In the second section a detailed description of the HEV model is given. The concept of ECMS is explained in the third section along with simulation results regarding fuel consumption comparisons to conventional IC engine vehicle. Finally the concept of SAGA system is investigated as applied to the HEV powertrain in fourth section followed by a short conclusion.

2 Vehicle Model

Since hybrid electric vehicles have two different power sources, various topologies are possible. The series hybrid vehicles completely isolate the engine from the task of actual vehicle propulsion. Hence the engine may run at the operating point of highest efficiency and the motor will produce the actual propulsion torque. This means that the motor must be powerful and oversized in the series topology. In the parallel hybrid, the relatively smaller motor assists the engine in order to avoid operating points of very low efficiency. While it is simpler, it does not provide the efficiency levels of series topology. A combination of series and parallel topologies can be realized in the power-split hybrid. This topology uses a planetary gear system to split the power requirement between two power sources. A through the road (TtR) hybrid may be classified as a sub-type of power-split hybrids which does away with the complex planetary gear device. Since the front and rear axle are mechanically isolated (except through the road), TtR hybrid offers more flexibility for engine and motor power split control. The Peugeot 3008 based on the PSA Hybrid4 technology is a TtR hybrid [7].

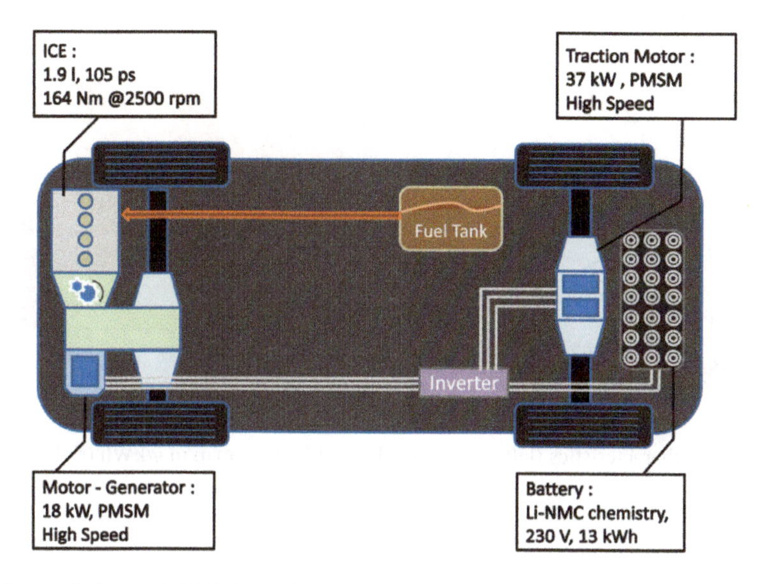

Fig. 1 Through the road (TtR) hybrid topology

The above figure is a schematic representation of the vehicle model used in this paper. It is mathematically modeled using Simulink®.

2.1 Energy Model and Power-Split Definition

A backward looking vehicle model is developed. It consists of two energy subsystems, namely, mechanical (combustion energy) and electrical. A power-split block demands part of total required power from each of the power source. The required power is calculated as follows.

$$F_x = \left(\frac{C_w.\rho \cdot A \cdot v^2}{2} + M_v \cdot g \cdot f_r + M_v \cdot g \cdot sin\beta + M_v \cdot a\right)$$

where, M_v is vehicle mass, A is front area, C_w is drag co-efficient, f_r is co-efficient of rolling resistance, r is wheel radius, a is vehicle acceleration, v is vehicle velocity, ρ is air density, β is grade angle and F_x is the required force at the wheel. The power-split is defined as follows.

$$u_{split} = T_{eM}/T_{total} \mid T_{total} = F_x \cdot r \mid T_{eM} = u \cdot T_{total} \mid T_{ice} = (1-u)\, T_{total}/g_{ice}$$

where, u_{split} or u is power-split ratio, T_{total} is the required torque, T_{eM} is torque demanded from electric motor, T_{ice} is torque demanded from I.C. engine and g_{ice} is the gearbox reduction ratio. The energy supplied by the I.C. engine comes from fuel spent. Fig. 2 presents the base efficiency engine map containing the static 'sfc' (specific fuel consumption) in g/kWh for a 1.9 litre, 105 ps gasoline engine. It is used to determine the final energy consumption in litres/100 km.

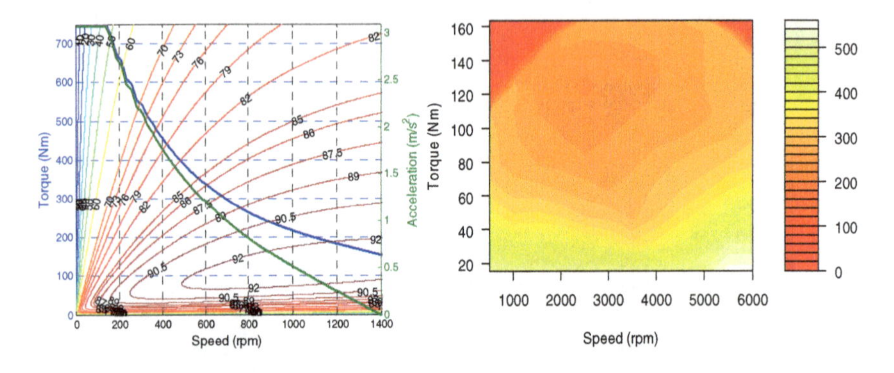

Fig. 2 Motor characteristics (left) and engine base efficiency map in g/kWh (right)

2.2 *eMotor*

An 18 kW PMSM e-Motor with 750 Nm peak torque is modeled. Front axle uses one motor as motor-generator whereas the rear axle uses two motors for propulsion. The motor model is based on look up tables which define the speed-torque characteristics and also the power losses corresponding to the operating point of the motor.

The motor model is supplied with speed (ω), torque request and battery voltage as inputs. The motor torque is calculated by adding the requested torque and power losses. Power required is the product of motor torque and speed. Since the voltage is known the current requirement for this power request is calculated. The calculated current value is supplied as an input signal to the battery. The torque speed characteristics for the e-Motor are shown in Fig. 2 along with the efficiency map and the maximum possible acceleration. Calculations are based on eFuture project [8] prototype vehicle described in [5].

2.3 Energy Storage System

The Energy Storage System (ESS) is a lithium-ion (NCM chemistry) battery developed as a part of the eFuture project by Miljøbil Greenland AS. From the energy calculation point of view, the battery is an energy storage device which supplies or accepts the requested current at the present battery voltage and efficiency. The model is based on a virtual circuit as defined in [9] which represents the battery capacity, internal resistance and transient behavior of the battery. It is sought to keep a balance between accuracy and model complexity in order to reduce the computational load. Thereby, a temperature dependent internal resistance variation is taken into consideration while calculating the power loss and efficiency. Energy equations for the battery charging and discharging are as follows. Inverter losses are assumed negligible.

$$E_{batt.dis} = P_{eM} dt / (\eta_{eM} \cdot \eta_{batt.dis}) \qquad E_{batt.chg} = P_{eM} \cdot \eta_{eM} \cdot \eta_{batt.chg} dt$$

Where $E_{batt.dis}$, $E_{batt.chg}$ denote energy discharged from and charged to the battery respectively, η_{eM} is the motor efficiency and $\eta_{batt.dis}$, $\eta_{batt.chg}$ denote the battery discharging and charging efficiencies respectively.

2.4 Transmission

A 5 speed manual transmission is modeled which complements the I.C. engine torque-speed characteristics. Since gear selection is controlled by the energy management algorithm, it may as well be classified as a semi-automatic transmission. The gear ratios in increasing respect are 3.250, 1.950, 1.423, 1.033 and 0.730. The final gear ratio is 4.063 and it yields the g_{ice} as a product with individual gear ratios.

3 Equivalent Consumption Minimization Strategy

The concept of ECMS (Equivalent consumption minimization strategy) was developed by G. Paganelli *et al.* in [6]. It is based on optimal control theory and can be derived from Pontryagin's Minimum Principle. In this paper, ECMS is used as a direct powertrain control algorithm in combination with the SAGA function for efficient autonomous vehicle control. The aim is to minimize the fuel consumption and sustain the battery charge level. It is explained as follows.

$$E_{total} = E_{ice} + E_{batt}$$

where, E is energy. But all the energy ultimately comes from the fuel. Even the electrical energy E_{batt} which is stored in battery is produced by the motor-generator powered by I.C. engine running at a specific operating point. If this operating point is known then there can be a direct comparison between fuel energy and electrical energy. This comparison factor is $s(t)$ and is called 'Equivalence factor'. For the rate of energy consumption it can be stated that,

$$\dot{m}_{fuel} = \dot{m}_{ice} + s(t) \cdot \dot{m}_{batt}$$

$$\dot{m}_{fuel} = \dot{m}_{ice} + s(t) \cdot \frac{E_{batt}}{Q_{lhv}} \cdot s\dot{o}c(t)$$

$$\dot{m}_{fuel} = \dot{m}_{ice} + s(t) \cdot \frac{P_{eM}}{Q_{lhv}}$$

where, $s\dot{o}c(t)$ is the change in battery state of charge, m is fuel mass and Q_{lhv} is the lower heating value of gasoline in MJ/g. The aim is to minimize the cost function 'J' i.e. the total fuel cost m_{fuel} subject to operational constraints as described below.

$$\text{Min } J\left(u_{split}, g_{ice}\right) = \int_0^T \dot{m}_{fuel}(t)dt$$

subject to,

$$P_{total} = P_{ice} + P_{eM}$$

$$P_{ice.min} \leq P_{ice}(t) \leq P_{ice.max}$$

$$P_{eM.min} \leq P_{eM}(t) \leq P_{eM.max}$$

$$P_{gen.min} \leq P_{gen}(t) \leq P_{gen.max}$$

$$SOC_{min} \leq SOC_{batt}(t) \leq \text{SOC}_{max}$$

In order to minimize the fuel consumption in the driving mission of time T, the global optimization problem is reduced to an instantaneous minimization problem by discretizing it into smaller time segments. For each segment the optimum values for u_{split} and g_{ice} must be determined such that optimum fuel consumption is obtained without violating the SOC constraints.

As stated earlier the equivalence factor allows a direct comparison between fuel and electrical energy for known engine operating point and overall system efficiency. Thus, if the $s(t)$ is high then the ECMS tries to penalize the use of electrical energy. On the other hand if $s(t)$ is low the use of electrical energy is encouraged. Since at any instant, the future operating point of the system is unknown, it is not possible to find the conversion between the electrical energy being spent which will be replenished by fuel energy at a later time or electrical energy being generated which will compensate fuel energy later. But if a driving cycle is known *a prori* then it is possible to find a lumped efficiency parameter for electrical and fuel energy conversion which is a characteristic of that particular velocity cycle. Such a method is described in [10] where two values of $s(t)$ are derived applicable to charging (s_{chg}) or discharging (s_{dis}) battery efficiencies. In [11], it is shown that only one equivalence factor suffices in ECMS. This $s(t)$ is always is located between s_{chg} and s_{dis} values. However, this $s(t)$ only offers optimum performance for the respective velocity cycle. Any deviation in the velocity cycle will result in sub-optimal and non-charge sustaining behavior. Pre-defined velocity cycles like NEDC, Artemis are hardly followed in real world situation. Hence a normal ECMS cannot be applied in real time.

In [11], an adaptive ECMS concept was introduced. Since then many different methods were introduced which used real time update for the equivalence factor. Such an algorithm facilitates constant adaptation of the strategy based on battery SOC status. One such method is described in [12].

$$s(t) = s_0 + s_0 \cdot K \cdot \tan\left(\frac{soc_{ref} - soc(t)}{2\pi} \right)$$

where, s_0 is the nominal equivalence factor. Since we know that equivalence factors lie between s_{chg} and s_{dis} the nominal equivalence factor may be derived as an average of the two. K is a feedback factor, it decides $s(t)$ correction intensity related to the soc deviation from soc_{ref}. K is tuned manually to obtain an acceptable performance of the system.

The adaptive ECMS (A-ECMS) is implemented and tested for both charge sustaining normal HEV and blended mode Plug-in HEV. While for normal HEV the soc_{ref} does not change, in a PHEV the aim is to use as much battery as possible. Hence the soc_{ref} must be adjusted according to the distance covered. For a PHEV,

$$soc_{ref}(t) = \left(\frac{soc_{final} - soc_{init}}{D_{total}} \right) \cdot d(t) + soc_{init}$$

3.1 Simulation and Results

The A-ECMS is developed as a Matlab s-function and implemented in the TtR hybrid Simulink model described in section 2. It acts as power-split and transmission gear controller. To test the assumption that the adaptive ECMS can be implemented in real time, the strategy is initially tuned for a WLTP

(Worldwide harmonized Light vehicles Test Procedures) class 3 cycle [13] and then tested with a completely different set of random velocity points which are recorded during real world driving [14].

3.1.1 Charge Sustaining (HEV) and Blended (PHEV) Strategy: WLTP

After an initial approximation using s_{chg} and s_{dis} values, the value of s_0 is finally manually tuned to 3.5. The value of K was manually tuned to 0.7. soc_{ref} is fixed at 60%. The performance of the A-ECMS for charge sustaining HEV operation is compared to that of a conventional I.C. engine vehicle.

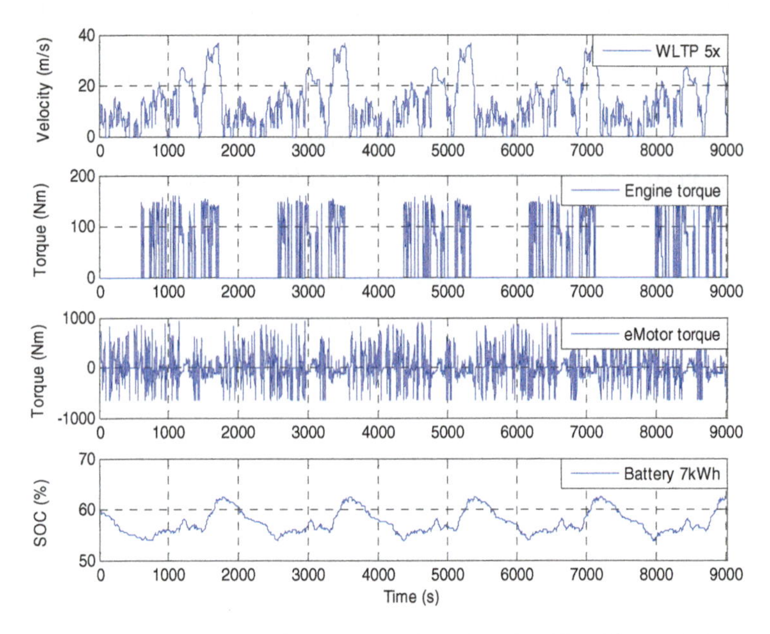

Fig. 3 ECMS controller behavior (Charge sustaining HEV, WLTP 116 km, 7 kWh battery)

In Fig. 3 and 4, behavior of ECMS controller is shown as applied to WLTP class 3 cycle which is repeated 5 times for a total distance of 116 km. It can be seen that ECMS successfully sustains SOC within certain limits for a charge sustaining pure HEV. Whereas, for a plug-in hybrid, full utilization of battery energy between SOC levels 0.8-0.2 is ensured. For this it is necessary that the trip distance must be known beforehand. Apart from SOC regulation the aim of ECMS is to reduce consumption. ECMS actively avoids the engine operation points of lower efficiency as seen in Fig. 5 where a single run of WLTP cycle is compared for ECMS hybrid and conventional ICE.

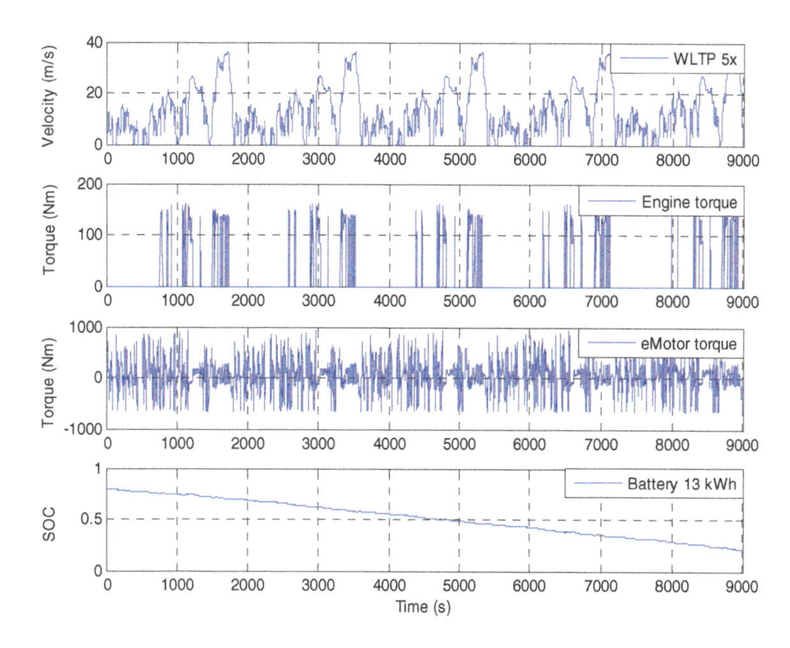

Fig. 4 ECMS controller behavior (Blended Plug-in HEV, WLTP 116 km, 13 kWh battery)

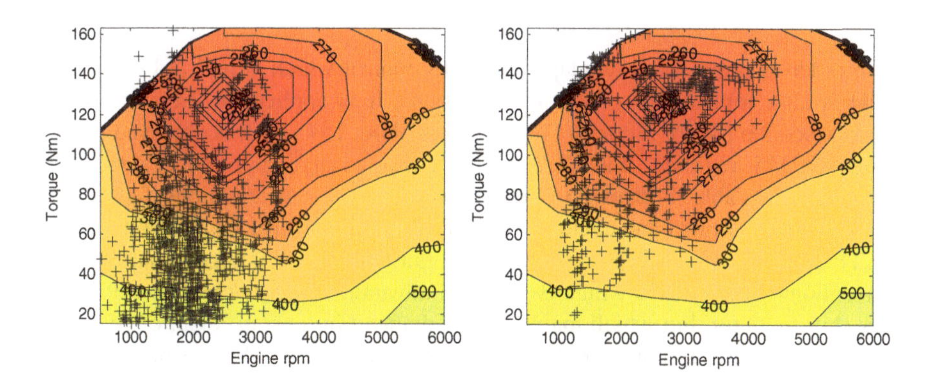

Fig. 5 WLTP class3 cycle engine operating points, conventional ice (left), TtR ECMS (right)

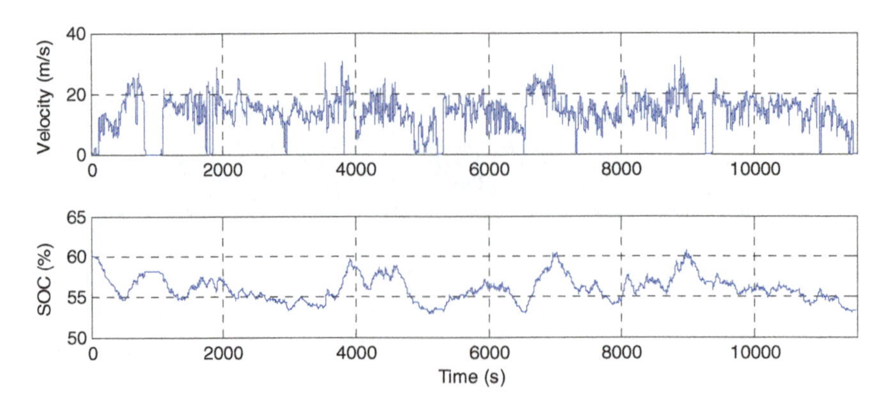

Fig. 6 ECMS controller behavior on a real world recorded velocity session

Table 1 Fuel consumption comparison for various powertrain configurations

Velocity cycle	*Powertrain*	*Consumption(l/100km)*
WLTP class3 (5x)	Conventional ICE	6.68
WLTP class3 (5x)	Pure HEV (CS)	5.52
WLTP class3 (5x)	Plug-in HEV (Blended)	3.23
Real world (A_M)	Conventional ICE	5.54
Real world (A_M)	Pure HEV (CS)	4.19
Real world (A_M)	Plug-in HEV (Blended)	1.96

In order to prove that the adaptive ECMS control which is realized on WLTP cycle is also applicable to real time situations, it is applied on actual recorded velocity session (Arco_Merano) [14] spanning about 157 km. It is observed in Fig. 6 that ECMS regulates the SOC within certain range and reduces the fuel consumption. Table 1 gives an overview of different consumptions figures for respective powertrains and cycles. A charge sustaining hybrid saves about 18% in WLTP cycle and about 24% in the Arco_Merano velocity session compared to the conventional ICE vehicle which is given an advantage of 200 kg and an efficient gear shifting algorithm which maintains the demand in most efficient engine map patch. It is not fair to compare the performance of plug-in hybrids since they use externally refillable electric energy.

4 SAGA as a Supervisory Controller

The SAGA which is basically a green cruise control function, operates in speed control and distance control (vehicle following) modes. In the speed control mode, the controller determines the acceleration depending on the difference in desired and actual speed. It is limited at 2 m/s^2. The distance control function is far more critical as it directly deals with the interaction between vehicles. The objective of the ACC is to regulate the error on the clearance e_d ($e_d = d - T_dV$) around 0 where, d is the headway spacing from front vehicle, V is the speed of ego vehicle and T_d is desired time headway. Depending on the sensors used to measure the distance, the algorithm may be more robust by integrating the error on the relative speed ΔV. The resulting acceleration is a function of these two errors. During the deceleration phase, the aim of SAGA is to regenerate as much energy as possible. For this, it tries to use exclusive motor braking taking into consideration the speed dependent limited braking torque offered by the motors.

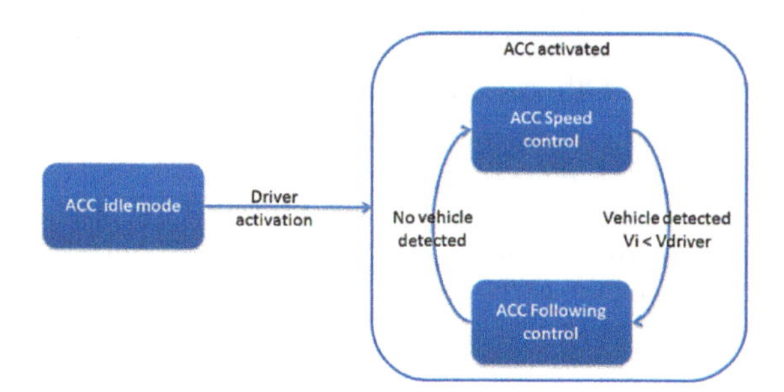

Fig. 7 Operation domains for a SAGA controller

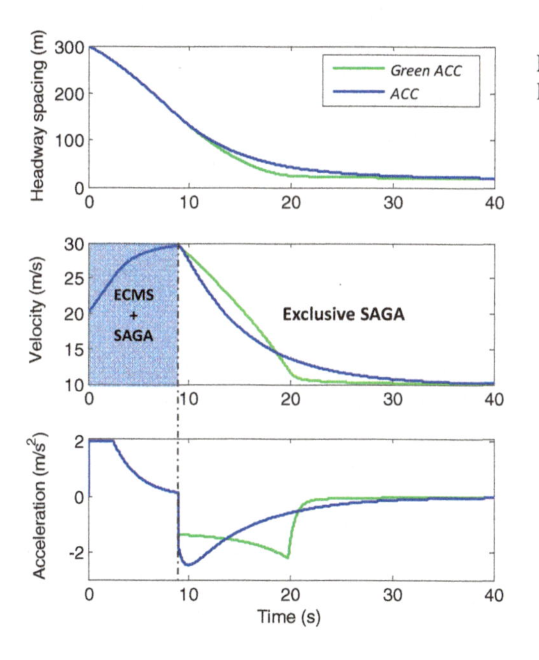

Fig. 8 Definition of SAGA and ECMS operation domains

Fig.8 demonstrates a maneuver where the desired set speed is 30 m/s and actual speed is 20 m/s. The front vehicle is 300 m ahead and sensors range is 150 m, so the SAGA function determines an acceleration value in speed control mode. The approximate amount of torque to achieve the specified acceleration is calculated and passed on to the ECMS function. Dependent on the system parameters, ECMS determines the power-split, transmission gear and amount of torque to be generated or consumed (stored as electrical energy) for each powertrain component. In short, for the acceleration domain SAGA acts as a surficial command generator whereas ECMS manages the actual powertrain activity.

At headway spacing of 150 m SAGA senses the front vehicle and shifts into distance control mode. The front vehicle is moving at 10 m/s. SAGA determines if the motor braking capacity is sufficient to achieve the required deceleration within safe headway distance. Otherwise the service brakes are engaged as seen in case of normal ACC. ECMS is held dormant during this time and SAGA directly controls the regenerative power components. When the speed of ego vehicle is equalized with front vehicle, the objective is to follow the front vehicle speed. In this operation domain characterized with constant speed and intermittent acceleration, SAGA again performs the supervisory control with ECMS as the active powertrain controller.

5 Conclusion

Autonomous driving may become reality sooner than projected predictions. While the main ounce for development is on safety, it presents a significant opportunity towards Eco-Driving. This paper presents an autonomous longitudinal control system where a combination of a Green ACC function and a powertrain management system is employed to control a TtR hybrid vehicle. For future development, real time simulator test runs are planned for an investigation of reliability of such a system.

References

[1] EUREKA Project E45 PROMETHEUS. EUREKA website, http://www.eurekanetwork.org/project/-/id/45 (retrieved January 21, 2014)

[2] "eCoMove Project" official website, http://www.ecomove-project.eu/ (retrieved January 21, 2014)

[3] "ecoDriver Project" official website, http://www.ecodriver-project.eu/ (retrieved January 21, 2014)

[4] Glaser, S., Akhegaonkar, S., Orfila, O., Nouveliere, L., Holzmann, F.: Smart and Green ACC, Safety and Efficiency for a Longitudinal Driving Assistance. In: Fischer-Wolfarth, J., Meyer, G. (eds.) Advanced Microsystems for Automotive Applications 2013 – Smart Systems for Safe and Green Vehicles (AMAA 2013). Lecture Notes in Mobility, pp. 123–133. Springer (2013)

[5] Akhegaonkar, S., Glaser, S., Nouveliere, L., Holzmann, F.: Smart and Green ACC series: A city and highway specific approach towards a safe and efficient eDAS, EVS-27, November 17-20, Barcelona, Spain (2013)
[6] Paganelli, G., Delprat, S., Guerra, T.-M., Rimaux, J., Santin, J.J.: Equivalent consumption minimization strategy for parallel hybrid powertrains. In: IEEE 55th Vehicular Technology Conference, VTC Spring 2002, vol. 4, pp. 2076–2081 (2002)
[7] http://en.wikipedia.org/wiki/PSA_HYbrid4 (retrieved January 21, 2014)
[8] "eFuture Project" official website, http://www.efuture-eu.org/ (retrieved January 21, 2014)
[9] Knauff, M., McLaughlin, J., Dafis, C., Niebur, D., Singh, P., Kwatny, H., Nwankpa, C.: Simulink model of a lithium-ion battery for the hybrid power system testbed. In: Proceedings of the ASNE Intelligent Ships Symposium (May 2007)
[10] Sciarretta, A., Back, M., Guzzella, L.: Optimal control of parallel hybrid electric vehicles. IEEE Transactions on Control Systems Technology 12(3), 352 (2004)
[11] Musardo, C., Rizzoni, G., Staccia, B.: A-ECMS: An Adaptive Algorithm for Hybrid Electric Vehicle Energy Management. In: 44th IEEE Conference on Decision and Control, 2005 and European Control Conference, CDC-ECC 2005, December 12-15, pp. 1816–1823 (2005)
[12] Friden, H., Sahlin, H.: Energy Management Strategies for Plug-in Hybrid Electric Vehicles. Master of Science Thesis, Report No. EX043/2012
[13] https://www2.unece.org/wiki/pages/viewpage.action?pageId=2523179 (retrieved January 21, 2014)
[14] Arco_Merano.mat, http://www.ecosm12.org/node/21 (retrieved January 21, 2014)

Layer-Based Multi-Sensor Fusion Architecture for Cooperative and Automated Driving Application Development

Maurice Kwakkernaat, Tjerk Bijlsma and Frank Ophelders

Abstract. Development of current ADAS is focused on single functionality and independent operation. Development of next generation cooperative and automated ADAS applications requires large amounts of information to be combined and interpreted. To operate efficiently and effectively, these applications should not operate in isolation, but share resources, information and functionalities. Furthermore, development, prototyping, real-life testing and evaluation of the applications in multiple-vehicle scenarios becomes more complex. In this paper iVSP, a scalable, multi-sensor fusion and processing architecture, is proposed for efficient development, prototyping, testing and evaluation of cooperative and automated driving applications in small to medium scale pilots.

Keywords: Advanced driver assistance systems, multi-sensor fusion, layer based architecture, vehicle automation, cooperative driving, sensing and perception, prototyping.

1 Introduction

Next generation intelligent vehicles will incorporate more complex functionalities, sensors and wireless communication to interact with drivers and the environment by performing automatic and cooperative driving tasks. This is driven by the continuing trends towards improved vehicle and traffic safety, supported by updated vehicle rating schemes [1], and the need for improved traffic flow and reduced CO2 emissions. To obtain a cost effective and reliable implementation these applications should not operate in isolation, but rather share processors, resources, information and functionalities. The development of these complex applications requires challenging solutions, e.g. multi-sensor information fusion, parallel real-time processing, security, functional safety and real-life evaluation.

M. Kwakkernaat(✉) · T. Bijlsma · F. Ophelders
TNO, Steenovenweg 1, 5708 HN Helmond, The Netherlands
e-mail: {maurice.kwakkernaat,tjerk.bijlsma,frank.ophelders}@tno.nl

J. Fischer-Wolfarth and G. Meyer (eds.), *Advanced Microsystems for Automotive Applications 2014*, Lecture Notes in Mobility,
DOI: 10.1007/978-3-319-08087-1_3, © Springer International Publishing Switzerland 2014

It is well recognized that the development of advanced automated driving and cooperative applications, or more generally advanced driver assistance systems (ADAS) applications, requires an architecture and platform with increased functionality, performance and interoperability compared to the current available sensor and processing platforms [2-6]. In [4] a server-client architecture was developed that consists of a local dynamic map (LDM) database server, a LDM or data fusion producer and client processes. The proposed architecture was developed to support rapid development and efficient use of resources, while decoupling algorithms and application functionality from low-level interfaces. In [5] a functional architecture was defined based on a perception layer, command layer, execution layer and HMI layer. This architecture allows effective and easy introduction of new functions compared to traditional architectures. In [6] an approach for the integration of ADAS functions was proposed based on an unified perception layer, a decision layer and an action layer. Although perception is integrated, parallel applications with independent reasoning are still allowed.

The work in this paper presents a scalable, multi-sensor fusion and processing architecture called Intelligent Vehicle Safety Platform (iVSP). The approach builds on previous work with the focus on prototyping, testing and evaluation of cooperative and automated driving applications in small to medium scale pilots. This includes the ability to efficiently develop and evaluate new applications in real-life. The objective is to define and implement a layer-based software architecture with appropriate interfaces and security that allow to collect, process and exchange information from multiple sources, such as external and internal sensors, wireless communication and digital-map data. The information is processed to generate real-time situational awareness (e.g. an LDM). This information can be made available to multiple parallel applications. The layer-based architecture decouples applications from low-level interfaces and enables developers to implement application logic relying on information and its confidence instead of low-level data from sensors. Based on the defined architecture an unique test and demonstration vehicle was instrumented for developing and evaluating cooperative and automated ADAS applications. The technology is integrated into a Toyota Prius and uses ITS-G5 communication, radar and camera sensors, information fusion, object tracking and control algorithms.

The paper is organized as follows. In Section 2, the iVSP system is described, including the general architecture and the layer-based structure. In Section 3, the results of two applications that are developed using the iVSP architecture are presented. Finally, Section 4 summarizes the work and provides future work.

2 System Description

2.1 Architecture

iVSP is an in-vehicle software architecture based on interconnected layers. The five layers are visualized in Figure 1 and consist of a sensor layer, communication layer, information layer, application layer and interface and authorization layer. The reason to adopt a layer based design is to adapt an object oriented approach in setting up ADAS. The layer functionalities can be reused when new ADAS applications are implemented in the application layer. The sensor and communication layer provide information to the information layer, which provides information to applications running in the application layer. All layers are interconnected using facilities from the interface and authorization layer.

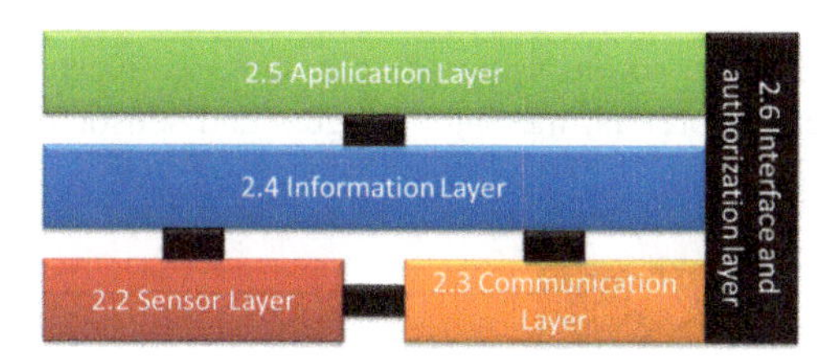

Fig. 1 iVSP layers

For the sensor layer a flexible approach has been chosen that enables the addition of sensors and interpreters for sensor data. The sensor layer performs preprocessing, filtering and low-level sensor fusion. A separate communication layer exchanges and interprets information coming from multimodal V2X wireless communication between other platforms or road-side units. Communicated sensor-based information will likely be provided to the sensor layer, while other types of information will be passed to the information layer. In the information layer the processed information can be stored in the form of *permanent static*, *transient static* and *transient dynamic* information. These different forms of information can be made available to authorized applications in the application layer at a low latency. The information layer also takes care of removing old non-relevant information using a garbage collector. In the application layer, different parallel applications can be executed.

In each layer specific functions can be defined, e.g. reading raw sensor data and interpreting the information should be handled by the sensor layer in order to pass it to the information layer. Some functions cannot be explicitly assigned to a specific layer, for example risk estimation can be assigned to either the information or application layer. In this specific example, the approach would be to run the function in the application layer when it is application specific, but when this function is requested by multiple applications it should run in the information layer to make it available for multiple applications.

Applications executed in the application layer are authorized to access certain information. To protect information, identification and security measures have been applied, such that not all applications have access to all collected information. Furthermore, for wireless communicated information certificates and basic encryption are used.

Some ADAS applications require a bounded latency for the retrieval of sensor values and predictable behavior. Therefore, iVSP has ongoing activities to apply real-time scheduling for critical parts of the layers.

In the next section the five iVSP layers and their functionalities will be detailed.

2.2 Sensor Layer

The sensor layer has been set up using sources, sinks between which information flows, via message interpreters, message mergers and sensor fusion processes, as depicted in Figure 2.

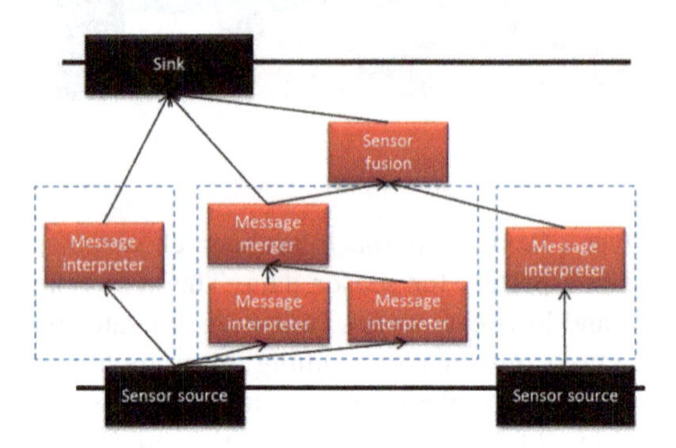

Fig. 2 Schematic drawing of the sensor layer

The sensor layer has been setup using a data flow structure. It is an acyclic event based system, where information flows from sources to the sink. This layer lends itself perfectly for parallel execution, for which the Intel thread library is used.

In the sensor layer, multiple sources are used, e.g. to receive CAN (Controller Area Network) frames and GPS NMEA string. The source receives the raw sensor information and forwards it to an available interpreter, e.g. for a CAN frame the source selects the interpreter corresponding to the ID of the CAN frame.

Interpreters extract sensor information from the raw sensor data and scale them according to the sensor specification. If sensor information is provided in multiple chunks, e.g. multiple CAN frames, a message merger is used. Additionally, it may be possible to significantly improve sensor information by combining it with information from other sensors, i.e. sensor fusion. An additional benefit of sensor fusion is that often the sensor values that serve as input for the fusion become obsolete, thus the amount of data in the system is reduced.

The sink receives sensor information from the interpreters, message mergers and sensor fusion processes. Currently a sink is used to forward information to the information layer. Additionally, sinks can be added to send information to different platforms or layers, e.g. to send information to external platforms.

The source, the sink, message interpreters, message merger and sensor fusion processes have been setup in a generic form. These can be easily adapted towards a new sensor or information. For example, the basic message interpreter class can easily be configured to add a new decoder for a specific CAN frame for which the values are scaled and decoded according to a dbc specification.

2.3 Communication Layer

The Communication layer has been set up using a communication source, message encoders and decoders and the observer pattern, as depicted in Figure 3.

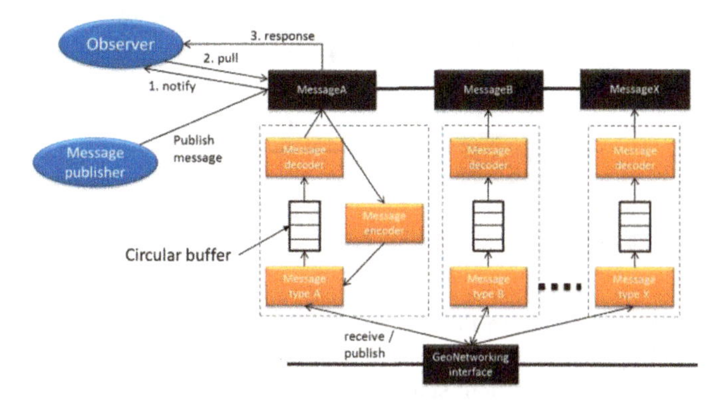

Fig. 3 Schematic drawing of the communication layer

In the current setup, communication is based on ITS-G5 with GeoNetworking, standards for vehicle to vehicle and road side communication. Multiple message types are supported and can be added to the communication layer. Processes in the information layer can register themselves as *observers* of a specific message type, meaning they will be notified if new messages of that type have arrived. Upon a notify the observer decides whether to pull the data, which forces the message decoder to parse the message and provide the observer with the decoded message. For efficiency reasons a message is only decoded when it is pulled by an observer, and the decoded message is kept in a cache to efficiently serve other observers.

Raw messages that are received through the GeoNetworking interface are put in the message buffer corresponding to the message type. In iVSP a circular buffer with a configurable size is used.

Processes in the information layer and application layer are allowed to publish messages through the communication layer. A message is provided at the interface of a specific message, which calls the message encoder to create the raw messages that are transmitted over the GeoNetworking interface.

The framework with message encoders and decoders and their source is setup in a generic form. New message types or network interfaces can easily be added. The architecture allows for addition of cellular communication.

2.4 Information Layer

The information layer stores and provides information for and to processes in all iVSP layers. Figure 4 depicts the information layer, which has been setup with the focus at three main activities: information storage, information provision and information management. These three main activities operate around the information storage that is centralized in iVSP.

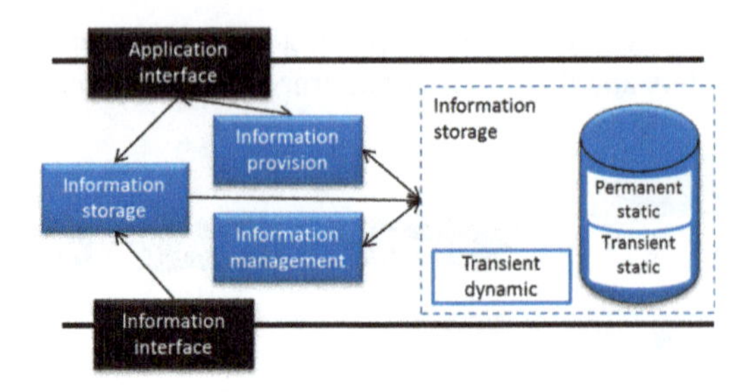

Fig. 4 Schematic drawing of the information layer

The information storage in iVSP uses a LDM, similar to the ETSI description in [7]. The iVSP LDM contains *permanent static*, *transient static* and *transient dynamic* information. Permanent static information includes information that rarely changes, e.g. map data and system configuration data. The transient static information has a life span from approximately a minute until hours or even days, e.g. signal phase and timing messages and road works warnings. Transient dynamic information has a short life-span, typically the most recent sensor layer information. Note that depending on the application of iVSP, the storage of information is determined. For example, the DPSI application in Section 3.1, requires sensor information to be analyzed for safety-critical events (SCEs), thus sensor information is stored as transient static, rather than transient dynamic.

The information storage activity stores the information received from the interfaces. The information provision retrieves requested information from the storage, for which it verifies the authorization with the interface and authorization layer. The information management process periodically reads the stored information, discards outdated information and in the future may be configurable to aggregate certain types of stored information and if necessary reallocate it. The information storage for transient dynamic information is implemented in memory, to minimize the latency and since volatile storage is sufficient. For both permanent static and transient static information a MySQL database is used. Considered alternatives include SQLite, Berkeley DB, Firebird and Redis. MySQL was chosen, due to its large user base and support. However, the choice may have to be reconsidered in case predictable behavior becomes desirable for the database.

2.5 Application Layer

In iVSP applications can be developed without knowledge of the specific sensor set used. Applications can obtain information from the information layer, which enables the application developer to use accurate information obtained by sensor fusion without having a detailed understanding of the applied algorithms.

Furthermore, applications execute independently of each other, and can be executed on different physical platforms (e.g. human machine interfaces and applications can be decoupled).

2.6 Interface and Authorization Layer

The interface and authorization layer provides means for the communication between layers and security, authentication and authorization mechanisms.

For the communication between layers and the message format specification, we use Apache Thrift, due to its flexibility. Data types and service interfaces are specified from which the Thrift framework can generate code, with which clients and servers can be setup that interact using remote procedure calls (RPCs). We use Thrift to generate code in C++, Java, Python and JavaScript. Thrift has been extended with support for the observer pattern. Alternatives considered besides Thrift were Lightweight Communications and Marshalling (LCM), Google Protocol buffers and the MPI framework, these were not selected, amongst others due to their maturity and absence of code generation for communication.

For the secure communication and authentication between layers, public key cryptography is applied. Layers and applications can register a certificate at the interface and security layer, which maintains a repository with certificates. Additionally, this layer manages an access control matrix, which lists the information for which a certificate is authorized. This matrix can be consulted by an information provider to determine if information can be shared. For the cryptography the Keyczar toolkit is used, which supports authentication and encryption and cab be used in multiple programming languages.

3 Results

3.1 Driver Behavior Monitoring Application

iVSP has been used to perform a study to monitor driver behavior [8], based upon basic vehicle sensors. iVSP was applied to monitor driver behavior, in order to determine the role of a driver during SCEs. The goal of this study was to quantify if a driver was distracted for SCEs and to determine if the event occurred due to a driver error or due to an external event.

For this study iVSP was applied in a passenger vehicle, where the vehicle sensors and a mono-camera where connected to iVSP. The sensor layer of iVSP was configured to receive and interpret the sensor information from the vehicle sensors and the mono-camera and forward it to the information layer. The information layer stored the information and provided it to the driver safety performance indication (DSPI) application running in the application layer. The DSPI application estimates the distraction of a driver, based upon the lane-keeping and car-following behavior and in case of a detected SCE sends the estimated distraction to the information layer for storage. The information layer has been configured to store relevant sensor information and the results of the DSPI application, which enabled replaying and analyzing scenarios with SCEs.

The DSPI application was evaluated in a simulator and applied in an instrumented vehicle. In a driving simulator experiment with 18 drivers the DSPI application was evaluated. The application correctly predicted distraction in 79% of the car-following detections and in 85% of the lane-keeping detections on highway sections. The application was demonstrated with an instrumented vehicle, where the DSPI application was executed by iVSP. In the instrumented vehicle, approximately 10 hours of driving behavior and corresponding SCEs were stored. On highways and rural roads, 84 SCEs were detected, from which 80% was due to distractions. Because the DSPI algorithms are based upon lane-keeping and car-following, the results of this study can be applied to reduce the number of false positive warnings for lane keeping assist and forward collision warning systems. iVSP has been applied to validate the DSPI algorithms in real-life at the road. For follow-up studies, it enables low-cost up-scaling to a large fleet for evaluation of DSPI algorithms.

3.2 Cooperative Automated Emergency Braking Application

Based on iVSP a cooperative AEB system (C-AEB) was developed with the focus on improved bicycle collision avoidance in crossing scenarios and beyond line-of-sight and beyond field-of-view [9]. The objective is to identify a bicycle object in the direct vicinity of the host vehicle and to warn the driver several seconds before a collision is expected (visual, audible, seatbelt pre-tensioner) and actuate the vehicle by automated braking to prevent a collision. To identify the bicycle the vehicle is equipped with two on-board sensors (radar, camera) and a "virtual" sensor (ITS-G5 wireless communication). The bicycle motion data is transmitted wirelessly after its state is estimated by information fusion and tracking in the so-called Host tracking function at the bicycle. The host vehicle is also equipped with a Host tracking function that is used to calculate the vehicle state, which is used by the Object tracking and Risk estimation. The radar, camera, communicated bicycle motion data and vehicle motion data are input to the Object tracking, which fuses these types of information in order to create reliable bicycle object information that holds the relative bicycle motion state with respect to the vehicle. The bicycle object and vehicle motion information, in turn, serve as input to the Risk estimation function that determines the risk level and time-to-collision, the parameters which are used in the actual C-AEB control algorithm.

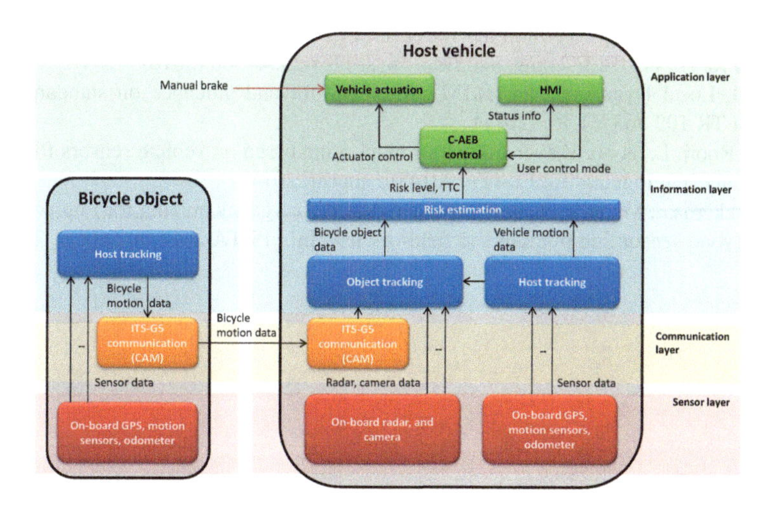

Fig. 5 Functional architecture of the C-AEB application based on iVSP

4 Summary and Outlook

This paper presents iVSP, a layer-based multi-sensor fusion architecture for rapid development of cooperative and automated driving applications. The approach collects, processes and exchanges information from different sources to generate real-time situational awareness, suitable for the execution of ADAS applications. iVSP has successfully been applied in two applications: driver behavior monitoring and cooperative automated emergency braking. An interesting addition considered for iVSP is the distribution of layers over multiple systems.

References

[1] Schram, R., Williams, A., van Ratingen, M.: Implementation of Autonomous Emergency Braking (AEB), the next step in Euro NCAP'S safety assessment. In: ESV, Seoul (2013)

[2] Valldorf, J., Gessner, W.: Advanced Microsystems for Automotive Applications. Springer, Heidelberg (2006)

[3] Meyer, G., Valldorf, J.: Advanced Microsystems for Automotive Applications. Springer, Heidelberg (2010)

[4] SAFESPOT consortium, Deliverable D1.3.4: Hardware and Software Platform Public Specification (2008)

[5] HAVEit consortium, Deliverable 12.1: Architecture (2009)

[6] Amditis, A., et al.: A Holistic Approach to the Integration of Safety Applications: The INSAFES Subproject Within the European Framework Programme 6 Integrating Project PReVENT. IEEE Trans. Int. Trans. Sys. 11(3), 554–566 (2010)
[7] ETSI, Local Dynamic Map (LDM); Rationale for and guidance on standardization. ETSI TR 102 863 V1.1.1 (2011)
[8] Van Rooij, L., et al.: Driver behaviour monitoring based on vehicle sensors for safety-critical event logging. In: FISITA 2014 (to appear, 2014)
[9] Kwakkernaat, M., et al.: Cooperative automated emergency braking for improved safety beyond sensor line-of-sight and field-of-view. In: FISITA 2014 (to appear, 2014)

Design of Real-Time Transition from Driving Assistance to Automation Function: Bayesian Artificial Intelligence Approach

Ata M. Khan

Abstract. Forecasts of automation in driving suggest that wide spread market penetration of fully autonomous vehicles will be decades away and that before such vehicles will gain acceptance by all stake holders, there will be a need for driving assistance in key driving tasks, supplemented by automated active safety capability. This paper advances a Bayesian Artificial Intelligence model for the design of real time transition from assisted driving to automated driving under conditions of high probability of a collision if no action is taken to avoid the collision. Systems can be designed to feature collision warnings as well as automated active safety capabilities. In addition to the high level architecture of the Bayesian transition model, example scenarios illustrate the function of the real-time transition model.

Keywords: Driving assistance, cognitive vehicle, safety, Bayesian, artificial intelligence, autonomous vehicle.

1 Introduction

Advances in information and communication technologies (ICT) in association with automotive technology developments have already resulted in automation features in road vehicles and this trend is expected to continue in the future owing to consumer demand, dropping costs of components, and improved reliability. Although autonomous vehicles have been developed to show-case the potential of technology and are being tested in controlled driving missions, the autonomous driving case is only a long term (i.e. 2025+) scenario [1,2]. According to a recent socio-technical study, self-driving vehicles offer potential benefits. But, there are also policy challenges for lawmakers [3].

A.M. Khan(✉)
Carleton University, 1125 Colonel By Drive, Ottawa, Ontario
K1S 5B6, Canada
e-mail: ata_khan@carleton.ca

J. Fischer-Wolfarth and G. Meyer (eds.), *Advanced Microsystems for Automotive Applications 2014*, Lecture Notes in Mobility,
DOI: 10.1007/978-3-319-08087-1_4,

High level automation has been under study for a number of years [4]. Research, development, and technology demonstrations have enabled major automotive manufacturers to initiate plans to add semi-autonomous features to their new vehicles in the near future [1,4]. While the present automation features are mainly in the form of information and driver warning technologies, future developments in the medium term (i.e. up to 2025) are expected to exhibit connected cognitive vehicle capability and encompass increasing degree of automation in the form of advanced driver assistance systems [2]. A number of original equipment manufacturers (OEMs) are accelerating their R&D in autonomous functionality and advanced technologies for implementation after 2025 [5].

Technological forecasters generally agree on the move toward autonomous vehicles in the long term and the availability of driving assistance in the short term. But there is no clear definition of how to integrate human and technology factors in order to make the human control and automation seamless and to overcome shared authority concerns in increasing automation in driving [6]. This paper advances a Bayesian Artificial Intelligence approach to the design of real-time transition from driving assistance to automation function.

2 Necessity of Cognitive Features in Driving Assistance System Design

Attributes of the cognitive vehicle have been of research interest [7,8,9]. Figure 1 presents a challenging set of cognitive features. These lead to the general specifications of a multifunctional advanced driver assistance system (ADAS) (Figure 2). In arriving at the suggested list of capabilities shown in Figures 1 and 2, current and recent developments in the use of artificial intelligence in vehicle design are taken into account.

Cognitive features

- Situational awareness (position, surroundings)
- Ability to gather data and send out data
- Ability to process data
- Ability to cooperate/collaborate
- Communication for active safety
- Informing driver about situations (warning, advice)
- Diagnostic capability
- In case of crash, capability to send and receive information
- Ability to provide non-distractive user interface for safe and efficient operation.
- Capability to perform user-requested infotainment tasks (not related to safety)

Fig. 1 Features of Cognitive Connected Vehicle

On the basis of industry analyses and research studies, cognitive vehicle features of high market potential are suggested here for inclusion in the design of the driver assistance system. Specifically, the following observations are noteworthy. First, for improving safety, driver workload and distraction should be reduced. It is necessary to provide a natural non-distracting driver-vehicle interface that reduces driver stress and workload. Second, selected active safety features are likely to gain favor with the users, provided that their designs are improved substantially. Third, advanced driver assistance (driver support) systems should take into account "driver intent". Fourth, automated non-distracted and non-aggressive driving feature, if activated by the driver for reasons of comfort, convenience and safety, will be a highly valuable design contribution. Fifth, ability to connect with other vehicles, infrastructure and devices is essential for future vehicles [10,11,12].

Multifunctional advanced driver assistance system (ADAS) design

- Open architecture & algorithms
- Natural interface of driver and automation features
- Interface with portable device
- Sensor network for data capture
- Integrated sensing for state estimation
- Communication systems
- Mechatronics/Microelectromechanical systems (MEMS)

Fig. 2 Driver Assistance Design Features

This paper defines the role of driving assistance in cognitive vehicle design that features a high level of automation. But, the driver remains in the loop unless the drive voluntarily relinquishes the driving task to automation or in circumstances that the driver does not take actions following the collision warning provided by the ADAS.

3 Bayesian Artificial Intelligence

The Bayesian methodology can serve as the foundation for system design and decision analysis in situations where uncertain "states of nature" are encountered and opportunities are available to refine knowledge of uncertain factors such as driving states, driver distraction and driver intention. Korb and Nicholson [13] have defined artificial intelligent as the "intelligence developed by humans, implemented as an artefact". More to the point, the Bayesian artificial intelligence integrates two facets of problem-solving in the design of cognitive features of the connected vehicle. The first one is the descriptive artificial intelligence which models a human action (e.g., non-distracted non-aggressive driving). The second is modelling our best understanding of what is "optimal" [11,13].

The Bayesian artificial intelligence was used in this research to produce algorithms. It is applied in three steps. First is the use of algorithms for Bayesian analysis of driving missions. Second requires the computation of expected gains/utilities. Finally, the third step involves the identification of the optimal course of action on the basis of maximum gain/utility.

Knowledge of evolving intelligent technology and human factors can be used in designing the cognitive vehicle's driver assistance system with the capability to provide seamless transition between human control and automation. The objective is to go beyond existing publicly reported driver assistance designs that simply rely on technology-assisted conflict projection method for alerting drivers about the driving state. The increasing literature on driver dissatisfaction with false alarms suggests that distracted driving and driver intent are not formally included in the model. The driver assistance system described in this paper has the capability to avoid rear as well as lateral crashes. The high level architecture of the system is shown in Figure 3.

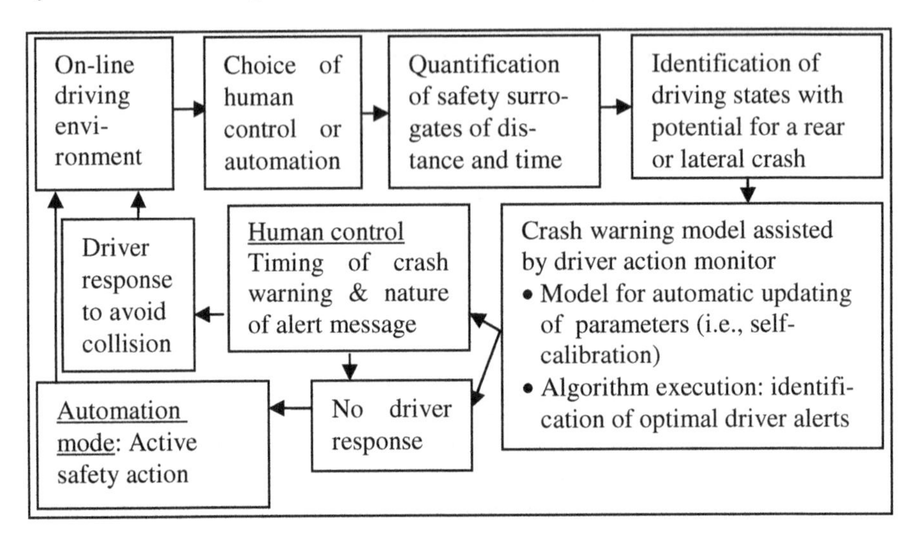

Fig. 3 High Level Architecture of Driver Assistance System's Advanced Safety Function

The models advanced in this paper for use in cognitive vehicle design go a step beyond existing work by developing more realistic probabilistic methods for quantifying driver distraction/driver intent and their use in issuing optimal driver alert message as well as the timing of providing this message.

4 Transition from Human Control to Automation

The seamless automated transition from human control to machine can be seen in the architecture presented in Figure 3. Figure 4 shows the transition model algorithm that can be built in the ADAS. The crash warning-active safety component is designed to work with the driver action monitor. For example, if a driver is in a distracted state with a high probability of crash, and does not show the intent to take corrective action, the driver monitoring part of the system will immediately modify the driver alertness parameter. The system can be designed to have the capability to automatically update key driving parameters, namely the probabilities of safety surrogates (i.e., distances or times), as well as the probability of driver's awareness of distance and time. This feature, in essence, is the adaptive self-calibration capability reported by Khan [11,12,14].

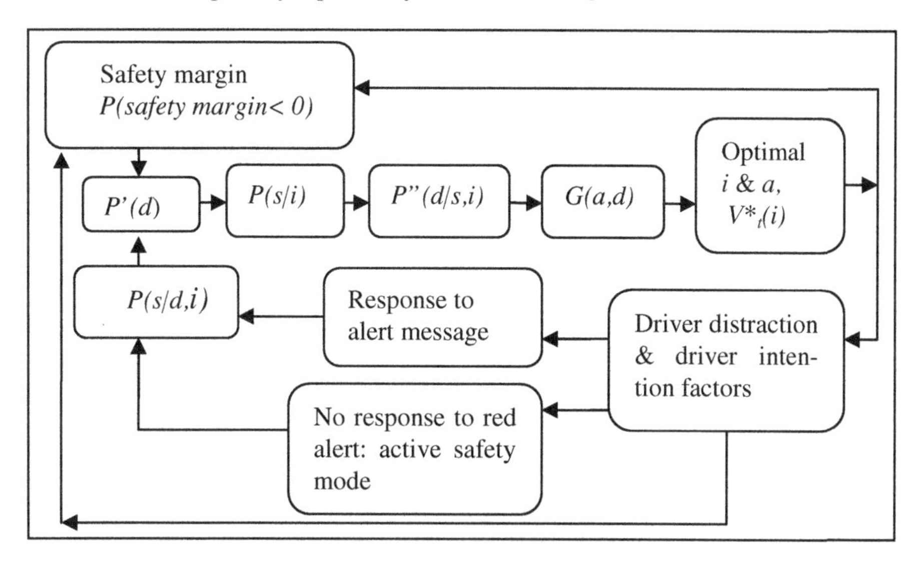

Fig. 4 Operation of the Collision Warning and Active Safety System

The transition model shown in Figure 4 can be used for the identification of optimal driver alerts in terms of the timing of rear or lateral crash warning (i.e., immediate or wait for an indication of driver intent) and the nature of alert message (e.g., no alert message, amber alert, red alert). If the driver does not respond, the system automatically initiates active safety action and informs the driver accordingly.

In the longitudinal direction, the distance between the subject vehicle and the leading vehicle and speeds of these vehicles are checked on a real time basis. If the longitudinal distance is less than or equal to 1.5 times critical distance, or if the lateral distance falls below the normal distance ND defined below, the algorithm presented in Figure 4 is launched. The components of the transition model are presented in Figure 5 and described briefly in this paper. For details, please see [10] and [11].

A Montecarlo simulation of safety margins enables the real-time estimation of vehicle location and distances within the driving environment. The prior probabilities $P'(d)$ and the conditional probabilities $P(s|d,i)$ are updated automatically as driving progresses. As noted in Figure 5, these reflect driver reliability. See Figure 6 for an illustration of the difference between a distracted driver and the

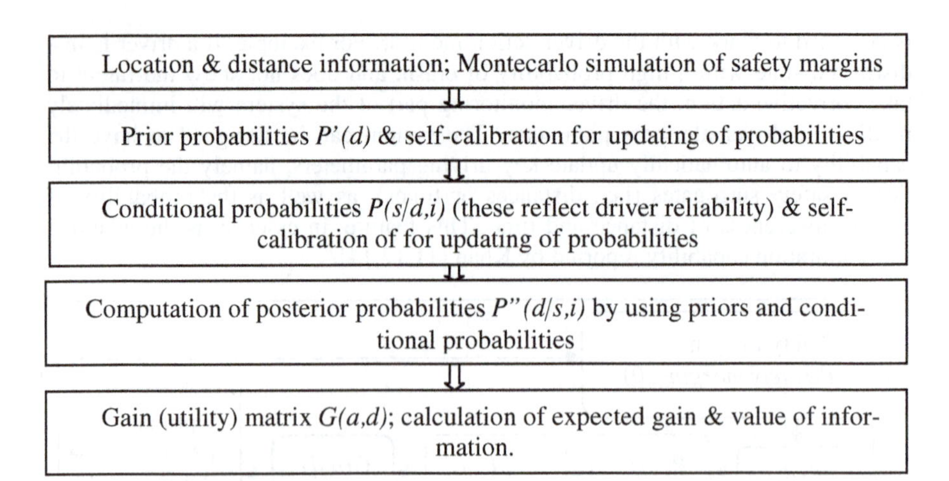

Fig. 5 Components of the Transition Algorithm [11]

automated system in terms of reliability. The other probabilities (i.e. marginal probability $P(s/i)$ and posterior probability $P''(d/s,i)$ are computed internally by the algorithm.

The Alternative actions of the crash warning system are: a_0 (no driver alert to be issued), a_a (amber alert -- calls for higher than normal deceleration in the longitudinal direction or steering action to increase transverse separation distance), a_r (red alert -- calls for serious emergency deceleration for avoiding a rear crash or steering action for preventing a lateral collision). In the case of rear crash, states of driving are defined by the distance between the following and the leading vehicles: d_c, $d_{1.25c}$, $d_{1.5c}$:

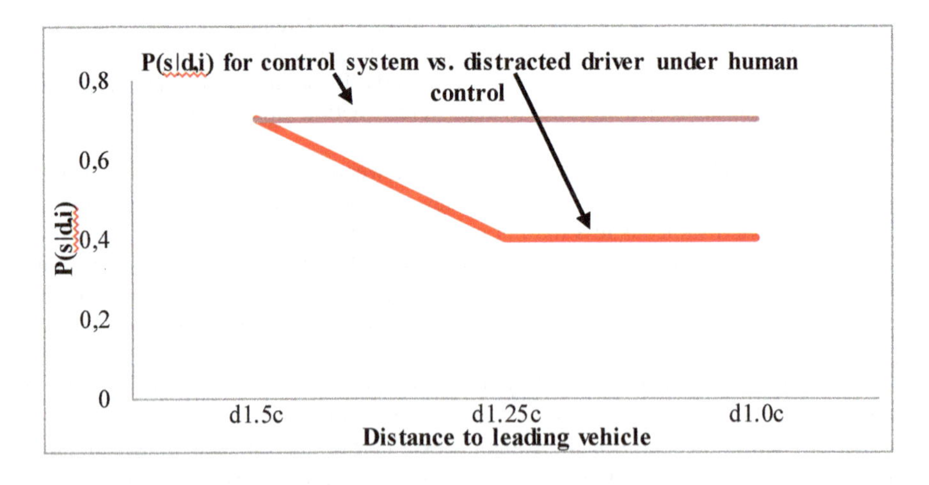

Fig. 6 Comparison of Distracted Driving and Automation [14]

d_c (critical distance that would lead to a rear crash -- but can be avoided if the required action is taken by the driver), $d_{1.25c}$ represents 1.25 times the critical distance, and $d_{1.5c}$ is 1.5 times the critical distance.

The number of states as well as the 1.0, 1.25, 1.5, etc. multipliers can be changed by the designer.

In the case of lateral driving states, the distance between envelopes of two vehicles in the transverse direction is tracked on the basis of the following thresholds: d_c (critical distance is set equal to 0.5(lane width L_w of the adjacent lane - width of the vehicle L_v in the adjacent lane). This definition implies that the subject vehicle is at the edge of the travel lane and should be prevented from encroaching on the adjacent lane. The normal distance *ND* is set equal to 0.5(L_w of the travel lane – L_v of the subject vehicle) + 0.5(L_w of the adjacent lane - L_v of the vehicle in the adjacent lane). If the two lanes are of equal width, the normal distance between the envelopes of the vehicles will be equal to $2.0d_c$. The intermediate distance is set equal to $1.5dc$ [10].

Options of driver information (alerts) are: i_0 (early applicable warning issued on the basis of initial information), i_w (the waiting mode so as to acquire and analyze additional safety surrogate information on the dynamics of vehicle-following or lane migration/lane change/merge and then issue the appropriate warning, if applicable) [10].

Possible surrogate measures (i.e., readings) on the states of longitudinal driving condition are: s_0 (no new reading, if i_0 is selected), s_c (corresponds to d_c), $s_{1.25c}$ (corresponds to $d_{1.25c}$), and $s_{1.5c}$ (corresponds to $d_{1.5c}$). The corresponding multipliers for the lateral collisions are s_c (corresponds to d_c), $s_{1.5c}$ (corresponds to $d_{1.5c}$), and s_{2c} (corresponds to d_{2c}) [10].

In order to apply the algorithm for identifying optimal *i&a* and the associated value of information $V^*_t(i)$, on a real time basis, the utility or gain $G(a,d)$ matrix has to be defined by the designer following consultations with policy experts. For details, please see [10,11].

The optimal courses of action for avoiding rear crashes and transition to automation are shown in Table 1. As noted above, model results show that at $d_{1.5c,}$ the driver is driving in a non-distracted manner and that the warning system is in the monitoring mode (i.e., additional information is being gathered and that no alert message is required). As the driver reaches $d_{1.25c}$ in a somewhat distracted condition, the system is gathering additional information on driver intention and if confirmed, amber alert will be given. In spite of the amber alert, this distracted driver moves even closer to the leading vehicle (at $d_{1.0c}$). The system has decided to issue a red alert without waiting for further information on driver action. If no action is taken, the active safety mode is launched [10,11]. Table 2 presents results for avoiding a lateral collision.

Table 1 Optimal Courses of Action for Avoiding Rear Crashes and Transition to Automation

Location of vehicle & prior probabilities	$d_{1.5c}$ $P'(d_{1.0c}) = 0.1$ $P'(d_{1.25c}) = 0.2$ $P'(d_{1.5c}) = 0.7$	$d_{1.25c}$ $P'(d_{1.0c}) = 0.15$ $P'(d_{1.25c}) = 0.7$ $P'(d_{1.5c}) = 0.15$	$d_{1.0c}$ $P'(d_{1.0c}) = 0.7$ $P'(d_{1.25c})= 0.2$ $P'(d_{1.5c})= 0.1$
Driver distraction	Not distracted	Somewhat distracted	Distracted
Optimal course of action	i_w & a_0	i_w & a_a	e_0 &a_r. *If no action is taken, launch automated braking*

Should the driver prefer to continue the use of the automation feature, the system can accelerate, decelerate, and make optimal lateral manoeuvres in order to travel within the traffic stream and avoid collisions. Table 3 presents optimal actions while under machine control. In this driving environment, it is assumed that some vehicles in the traffic stream may not be equipped with automation system and therefore unsafe lane changes and short headways be encountered.

Table 2 Optimal Courses of Action for Lateral Collision Cases

Separation distance & prior probabilities	s_{2c} $P'(s_{1c})$=0.1 $P'(s_{1.5c})$= 0.2 $P'(s_{2c})$= 0.7	$s_{1.5c}$ $P'(s_{1c})$ =0.15 $P'(s_{1.5c})$= 0.7 $P'(s_{2c})$= 0.15	s_c $P'(s_{1c})$ =0.7 $P'(s_{1.5c})$=0.2 $P'(s_{2c})$= 0.1
Driver Distraction	Not distracted	Somewhat distracted	Distracted
Optimal course of action	i_w & a_0	i_w & a_a	i_0 &a_r. *If no action is taken, launch automated steering action.*

Table 3 Driving Environment and Optimal Actions under Automation

Deceleration case		*Acceleration case*	
$d_{1.0c}$	i_0 & a_E	$d_{1.5c}$	i_W & a_0
$d_{1.25c}$	i_w & a_N	$d_{1.75c}$	i_W & a_N
$d_{1.5c}$	i_w & a_0	$d_{2.0}$	i_W & a_H

NOTES: a_0 is no action. a_E is emergency deceleration. a_N is normal speed change. a_H is high acceleration.

5 Conclusions

The contributions of this paper to knowledge on future advanced technology vehicle design are two-fold. First, research attention is drawn to the complexity of modeling the transition from human control to machine control under traffic states that involve high degrees of collision risk. Second, the Bayesian approach to meeting the requirements of the emergency transition is illustrated in the form of the developed model and its application in example scenarios.

This paper highlights the importance of a well-designed transition, the merits of the Bayesian approach in model development, and characterization of driving states that require real-time transition from driver-in-the loop to the automated function. The driver assistance example cases illustrate the integration of intelligent technology, Bayesian artificial intelligence, and abstracted human factors.

Acknowledgements. This paper is based on research that was sponsored by the Natural Sciences and Engineering Research Counsel of Canada (NSERC) and Ministry of Transportation of Ontario (MTO). The views expressed are those of the author.

References

[1] Smart Planet. When will cars be driverless? The Bulletin (January 15, 2014)
[2] Khan, A.M., Bacchus, A., Erwin, S.: Policy challenges of automation in driving. IATSS Research 35, 79–89 (2012)
[3] Anderson, J.M., Kalra, N., Stanley, K.D., Sorensen, P., Samaras, C., Oluwatola, O.A.: Autonomous Vehicle Technology, A Guide for Policy Makers. RAND Corporation (2014)
[4] HAVEit. Highly automated driving for intelligent transport. An EU Project
[5] Telemetics Update. Weekly Brief: No stopping the self-driving car (January 27, 2014)
[6] Cummings, M.L., Ryan, J.: Shared Authority Concerns in Automated Driving Applications, http://web.mit.edu/aeroastro/labs/halab/papers/cummingsryan_driverless2013_draft.pdf
[7] Heide, A., Henning, K.: The "cognitive car": A roadmap for research issues in the automotive sector. Annual Reviews in Control 30, 197–203 (2006)
[8] Stiller, C., Farber, G., Kammel, S.: Cooperative Cognitive Automobiles. In: Proceedings of the 2007 IEEE Intelligent Vehicles Symposium, Istanbul, Turkey, June 13-15, WeC1.1, pp. 215–220 (2007)
[9] Hoch, S., Schweigert, M., Althoff, F.: The BMW SURF Project: A contribution to the Research on Cognitive Vehicles. In: Proceedings of the 2007 IEEE Intelligent Vehicle Symposium, Istanbul, Turkey, June 13-15, ThB1.26, pp. 692–697 (2007)
[10] Khan, A.M., Bacchus, A., Erwin, S.: Surrogate safety measures as aid to driver assistance system design of the cognitive vehicle. IET Intelligent Transportation Systems (October 2013)
[11] Khan, A.M.: Cognitive Connected Vehicle Information System Design Requirement for Safety: Role of Bayesian Artificial Intelligence. Systemics, Cybernetics and Informatics 11(2), 54–59 (2013)

[12] Khan, A.M.: Bayesian-Monte Carlo Model for Collision Avoidance System Design of Cognitive Vehicle. International Journal of Intelligent Transportation Systems Research 11(1), 23–33 (2013)
[13] Korb, K.B., Nicholson, A.E.: Bayesian Artificial Intelligence. Chapman & Hall/CRC, UK (2004)
[14] Khan, A.M.: Design of Adaptive Longitudinal Control for Cognitive Connected Vehicle. In: Proceedings of the ITS World Congress, Orlando, USA (2011)

Enhancing Mobility Using Innovative Technologies and Highly Flexible Autonomous Vehicles

Timo Birnschein, Christian Oekermann, Mehmed Yüksel, Benjamin Girault, Roman Szczuka, David Grünwald, Sven Kroffke, Mohammed Ahmed, Yong-Ho Yoo and Frank Kirchner

Abstract. The combination of automobiles and typical robotic technologies such as high computational power, advanced exteroceptive sensors, and complex control algorithms can lead to a new kind of mobility. Features like extended maneuverability, autonomous driving systems, and new safety features become imaginable. To create the best combination of both fields, a new design philosophy is required. At DFKI, we have developed two generations of innovative concept cars (EO smart connecting car 1 and 2) with the intention to build the bridge between robotics and cars. The development of key parts like suspension, drivetrain, electrical steering, and braking system is more interconnected than in conventional vehicles. In this paper, we describe the development process of EO smart connecting car 2 - a highly innovative and fully functional robotic electric vehicle with double Ackermann steering, the ability to turn on the spot, go sideways, drive diagonally, change ride height, and shrink by adjusting the position of its

---

T. Birnschein(✉) · C. Oekermann · M. Yüksel · B. Girault · R. Szczuka ·
D. Grünwald · S. Kroffke · M. Ahmed · Y.-H. Yoo · F. Kirchner
German Research Center for Artificial Intelligence,
DFKI GmbH - Robotics Innovation Center,
Robert-Hooke-Straße 5, 28359, Bremen, Germany
e-mail: {timo.birnschein,christian.oekermann,mehmed.yueksel,genjamin.girault,
roman.szczuka,david.gruenwald,mohammed.ahmed,Yong-Ho.Yoo}@dfki.de,
sven.kroffke@uni-bremen.de

F. Kirchner
DFKI GmbH - Robotics Innovation Center,
Department of Mathematics and Computer Science, University of Bremen
Robert-Hooke-Straße 1, 28359, Bremen, Germany
e-mail: frank.kirchner@dfki.de
http://www.dfki.de/robotik

J. Fischer-Wolfarth and G. Meyer (eds.), *Advanced Microsystems for Automotive Applications 2014*, Lecture Notes in Mobility,
DOI: 10.1007/978-3-319-08087-1_5,

rear axle and tilting the cabin as well as docking at charging stations and extension modules. Detailed information of the utilization of Rapid Control Prototyping, and optimization strategies will be given as well as the design constraints for these technologies. Finally, the conclusion section will cover problems and challenges that had to be overcome and future work that will follow.

Keywords: Electric Vehicle, Autonomy, Artificial Intelligence, Drive by Wire, Hardware-in-the-loop, Software-in-the-loop, real-time control, path following, simulation, construction.

1 Introduction

Today's market of electric vehicles is still in a very early stage. For many car manufacturers it is a leap of faith when they actually start developing a completely new chassis around the electric power system - which is why most of them don't do it. Instead, most electric or partly electric (hybrid) vehicles are refits of their combustion engine driven models. This reduces financial risk and increases the chance of the models success - but not specifically the e-models success.

Truth is, cars don't evolve the way they could when designed completely around the fully flexible electric drive and battery system. Unlike e-cars that are based on a combustion driven model, a completely new design is not restricted by axels, gearbox position, differentials, main engine, and the gas tank. Instead, motors can be integrated into a wheel, motor controllers can be anywhere in the car and, for more stability, batteries help to get the center of gravity as close to the ground as possible. What can we do with this new category of flexibility in design? We can start from scratch and build a car that does not follow the rules everybody is used to for the last 125 years. Instead, we can build a car that is able to turn on the spot, go sideways and drive diagonally as well as change its ride height. We can even shrink the car because there is no need for lengthy solid parts running through the middle of the car because there is neither an exhaust system nor a center drive shaft.

Here at DFKI - Bremen we therefore decided to go the unknown and complex path of creating a completely new, fully electric vehicle with all of these features plus the ability to automatically dock at power outlets and autonomously drive in a virtual road train configuration: EO smart connecting car 1 and 2 (EOscc1 and 2).

This concept, realized in an actual fully functional and well-made concept car, is the basis for a new mobility concept of cars parking themselves at power outlets of regular parking spaces, automatically picking up passengers at their front door or drive in large platoons of road train driving cars on the highway or inside of big cities. This helps releasing the tension of heavy traffic today, reduces rush-hour traffic and enhances energy efficiency by having a smoother ride and the slipstream of other drivers without compromising safety.

The project started in October 2011 following the initial design of "EOscc1" (see [1] for a comprehensive state of art). The goal of optimizing this car and building a successor was a tricky and hard task considering the projects runtime of just 30 months.

Fig. 1 Folding EOscc2 for reduced parking space and higher maneuverability

After thoroughly reviewing the cars features as well as construction and its pros and cons we decided to remove some features and change major design decisions. The core decisions that were made in the process, are:

- Solid but light weight construction
- Wheel hub motors with integrated brakes
- Secure CAN-bus communication for all control units
- Simplified construction of major chassis and body parts
- Multiple sensors for autonomous parking and driving as well as comprehensive telemetry
- Docking interface for autonomous charging

After building the technology prototype EOscc1 in just twelve month with additional six month of software development and debugging the schedule for the development of EOscc2 was tight. Thanks to the deployed RCP system for hardware as well as software in the loop development, development of control algorithms, CAN-protocols and graphical user interfaces could be started immediately, even without the target hardware.

2 Designing a Robotic Car

Nowadays, most modern cars can be interfaced with a computer to control the car replacing the driver, but only for development or science purposes. A car for individual transportation usually is a front, rear, or all wheel drive vehicle with an electro-mechanical or hydraulic-mechanical steering and a direct link between the steering wheel and the tires.

EOscc2 is designed with the opposite approach. It is a car with all-wheel drive electric wheel hub motors, electrical servo motors directly controlling each of the four wheels individually and no mechanical link to any of the controlling systems, except for the hydraulic brake - for safety reasons. Every system is exclusively controlled by a computer and the driver may override the computer while driving manually. In normal manual drive mode, the car almost behaves like a regular car, except that four wheel steering is slowly activated at lower cornering speeds. It is also possible to turn on the spot, drive sideways into a parking space, and shrink the cars' length by 80 cm to safe even more parking space. In automatic drive mode it can move diagonally to change lanes, it can drive around corners sideways and park autonomously using sensors that are positioned all around the cars body. The car can also dock at charging stations. Basically, it is a wheeled mobile robot with two comfortable seats and a steering wheel as well as a gas and brake pedal.

When designing a car that has independent four wheel steering with a maximum steering angle of 92° as well as folding capabilities to shrink its size, designing an attractive body is a very challenging task. Big industrial manufacturers have many people working on different body parts, design, prototypes, clay models and final CAD construction using specialized software. The same is true for electronics, software, chassis construction, docking interface and wheel suspension. At DFKI the team working on the car contained between seven and 13 motivated scientists. A designer, two constructing engineers, two software engineers, an electronics engineer, a simulation engineer, and the project manager/software engineer as well as two students. A small team like this has to concentrate on all topics at the same time. In consequence, parallel and interconnected development was key and unavoidable as well as using as many readily available components as possible. Unfortunately, the latter didn't work out as planned as can be seen in the electronics chapter, later.

In the following chapters different design areas are described in more detail giving insights into the development as well as problems and challenges that the team dealt with during the last 27 month.

3 X by Wire Suspension

Technically, for an electric vehicle with wheel hub motors, the suspension system including the motors is the most important part.

Since EOscc2 uses four wheel drive as well as four wheel steering new motors with built-in brakes had to be designed to be used with standard rims. It was important that the motor mount and the kingpin lies inside of (or very close to) the motor and therefore close to the middle of the tires' contact area. This minimizes the scrub radius and therefore the necessary steering forces during turn, acceleration and braking. A 90° steering angle also requires the kingpin to be straight up vertical - otherwise the tire would tilt when going sideways. It also requires a solution for the steering mechanism so that the steering knuckle cannot lock-up with the steering gear in a force singularity when steering into the extreme positions. The steering knuckle is equipped with a complex innovative link mechanism (see fig. 2) to provide an almost linear force distribution throughout the unusually wide steering range of 122 degrees. Position and length of these links was determined using genetic algorithms within a precise computer simulation [2].

A double wishbone setup with two electric servo motors to control the steering angle and ride height was constructed as a module for each wheel in a symmetrical setup. Both servos utilize a worm-gear in combination with a spindle drive to create fast and powerful linear movement and are controlled using a CAN-bus. The lower mounting point of the shocks is connected to the wishbone; the upper one is connected to a movable support link and the spindle. This way, force from the cars' weight and bumps is only coaxially applied to spindle (see fig. 2 for details). These modules are then attached to a lightweight central axle housing in the front and at the back which protects electromechanical parts and also connects the axles to the chassis. Additionally, the center housing provides space and mounting points for the foldable docking interface.

Fig. 2 Shock absorber connected to ride height changing spindle and the new motor housing

At the same time, CAD design began on the construction of the BLDC motors. As a starting point, E-Max scooter wheel hub motors from Proud Eagle International ltd were disassembled and equipped with a new housing. The required internal mechanical brake as well as the constraint to be able to mount standard rims lead to a complete redesign of the motor housing (fig. 2), while the motor windings, hall sensors, and magnets remained untouched. The resulting motor is a compact direct drive 4 kW out-runner motor with internal brakes that fits into a 15" rim with its attachment flange inside of the rim.

4 Docking Interface

As advised from the requirements the car has to be able to dock at power outlets and drive with extension modules in a road train. Designing an invisible docking interface for an electric car is difficult because of its requirements:

- Foldable and invisible within the body
- Damping mechanism to absorb shocks
- Play free and stiff connection
- Strong enough to pull a car
- Provides communication and high current lines for charging.

Since the front tires of the car are further in front of the car than the front bumper, the docking interface has to be even longer when it is unfolded so that an extension module or another car can be docked while the car is making turns, similar to a regular drawbar of a trailer. Due to the very confined space, simply folding the docking mechanism was not enough. Hence, we designed an interface that can be elongated while unfolding. The designed mechanism also includes a shock absorber which is integrated into the elongation mechanism (fig 3).

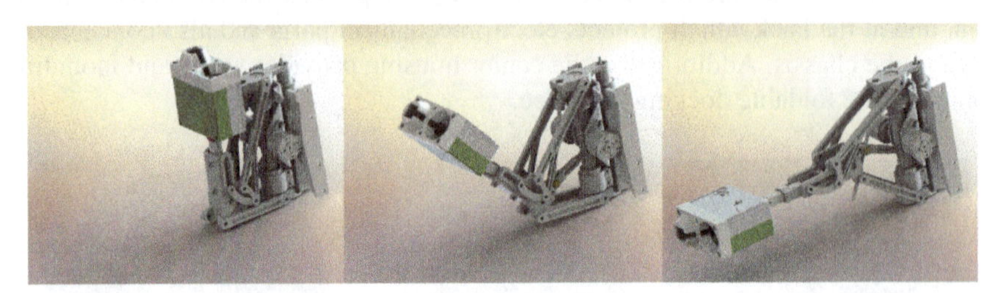

Fig. 3 EOscc2 Docking Interface in different folding positions

5 Morphological Chassis

Designing a chassis for a car with the ability to go sideways and change its size is a complex challenge - the same is true for the body design, as explained later. The cars' size was supposed to be roughly the size of EOscc1 which in turn is roughly the size of an older MC01 Smart ForTwo with a wheelbase of ~1800 mm when unfolded. When folded, EOscc2 shall reduce its size by 80 cm without compromising usability, except for the maximum allowed speed. A completely new tube frame chassis was designed with attachment points at the front for the axle. The rear was equipped with a rail system to provide the axle with a mechanical guide when sliding to the front while folding the car. During folding neither the front nor the rear axle changes its rotation relative to the road surface, therefore, except for the wheelbase, none of the steering parameters change (fig. 1). The rotation of the front axle is realized using a servo actuator.

The frame itself is built using welded 25CrMo4 steel tubes that were CNC cut and CNC bent with high accuracy. All joints were strengthened using welded triangles, more statically loaded parts are made out of thicker material with larger diameter. The chassis offers many mounting points for lights, body panels, electronics, seatbelts, seats, sensors, and so on.

6 Body Design

Similar to the chassis design, the body has to fulfill the same requirements and is therefore very restricted in terms of length as well as front and rear design caused by the 92° steering and the docking interface. From very early on, it was decided that the tires shall not be covered up by the body and have individual fenders instead. While in the car, the driver must have the best possible overview over the surroundings to be able to safely navigate through dense traffic which led to a massive windscreen going from the headlights right onto the roof. This also helps when driving in the folded pose. To enter and to leave the car without using additional space, the car is equipped with scissor style doors that rotate around the front wheel center which works perfectly in both poses. In order to maintain an ergonomic seat position at any cabin inclination angle the seats are mounted on a mechanism that varies the seats pitch and position. In both positions, all controls need to be reachable without a stretch, especially the steering wheel, both gas and brake pedal, as well as the drive mode selector and the manual hand brake. The seat mounting plate therefore utilizes an additional actuator with a spindle drive to move.

Since the rechargeable batteries are swappable a hatch was needed near the door sill under the car to be able to slide them in and out.

Figure 1 showcases the final design of the body shell which was fabricated using CNC milled molds for each panel and glass fiber. Finally, the finished part will be laminated to the chassis where necessary using specially shaped frames.

7 Electronics and Sensors

Within the project, a set of electronic control units has been developed to support and control all the necessary actuators, motors, sensors, and batteries in the car. Namely, these are:

- Vehicle Control Unit (VCU)
- Peripheral Control Unit PCU)
- Power Supply Unit (PSU)
- Analog Sensor Array (ASA)
- High Current Battery Management System (BMS)
- High Current Solid State Relay (SSR)
- High Current Ultra Cap Charger (UCC).

The VCU acts as the main controller. In the final stage, there will be three VCUs comparing results of calculations to rule out any possible error and improving safety for the drive by wire system. It is equipped with five CAN bus drivers, three of which are realized using additional CAN controllers, the two main busses are driven directly using two internal CAN controllers of the STM32F4 microcontroller. The VCU, as well as the PCU and PSU, also provides an extension interface to connect several analog and digital signals directly to the board.

To control all peripheral units such as the ASA boards, lights, and SSRs, the PCU is connected to the VCU via low speed CAN bus. It also communicates with the ASA board via RS485 and collects all its sensor information from front and rear axle. Two ASA boards are integrated into the axles housing themselves to be as close as possible to the analog sensors of the suspension system. Shielded wires needed to be used on these sensors because EM radiation from high current switching of the motor controllers became visible on the sensor feedback data. Each axle module contains two of each sensor: linear potentiometer for wishbone angle, rotatory sensor for lift actuator position, rotatory sensor for steering angle.

In the front part of the main chassis, covered by the dashboard, an electronics rack contains power electronics and control systems. Two motor controllers, two SSRs, a VCU, PCU, and PSU, a high power DC/DC converter and a high as well as a low current fuse box are bolted into place. The SSR is an additional safety device which works in series with a mechanical high current Schuetz-relay.

Each of the four battery packs contains four $LiFePo_4$ rechargeable batteries and a complex battery management system. The BMS is able to balance the packs under load with a maximum current of five amps. It was developed with the need for CAN-communication and active high current balancing with the additional option to switch itself off completely. This way, the battery cannot discharge itself by just balancing. In addition to the master-slave communication and safety features it also supports active balancing from cell to cell to further enhance efficiency.

When braking with an electric vehicle, like with a conventional vehicle, energy is wasted by heating up the brakes. Some electric cars support recuperative braking where a small part of the kinetic energy of the car is fed back into the batteries. However, this technique is limited in several ways, like the maximum C-rating for charging a battery and the maximum capacity of the system as well as the number of available recharge cycles. For example, our battery system can be charged with 100 A current. While normal braking from 75kp/h (~ 47mph) the car produces almost 790 A. This process can lead to the destruction of the batteries molecular structure and is dangerous for the battery module and its ability to store energy. A stronger system with the ability to store energy with a much higher current was needed, so we decided to design an UltraCapCharger. Four of these modules where integrated and tested. Connecting these in parallel results in 650 A brake current available for four seconds. This energy almost completely bypasses the vehicles batteries and is stored into high current capacitors instead and can be used for acceleration.

8 Software and Hardware in the Loop

A drive by wire car of this complexity needs a safe and free of bugs software (ISO 26262 - *Functional Safety*). In addition, debugging should be easy and comfortable. To accomplish this, we implemented a development environment that includes several hardware and software components based on RCP and backed with precise and detailed simulation tools. Every component of the control process could therefore be designed and implemented within this environment (for further details check M.Yüksel et al. [3]). We built a functional test platform for EOscc2 named SujeeCar. Basically, it is made of a chassis equipped with most of the final electronics and actuators to make it fully drivable, hence, all available hardware and software in the loop. Graphical user interfaces were designed and used for data logging and online display of telemetry data. To enhance productivity, RCP tools offer several ways to visualize available data like graphs, gauges, sliders and numbers, all of which can be added and removed during runtime making it easy to perform online experiments. Next in the process, all of the developed control models were implemented into the VCU. With its five CAN busses and a faster main loop frequency, especially with operating system out of the way, all required drive modes, kinematic calculations and user interface functions could be implemented efficiently and fast.

9 Simulation

Our simulation environment (e.g. Adams/View, Matlab Simulink, and Mars Simulation) is used to simulate, optimize and build extremely complex robots and cars as well as combinations of the two in a relatively short time. Within this project, simulation was used to evaluate the dynamic properties of the design, implement and test path and motion planning algorithms based on 3D-maps scanned from the test environment. Also, it was used to optimize the links for the steering mechanics. To accomplish the required wide steering range (-32° up to 92°), an innovative steering mechanism was constructed. This mechanism is designed to minimize the steering forces and to attain almost linear forces. Evolutionary algorithms where used to optimize the link length and mounting. Additionally, all forces for spring rates, steering, ride height, and folding could be estimated to define the necessary gears and servo motors (further details can be found in Ahmed et al. [2]).

10 Conclusion

In this paper, we presented a comprehensive overview of the development of EOscc2, the second generation EOscc, with the ability to turn on the spot, drive sideways, diagonally, normal, to shrink in size to save parking, space and a foldable docking interface.

Later, the car will be able to drive autonomously, search for parking spaces and will be able to dock to charging stations. During development, several problems had to be overcome. Most of these problems were solved with interdependent and modular design. While conventional car design starts with the design first and build the chassis afterwards – or vice versa – our modules were built simultaneously to reach the functionality and the looks on time. During specification phase it was advised to include as many of-the-shelf components as possible to minimize development effort and get faster results. Unfortunately, this turned out to be a naive approach because not even wheel hub motors with integrated brakes to be used with standard car rims were available on the market, not to mention force feedback steering wheels with a CAN interface or simple low force - high current connectors for a possible docking interface. For this reason, most of the EOscc2 components were designed and implemented from scratch.

At present, the final assembly of the car takes place and will be finished within the next three months. After several experiments with a test platform and the two fully functional axles we are confident that no major problems will occur and we can finalize the project on budget and on time.

Acknowledgement. The authors would like to thank the whole EO smart connecting car team for their incredible work, motivation, imagination and willpower. In addition, Janosch Machowinski, Pierre Willenbrock, Sujeef Shanmugalingam, and Robert Thiel supported and enriched the project to bring the vision to life. This project is funded by the German Federal Ministry of Transport, Building and Urban Development. The program coordination is carried out by the NOW GmbH National Organization Hydrogen and Fuel Cell Technology (Grant Nr. 03ME0400G).

References

[1] Birnschein, T., Kirchner, F., Girault, B., Yueksel, M., Machowinski, J.: An innovative, comprehensive concept for energy efficient electric mobility - eo smart connecting car. In: IEEE International Energy Conference and Exhibition (EnergyCon-2012), Towards User-Centric Smart Systems, Florence, Italy, pp. 1028–1033 (September 9, 2012)

[2] Ahmed, M., Oekermann, C., Kirchner, F.: Simulation environment for mechanical design optimization with evolutionary algorithms. In: World Symposium on Computer Applications & Research WSCAR 2014, International Conference on Artificial Intelligence (ICAI 2014), Sousse, Tunisia, January 18-20. IEEE (2014)

[3] Yüksel, M., Ahmed, M., Girault, G., Birnschein, T., Kirchner, F.: A Framework for Design, Test and Validation of Electric Car Modules. In: Fischer-Wolfarth, J., Meyer, G. (eds.) Advanced Microsystems for Automotive Applications 2014 – Smart Systems for Safe, Clean and Automated Vehicles (AMAA 2014). Lecture Notes in Mobility. Springer (2014)

Part II
Networked Vehicles, ITS and Road Safety

Assessing the Evolution of E/E Hardware Modules with Conceptual Function Architectures

Stefan Raue, Markus Conrath, Bernd Hedenetz and Wolfgang Rosenstiel

Abstract. Original Equipment Manufacturers (OEMs) use hardware module strategies to reduce hardware costs and development time. They prefer stable and cost efficient design concepts but the introduction of innovative functions requires agile methods. Therefore, it is still a challenge to transfer product features to the design of the E/E architecture in the context of a modularised E/E hardware. Hence, this paper presents a novel approach for the E/E architecture modelling using a conceptual function architecture layer as further abstraction to allow design decisions. The approach improves the transparent description of E/E systems in the E/E architecture development, the comprehension of the complex systems of modern premium cars, and the traceability from product features to E/E hardware.

Keywords: Conceptual function architecture, modularisation, evolution, E/E architecture, driver assistance, abstraction, E/E architecture modelling.

1 Introduction

The E/E architecture of a vehicle describes the interfaces, the structure and the interaction of the networked E/E components, the power distribution and the wiring harness. In the vehicle development, E/E architectures are currently heavily influenced by the use of modular construction kits and thus by the modularisation

S. Raue(✉) · M. Conrath · B. Hedenetz
Daimler AG, Architecture & Body Controller,
HPC G007-BB, 71059 Sindelfingen, Germany
e-mail: {stefan.raue,markus.m.conrath,bernd.hedenetz}@daimler.com

W. Rosenstiel
University of Tübingen, Computer and Engineering Department,
Auf der Morgenstelle 8, 72076 Tübingen, Germany
e-mail: rosenstiel@informatik.uni-tuebingen.de

J. Fischer-Wolfarth and G. Meyer (eds.), *Advanced Microsystems for Automotive Applications 2014*, Lecture Notes in Mobility,
DOI: 10.1007/978-3-319-08087-1_6,

of the hardware components. Modularisation is performed by OEMs (Original Equipment Manufacturers) to reuse the once developed hardware components in several vehicle model series. The objectives of this modularisation are cost reduction while reducing variants [2]. In general, a modularisation allows the design of modular product architectures [3] and is performed in industry in different ways [4]. In the automotive industry, modularisation is based on assembly aspects, i.e. those hardware components, which are installed together into the vehicle in the production, form a module. Regarding the E/E, an E/E module thus contains sensors, ECUs (Electronic Control Units) and actuators that are associated to an assembly unit, e.g. door.

In addition to the modularisation, E/E architectures are heavily influenced by distributed functions which e.g. result from innovations. Thus, both different function components and hardware components realise a functionality. The part of a function that is responsible for the realisation on a hardware component is referred to as a function contribution (FC). The FCs of various functions on an ECU are implemented by AUTOSAR software components (SWCs) [5].

In the concept phase of the E/E architecture development, new product features, like innovative functions, are transferred and integrated into the E/E architecture by a refinement procedure in two phases: In the first phase, a new product feature is assessed from the E/E architecture point of view: Technical aspects like the distribution of the function, the required bandwidth and the impact on the vehicle are examined. Also, organisational aspects like the distribution of the involved persons in the organisation and their responsibilities are regarded. Finally, estimations of efforts are supplied and management decisions are driven. Subsequently in the second phase, the involved specialist departments are supported in the change or elaboration of interfaces and the functionality. Additionally, resulting changes or new requirements can be introduced into the networking concepts of the E/E architecture development and the hardware module strategy processes.

Hardware modularisation on the one hand and the introduction of innovative functions on the other hand drive opposing requirements for the E/E architecture development. The hardware module strategy prefers stable and cost efficient design concepts and the introduction of innovative functions agile methods. Therefore, it is still a challenge to transfer product features like new innovative functions to the design of the E/E architecture in the context of a modularised E/E hardware. At the same time, a well-known trend in the automotive industry is the further increase of the function range and therefore E/E complexity [6].

These aspects motivate to develop an approach coping with these issues and allowing a transparent and traceable description of the functions of a module. This approach should help in a module evolution to assess the impact of function changes to the affected module variant and vice versa. Additionally, this approach should provide a basis for design decisions which have to be taken in the concept phase of the E/E architecture development.

2 Previous Work

For the concept phase of an E/E architecture family, Daimler AG decided very early to perform the development of the E/E architecture using a model-based approach [6]. In subsequent years, the methods were improved steadily. From the perspective of today, a model-based design is thus a state-of-the-art and is also applied by most competitors [7]. In addition to the function network, on the levels hardware networking and topology is modelled, each representing only one aspect of the E/E architecture (Figure 1a). The hardware networking level includes the wiring harness and the power distribution. The connection of the different levels is established by a mapping relating the model objects of the different levels [6].

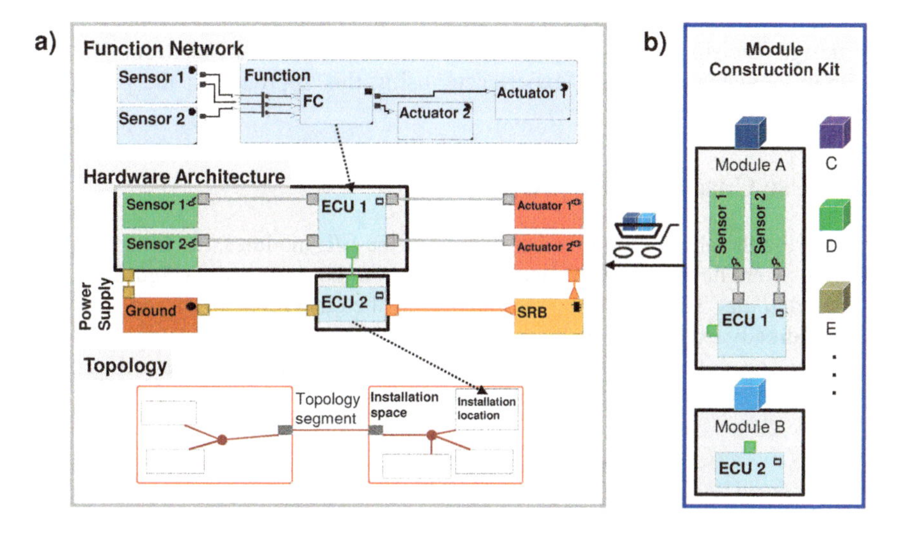

Fig. 1 Model based E/E architecture design. a) Modelling in levels. b) Module construction kit according to [8].

In the recent years, the methodology for the E/E architecture design has been significantly enhanced to allow the modelling of hardware modules of the modular strategy [8] by integrating modules in the model-based E/E architecture design [9] using a product line approach [10]. These hardware modules, modelled in the modular construction kit, can be evolved there and reused in different E/E architecture models (Figures 1b). Thus, the modelling effort was reduced significantly, the quality of the model increased and user support for the modelling of complex E/E modules achieved.

The hardware modules may have variants that arise e.g. out of the scaling. Thus, one ECU and one sensor, for example, could form a basis variant of a module while a premium variant of this module could contain the ECU with 3 sensors. This reuse of hardware components in the E/E architecture design is covered by the already existing modelling approach [9].

In the field of software development for distributed automotive systems, abstraction was proposed for the modelling of the development of software for ECUs [11]. In this approach, the structural information of automotive functions is specified in a function net (also denoted as logical architecture). Several functions can be realised by one SWC and no design decisions are taken on the function net.

Unlike the conventional software development, the E/E architecture modelling allows to capture the widespread heterogeneity of an E/E architecture. Previous approaches allow the modelling of the hardware layers of hardware modules and of a whole function net as abstraction of the software architecture. Nonetheless, these approaches neglect the necessity for an abstraction of the large amount of functions to support evolving hardware modules.

Thus, the contributions of this paper are threefold. First, use cases are provided which allow a description of the hardware module evolution. Second, a novel approach of a conceptual function architecture is presented which provides a basis for design decisions that can be used in the E/E architecture concept phase to integrate new features. Last, experiences using this approach in the productive architecture work are shared.

3 Use Cases and Requirements

To describe the above mentioned evolution of the E/E modules, the following four use cases were derived from an analysis of the changes in the past (Figure 2):

New and die: This trivial use case describes the emergence or disappearance of FCs on already existing modules

Move: The shift of FCs to other, already existing modules is addressed by this use case, which occurs especially in the context of optimisations.

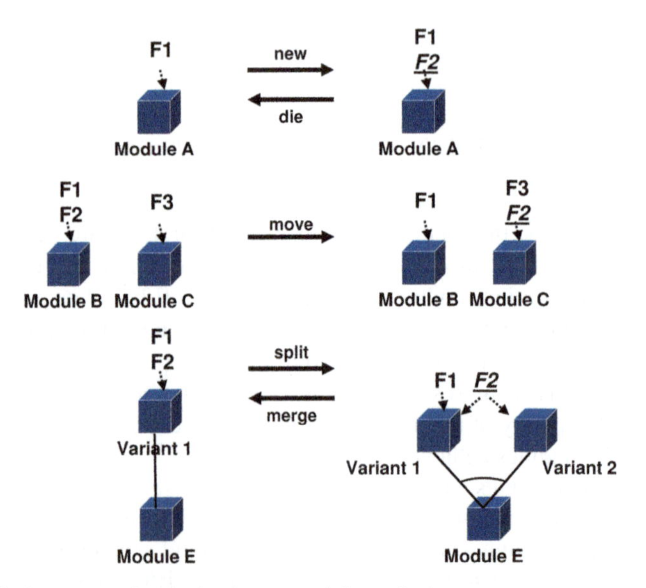

Fig. 2 Practical use cases for the hardware module evolution

Split: Creating new module variants, often only a part of the function, already implemented on an existing module variant, is adopted in the new variant. For example, a new basis variant for the compact segment containing a smaller function range is generated.

Merge: On the contrary, functions can meet on a module where they have not yet been implemented together when, for instance, module variants are reduced.

An approach must allow an integration of new features and thus must provide a basis for design decisions. A difficulty in deriving an approach coping with these aspects arises from the parallel development of the E/E architecture on the one side and the hardware and innovation development on the other side. An approach thus must allow the evolution use cases and be robust against flexible changes of the function net. Since a module contains several variants (Figure 3a), an 150% representation should be allowed.

A further difficulty emerges from the function net which describes the signal flow of a function from the signal generation of a sensor to the commands operating an actuator. Usually, FCs of several functions are present on an ECU. To optimise the implementation by AUTOSAR software components, communalities like a central coordinator might be implemented in a common SWC. This leads towards a m:n relationship between FCs and SWCs. As result, the FCs appear to be independent although they have a hidden dependency through this common SWC (Figure 3b). Moreover, in case many FCs are present on one ECU, the complexity makes an understanding difficult. Thus, an approach must allow a transparent description of the hidden communalities and easy understanding of the complexity of a large number of functions.

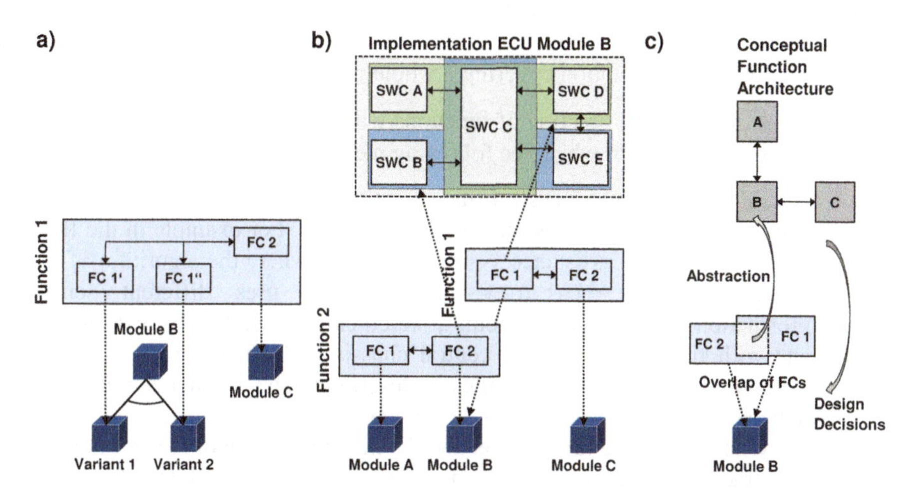

Fig. 3 a) Module Variants, b) Communalities of FCs in the implementation: The function contribution FC2 of function 2 is implemented by the SWCs B, C, and E and the function contribution FC1 of function 1 is implemented by the SWCs A, C, and D. Thus, both function 1 and 2 share SWC C as communality in their implementation. c) Abstraction to conceptual function architecture.

From E/E architecture point of view, an E/E module evolution requires an assessment of the affected interfaces and other E/E architecture criteria which are analysed by existing metrics. Using these, the expected bus load impact can be determined:

- Affected interfaces: It must be determined which interfaces are affected or required by the evolution. Due to a new function distribution, it is possible that ECU-internal interfaces are now ECU-external interfaces and sent via the bus system. By moving functions or FCs, interfaces might change on the bus system or are required there also.
- Bus load impact: After recognising the affected interfaces, it must be determined whether the affected bus systems have sufficient reserves to hold the new or changed interfaces.

Partly, these analyses must be possible on the basis of a rough concept and without a formalised E/E description since either an implementation is unavailable or will be developed in later stages.

4 Proposed Approach: Conceptual Function Architecture

To perform this assessment, our approach uses an abstraction of the function net. This is the basis and the enabler for an improved functional understanding and thus E/E module evolution. The abstraction describes

- the distributed functions implemented by E/E modules,
- their complex relationships, and
- their hidden dependencies
- in a 150% representation covering a whole module with all its variants.

In order to derive the abstraction of functions with an existing implementation, a reengineering is required for which the following method is proposed (Figure 3c):

- Several detailed interfaces of the function net that are considered as coupled are summarised as one abstract interface. For example in the field of advanced driver assistance systems (ADAS), where the algorithm of the costumer feature `distance-regulation` uses different sensor information from the bus system, a stereo-camera sends information of a lane with five interfaces to a central chassis ECU. These five interfaces containing the numerous attributes to describe a lane are then summarised in the abstract interface `stereo-camera_lane`.
- Every function of the function net and the implementation of the function net is abstracted by describing the elements of the function which transforms the abstract inputs to an output. This abstraction must provide an intuitive and clear understanding of the functionality. Implementation relevant information is ignored unless it is E/E architecture relevant. For example, if the implementation of several functions has a central coordinator SWC coordinating these functions, then a corresponding element is created in the abstraction.

In case a novel functionality should be described by the above mentioned refinement, the following method is proposed:

- The modelling of the first phase is done on the abstract level. For the important elements of the function, an abstract element is created. The abstract elements are connected over abstract interfaces describing the information flow. A size for each abstract interface is assigned. For new abstract interfaces, the size is approximated, while the size of the existing abstract interfaces is received from the detailed interfaces. After a mapping of the abstract elements, this size can be used for a model based evaluation of the expected busload. Stepwise optimisation in dialogue with the function and E/E component experts will thus provide a conceptual design which is then further concretised.
- In the second phase, the function and their interfaces are elaborated. These detailed information are modelled in the function net. Some findings might require adapting the abstraction.

As consequence of this approach, the abstract and detailed interfaces have to be related in the modelling (Figure 4a). Here, the relation between abstract interfaces and detailed interfaces should be 1:n.

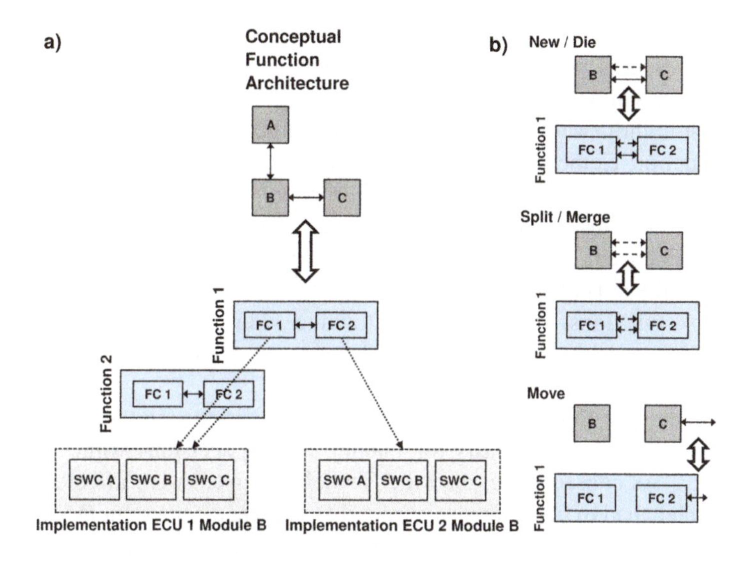

Fig. 4. a) Relation of abstract and detailed interfaces. b) Interface evolution use cases.

Evolution use cases also apply to the abstract interfaces and their relation to the detailed interfaces (Figure 4b). Also here, new abstract interfaces can be created (new), deleted (die), shifted to another abstract element (move), separated (split) or united (merge).

5 Experiences

The proposed approach was evaluated in a chassis E/E architecture development project at the Daimler AG. The abstraction of the detailed interfaces, the reengineering and the abstraction of the function net was done for the ADAS module of the current Mercedes-Benz S-class to support the evolution towards the next generation Mercedes-Benz S-class.

Using the abstraction, the overlap of the function net could successfully be resolved. Furthermore, a generation spanning description of the functionality was received which is robust against changes of the function net. An update of the function net layer with its detailed interfaces did not have an impact on the abstract layer. However, new or changed detailed interfaces had to be assigned to the corresponding abstract interfaces. To automatize the modelling of the connection between the abstract and detailed interfaces, and of the interface evolution use cases, an extension of the modelling tool was specified and is currently in implementation.

The derived abstract description was furthermore used as basis for the first discussion with function experts. At the beginning of the discussions, it was necessary to motivate the idea of an abstract 150%-representation of the functions as an essential basis for the integration since the function developers usually just regard their own function. After creating a common mind-set and wording, the abstract representation was an easy to understand method to discuss complex functionalities. On this basis, changes resulting from new features could be discussed straightforward. However, a small number of interfaces between the abstract function elements were left detailed to test this description in such a discussion. These detailed interfaces disturbed reading and comprehension in comparison to the abstract interfaces. Moreover, the order of the abstract function elements influenced the comprehension and discussion. A flow from sensor elements (on the left) to actuator elements (on the right) was crucial to understand the complex functions. Parts of the abstraction with an inconsistent functional flow in this sense hindered the comprehension or were even overlooked and missed by the experts.

In contrast to the function net where no design decisions are taken, the abstract representation was used as basis for making design decisions while integrating new features. As consequence, the other modelling layers were affected by these decisions. They had to be adapted accordingly to create a consistent model.

It is further possible to enrich the conceptual function architecture with cost details. Therefore, it is also suited for product management assessment, discussions and decision making.

6 Conclusion

This paper presented a novel approach to transfer new features to the design of an E/E architecture with a modularized hardware utilizing a conceptual function architecture layer as further abstraction allowing design decisions. The new conceptual function architecture is the most abstract view on a hardware module and allows a first assessment in the first refinement phase defining functional requirements and impacting the other modelling layers. The evaluation in productive module evolution at Daimler AG prove the conceptual function architecture to be robust against updates of the function architecture, supportive in understanding the complexity of a large number of functions, and utilisable for product management decisions.

References

[1] Ringler, T., Simons, M., Beck, R.: Reifegradsteigerung durch methodischen Architekturentwurf mit dem E/E-Konzeptwerkzeug, 13. Internationaler Kongress Elektronik im Kraftfahrzeug, Baden-Baden, VDI-Berichte Nr. 2000. VDI-Verlag GmbH, Düsseldorf (2007)
[2] Greiner, E., Steuer, C., Schaser, J.: Variabilität durch Standardisierung. ATZextra 17(4), 24–27 (2012)
[3] Ulrich, K., Seering, W.: Function sharing in mechanical design. Artificial Intelligence in Engineering Design 2, 185–213 (1992)
[4] Florent, C.: Strategic perspectives on modularity. in: The DRUID Tenth Anniversary Summer Conference on Dynamics of Industry and Innovation: Organizations, Networks and Systems, pp. 27–29 (2005)
[5] AUTOSAR. AUTomotive Open Systems ARchitecture, http://www.autosar.org
[6] Belschner, R., Frees, J., Mroßko, M.: Gesamtheitlicher Entwicklungsansatz für Entwurf, Dokumentation und Bewertung von E/E-Architekturen, 12. Internationaler Kongress Elektronik im Kraftfahrzeug, Baden-Baden, VDI-Berichte Nr. 1907. VDI-Verlag GmbH, Düsseldorf (2005)
[7] 7. Vector Congress, Stuttgart (November 2013)
[8] Jaensch, M., Prehl, P., Schwefer, G., Müller-Glaser, K., Conrath, M.: Model-Based Design of E/E Architectures 2.0. ATZelektronik Worldwide eMagazine 6(4), 16–21 (2011)
[9] Jaensch, M., Conrath, M., Hedenetz, B., Müller-Glaser, K.: Integration of Electrical/ Electronic-Relevant Modules in the Model-Based Design of Electrical/Electronic-Architectures, 1. Energy Efficient Vehicle Conference (EEVC), pp. 84–92 (2011)
[10] Jaensch, M., Hedenetz, B., Conrath, M., Müller-Glaser, K.: Transfer von Prozessen des Software-Produktlinien Engineering in die Elektrik/Elektronik-Architekturen twicklung von Fahrzeugen, 8. Workshop Automotive Software Engineering, pp. 497–502 (2010)
[11] von der Beeck, M.: Development of logical and technical architectures for automotive systems. Software & Systems Modeling 6(2), 205–219 (2007)

Increased Consumption in Oversaturated City Traffic Based on Empirical Vehicle Data

Peter Hemmerle, Micha Koller, Hubert Rehborn, Gerhard Hermanns, Boris S. Kerner and Michael Schreckenberg

Abstract. Congestion of urban roads causes extra travel time as well as additional fuel consumption. We present an approach to determine this additional fuel consumption on the basis of empirical vehicle data. We study probe vehicle data provided by TomTom to find the various traffic patterns of urban congestion. We use simulations of these urban traffic patterns based on a stochastic Kerner-Klenov model as input for an empirical fuel consumption matrix compiled from empirical CAN bus signals from vehicles. Our results confirm that in certain congested city traffic patterns vehicles consume more than twice as much fuel as in free city traffic.

Keywords: Urban traffic management, urban congested traffic, traffic signal, empirical fuel consumption, consumption matrix, eco-routing, oversaturated traffic, probe vehicle data, navigation, traffic simulation.

1 Introduction

The commonness of urban congestion and the growing importance of fuel saving create new needs for eco-routing and green traffic management. On the other hand, probe vehicle data from personal navigation devices contribute to a thorough understanding of urban traffic as they allow for the identification of spatiotemporal

P. Hemmerle(✉) · H. Rehborn
Daimler AG, RD/RTF, HPC: 059-X832, 71063 Sindelfingen, Germany
e-mail: {peter.hemmerle,hubert.rehborn}@daimler.com

M. Koller
IT-Designers GmbH, Entennest 2, 73730 Esslingen, Germany
e-mail: micha.koller@it-designers.de

G. Hermanns · B.S. Kerner · M. Schreckenberg
Universität Duisburg Essen, Physik von Transport und Verkehr,
Lotharstr. 1, 47057 Duisburg
e-mail: {gerhard.hermanns,boris.kerner,michael.schreckenberg}@uni-due.de

J. Fischer-Wolfarth and G. Meyer (eds.), *Advanced Microsystems for Automotive Applications 2014*, Lecture Notes in Mobility,
DOI: 10.1007/978-3-319-08087-1_7,

traffic patterns. In this article, we present an overview of the various urban traffic patterns and an approach to determine the additional fuel consumption associated with these traffic patterns. It is important to obtain an algorithm for calculating the fuel consumption due to traffic congestion as an input for traffic management and individual navigation. Our results will be used in an energy efficient routing strategy which is being developed in the project "UR:BAN – Urbaner Raum: Benutzergerechte Assistenzsysteme und Netzmanagement" (Urban Space: User Oriented Assistance Systems and Network Management) funded by the German Federal Ministry for Economic Affairs and Energy [1].

The article is organized as follows. In section 2, we propose a classification method for urban traffic patterns using anonymized probe vehicle data provided by TomTom. Then, in section 3, we explain our approach to compiling a fuel consumption matrix using an archive of recorded CAN (Controller Area Network) bus signals of probe vehicles with combustion engines moving on German highways and in cities. Finally, in section 4, we present some first results for the fuel consumption associated with the various urban traffic patterns.

2 Classification of Vehicle Trajectories

We study anonymized map-matched GPS (Global Positioning System) probe vehicle data provided by the company TomTom. There are both online data (sent directly by the navigation devices) and offline data (the user later connects the TomTom device to a computer), differing in their temporal resolution; offline data is available in steps of one second, and online data in steps of five seconds. However, we do not differentiate between the different types of data in our analysis. The data for each time step consist of timestamp and position (in meters) relative to the beginning of the road section. We use these data to reconstruct the vehicle trajectories and to derive the speed profiles along these trajectories.

The road section in question is a 630m long section of Völklinger Straße in the city of Düsseldorf with a traffic signal at the end and a speed limit of 60 km/h (Fig. 1). We regard this section as suitable because congested traffic is observed there on many days in the data of a stationary video detector [2]. Additionally, no junctions disturb the traffic flow between its beginning and end.

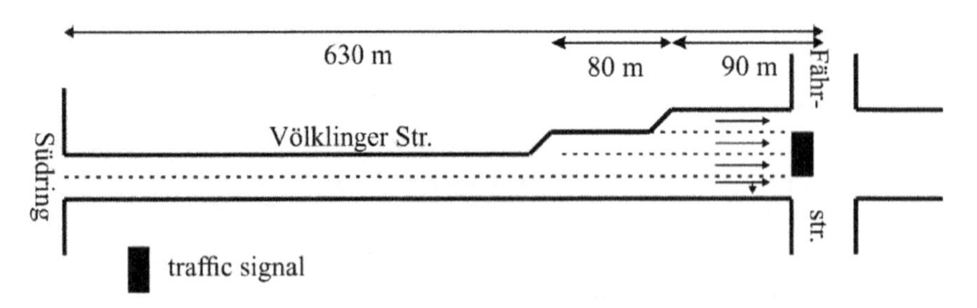

Fig. 1 Road section of Völklinger Straße in Düsseldorf. The speed limit is 60 km/h.

The data basis of our analysis is probe vehicle data from selected days in November and December of 2011 and in February, March and April of 2013. On all of these 19 days traffic breakdowns have been observed by means of a video detector measuring traffic flow and aggregated speed at the beginning of the section. All in all, data from some thousands of vehicles traced on the road section in question form our data basis.

We set up a classification scheme for vehicle trajectories with the average speed of a vehicle and its number of stops on the road section as the criteria for belonging to a class. The relevance of the average vehicle speed for navigation and routing is obvious due to the drivers' common wish to know the fastest route. The number of stops of a vehicle, on the other hand, plays a part in the fuel consumption of a vehicle. The more often a vehicle stops, the more often it accelerates from speed zero, consuming additional energy in each of these acceleration processes. In our scheme, there are a total of 15 classes, as listed in Table 1.

Table 1 Overview of the 15 vehicle trajectory classes

	Average Speed [km/h]		*Number of Stops*	
Class	*Minimum*	*Maximum*	*Minimum*	*Maximum*
1	30	75	0	0
2	30	75	1	1
3	15	30	0	1
4	15	30	2	4
5	10	15	0	1
6	10	15	2	4
7	10	15	5	6
8	8	10	5	6
9	8	10	7	10
10	6	8	5	6
11	6	8	7	10
12	6	8	11	13
13	3.5	6	7	10
14	3.5	6	11	13
15	0	3.5	13	30

Free flow on an urban road with a traffic signal means that the vehicles move with a speed that corresponds approximately to the local speed limit unless they have to stop at the signal. Whenever a queue is formed upstream of the traffic signal during a red phase, all cars in that queue can pass the signal during the next green phase; the traffic signal is said to be *undersaturated*. Therefore, in urban free flow a car stops zero times or once. Classes 1 and 2 are associated with urban free flow, where we distinguish between drives without a stop (class 1, Fig. 2) and drives with one stop at the traffic light (class 2, Fig. 3). In the case of green wave control we can assume that vehicles can pass road sections without any stops.

In such cases vehicles have almost constant speeds and no stops in undersaturated traffic. From Fig. 2 to Fig. 8 the position of the traffic signal at 630m is marked by a horizontal line in the left-sided figures; the vertical line in the right-sided figures indicates the point in time when the vehicle passes the light signal.

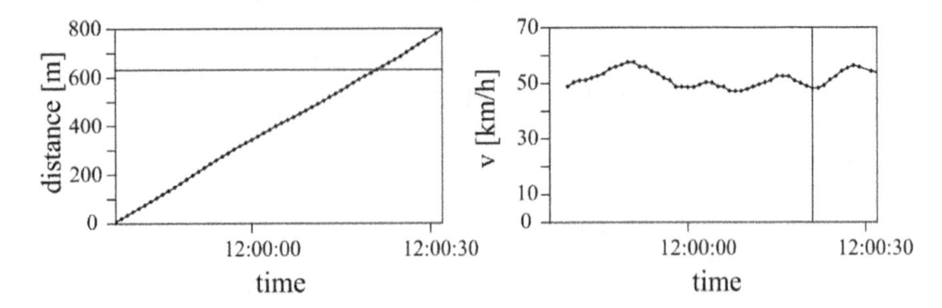

Fig. 2 Representative of class 1 (undersaturated traffic, no stop); vehicle trajectory and speed profile from February 2013

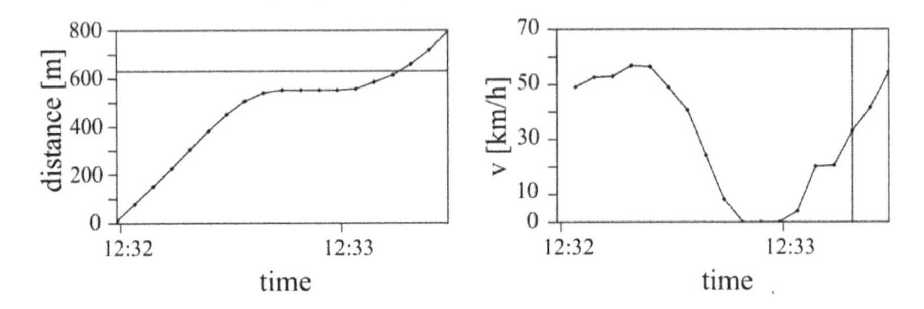

Fig. 3 Representative of class 2 (undersaturated traffic, one stop); vehicle trajectory and speed profile from November 2011

Classes 3 to 15 cover the various traffic patterns associated with urban congested traffic. For instance, class 3 trajectories have zero or one stops, but the average speed is much lower than in free flow (Fig. 4). Class 4 trajectories lie in the same speed range, but there are two to four stops, as in the example shown in Fig. 5. These additional stops occur due to the emergence of *moving queues* that move upstream from the location of the traffic light. Moving queues are a well-known phenomenon of *oversaturated* traffic signals where not all of the cars in a queue can pass the traffic signal during one green phase. These examples demonstrate how empirical vehicle trajectories can display considerable qualitative differences despite similar average speeds.

It should be noted that the distribution of the vehicle speed shown in Fig. 4 related to class 3 is an empirical example of *synchronized flow* in oversaturated urban traffic the theory of which has been developed recently [3]. In synchronized flow, the speed of a vehicle is considerably lower than the free flow speed the maximum value of which is limited to 60 km/h via traffic regulations on the street

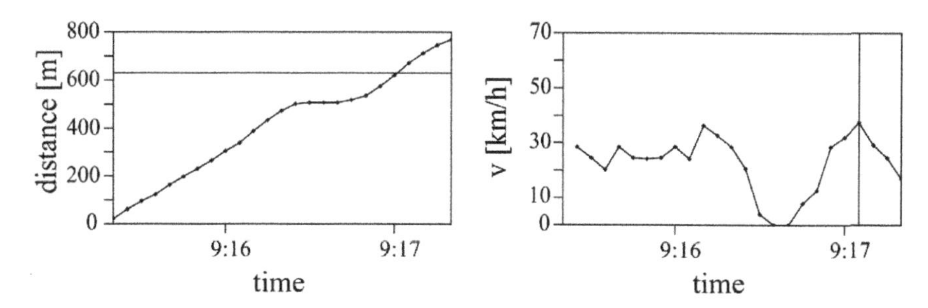

Fig. 4 Representative of class 3; vehicle trajectory and speed profile from February 2013

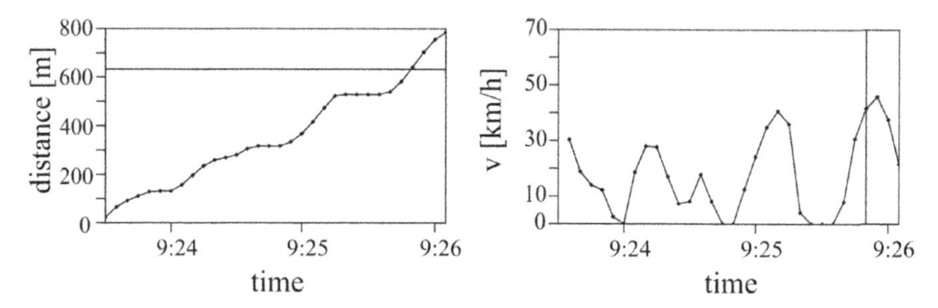

Fig. 5 Representative of class 4; vehicle trajectory and speed profile from February 2013

under consideration. However, in contrast to a vehicle queue within which vehicles are in a standstill and, therefore, traffic flow is interrupted through the queue, there is no flow interruption in synchronized flow [3-9].

Classes 8 and 9 repeat the finding that trajectories with average speeds within the same range can differ with regard to their stops. In the examples depicted in Figs. 6 and 7, not only the numbers of stops are different. Also, the stops are distributed along the road section differently. The representative of class 9 stops three times in the first 200 meters of the road section. In contrast, there is only one short stop in the first 200 meters for the class 8 example, but the vehicle moves with a low speed. On the remaining part of the road section, the stops are considerably longer. This speed profile can be explained with the dissolution of moving queues upstream of the traffic signal and the forming of synchronized flow [3].

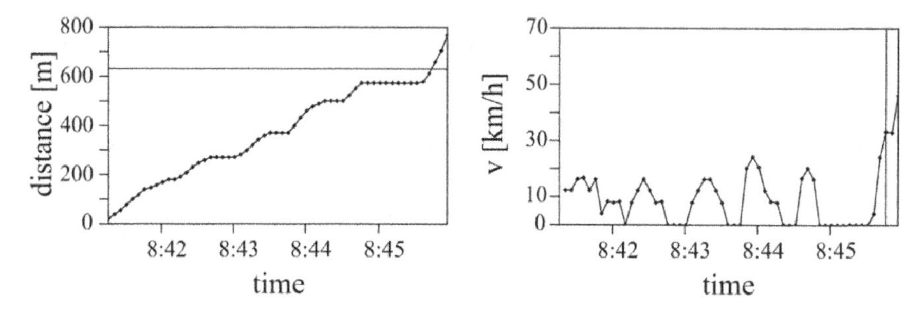

Fig. 6 Representative of class 8; vehicle trajectory and speed profile from November 2011

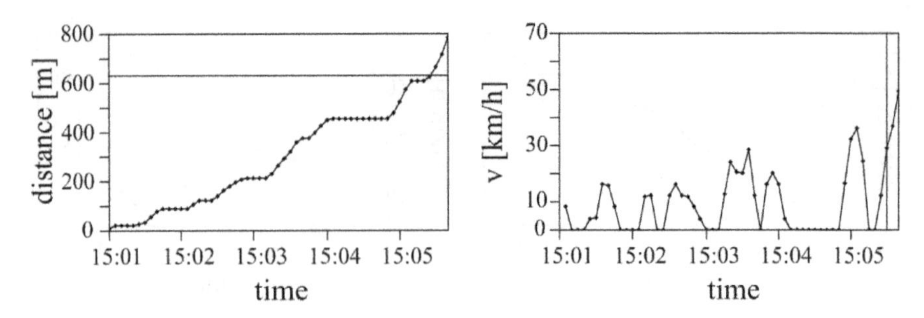

Fig. 7 Representative of class 9; vehicle trajectory and speed profile from February 2013

The highest degree of urban traffic congestion observed in the available data is associated with class 15. An example can be found in Fig. 8, where a sequence of numerous stops and short passages between these stops can clearly be seen.

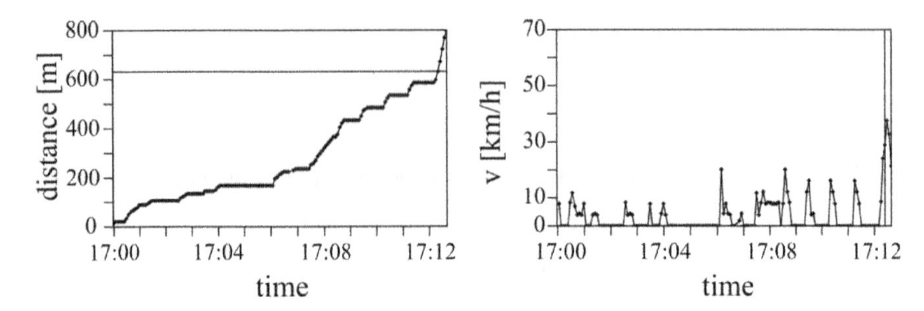

Fig. 8 Representative of class 15; vehicle trajectory and speed profile from February 2013

3 Empirical Fuel Consumption

We examine a recorded archive containing empirical data read from the CAN bus of vehicles driving on German freeways and urban areas. Current velocity, acceleration and instantaneous fuel consumption were recorded multiple times per second. We investigate data of about 90 hours from one vehicle to build a vehicle specific consumption matrix which is shown in Fig. 9. The resolution for velocity is 2 km/h, and the resolution for acceleration is 0.1 m/s². In total, more than 10 million measured data points have been assigned to their corresponding matrix element and aggregated by applying the statistical median operation for each matrix element. Blue pixels indicate consumption values near zero; red pixels indicate high fuel consumption values. A white element in the consumption matrix means that this combination of velocity and acceleration never appears in the data from 90 hours of measurements. As we intend to use the consumption matrix for the determination of the additional consumption caused by specific traffic situations, for instance different patterns of congestion, a normed scale for the fuel consumption is sufficient.

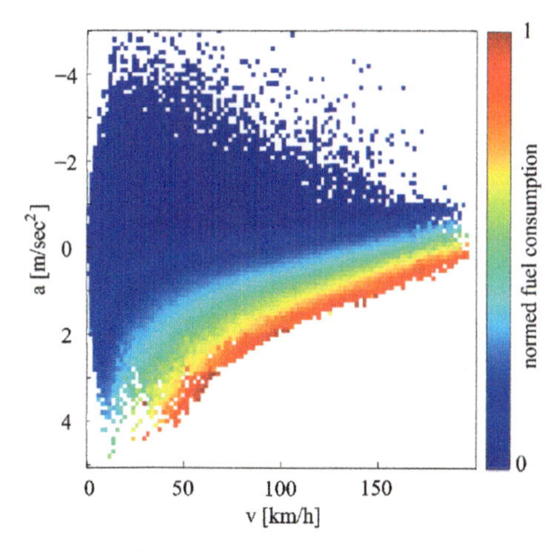

Fig. 9 Visualization of an empirical consumption matrix with the parameters acceleration a and velocity v

4 Increased Consumption in Urban Oversaturated Traffic

Empirical TomTom data do not easily allow a precise determination of the acceleration profiles of vehicles because the devices do not measure acceleration directly (it must be derived from the GPS signal with its impreciseness). Thus, the empirical trajectories and speed profiles described in section 2 cannot directly be used as input for the consumption matrix introduced in section 3.

To close this gap, obtain realistic acceleration values and make a large variety of vehicular congested traffic patterns available, we use a microscopic simulation of vehicular traffic based on a stochastic Kerner-Klenov three-phase traffic flow model [4-6]. The choice of simulation parameters aims at an accurate qualitative reproduction of the empirical vehicle trajectories. Therefore, the classification scheme for empirical vehicle trajectories can be applied to the simulated vehicle trajectories as well. We calculate the fuel consumption for each trajectory by means of the consumption matrix. We denote the average fuel consumption per traffic class N as C_{classN}. The average additional fuel consumption C_{classN}^{RA} relative to class 1,

$$C_{classN}^{RA} = \frac{C_{classN}}{C_{class1}}, \tag{1}$$

is depicted for the first 11 classes in Fig. 10. A gap between classes 1 and 2 of undersaturated traffic can clearly be seen, hinting at the additional fuel consumption caused by a stop at the traffic light. However, the gap between classes 2 and 3 is much larger. The latter gap and the increase in consumption from class 3 to class 11 illustrate the extent to which oversaturated city traffic causes additional fuel consumption. In an oversaturated traffic situation corresponding to class 11, a vehicle with a combustion engine consumes more than twice as much fuel as in the ideal traffic situation corresponding to class 1.

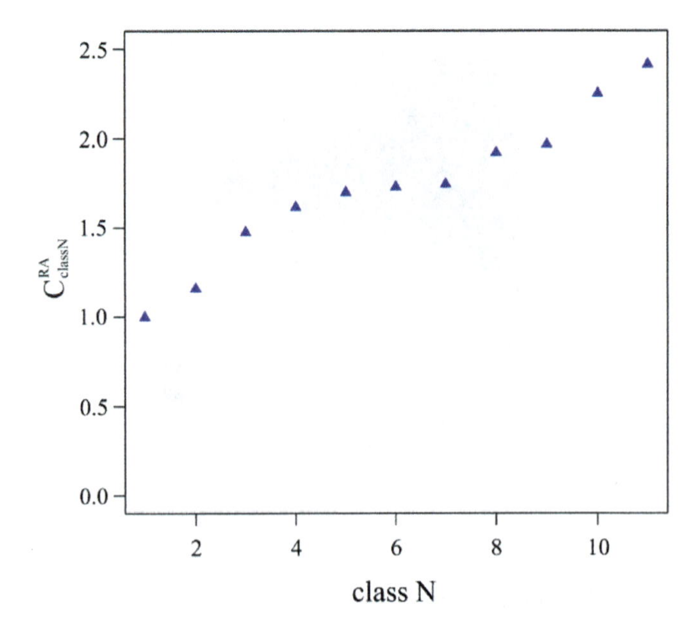

Fig. 10 Average fuel consumption for the first 11 traffic pattern classes as a multiple of the average fuel consumption of class 1

Acknowledgements. The authors thank the TomTom Content Production Unit Berlin for providing anonymized probe vehicle data and the German Federal Ministry for Economic Affairs and Energy for support in the project UR:BAN – Urbaner Raum: Benutzergerechte Assistenzsysteme und Netzmanagement (Urban Space: User Oriented Assistance Systems and Network Management).

References

[1] Rehborn, H., Schreckenberg, M., Kerner, B.S., Hermanns, G., Hemmerle, P., Kulkov, I., Kannenberg, O., Lorkowski, S., Witte, N., Böhme, H., Finke, T., Maier, P.: Eine methodische Einführung zur antriebsartabhängigen Routensuche in Ballungsräumen, Straßenverkehrstechnik 3, Kirschbaum (2014)
[2] Koller, M., Rehborn, H., Kerner, B.S., Hermanns, G., Schreckenberg, M.: Wahrscheinlichkeit des Verkehrszusammenbruchs an Lichtsignalanlagen. Straßenverkehrstechnik 11, Kirschbaum (2013)
[3] Kerner, B.S., Klenov, S.L., Hermanns, G., Rehborn, H., Hemmerle, P., Schreckenberg, M.: Synchronized flow in oversaturated city traffic. Phys. Rev. E 88, 054801 (2013)
[4] Kerner, B.S.: Three-phase theory of city traffic: Moving synchronized flow patterns in under-saturated city traffic at signals. Physica A 397, 76–110 (2014)
[5] Kerner, B.S.: The Physics of Traffic. Springer, Berlin (2004)
[6] Kerner, B.S.: Introduction to Modern Traffic Flow Theory and Control: The Long Road to Three-Phase Traffic Theory. Springer, Berlin (2009)

[7] Kerner, B.S., Rehborn, H.: Experimental Properties of Complexity in Traffic Flow. Phys. Rev. E 53, R4275–R4278 (1996)
[8] Kerner, B.S., Rehborn, H.: Experimental Properties of Phase Transitions in Traffic Flow. Phys. Rev. Lett. 79, 4030–4033 (1997)
[9] Kerner, B.S.: Experimental Features of Self-Organization in Traffic Flow. Phys. Rev. Lett. 81, 3797–3800 (1998)

An Active Vulnerable Road User Protection Based on One 24 GHz Automotive Radar

Michael Heuer, Marc-Michael Meinecke, Estrela Álvarez, Marga Sáez Tort, Francisco Sánchez and Stefano Mangosio

Abstract. Among other initiatives to improve safety of Vulnerable Road Users (VRUs), the European Commission is funding a research project called ARTRAC: "Advanced Radar Tracking and Classification for Enhanced Road Safety" aimed to develop an integrated safety concept for pedestrians. ARTRAC investigates the possibility to use a stand-alone radar sensor especially for the detection of VRUs in order to generate the appropriate supporting actions to the driver with the goal to avoid or mitigate a possible collision. The supporting actions comprise driver warning, automatic braking and steering recommendation. This paper provides an overview of the definition of the system solutions for the warning and vehicle automatic actuations in case of a potentially dangerous situation with respect to a VRU occurs; and of the development of specific software for the actuation over the vehicle braking and steering with the goal to perform an automatic response devoted to avoid an accident involving a VRU or, in the worst case, to mitigate the impact. The functional performance of the systems is analyzed with a specialized pedestrian test facility.

M. Heuer(✉)
Otto-von-Guericke-University of Magdeburg,
Universitaetsplatz 2, Magdeburg, Germany
e-mail: michael.heuer@ovgu.de

M.-M. Meinecke
Volkswagen AG, Brieffach 1777, Berliner Ring 2, Wolfsburg, Germany
e-mail: marc-michael.meinecke@volkswagen.de

E. Álvarez · M.S. Tort · F. Sánchez
Centro Tecnológico de Automoción de Galicia, Pol. Industrial A Granxa, Calle A,
Parcela 249-250, 36400, Porriño (Pontevedra), Spain
e-mail: {estrela.alvarez,marga.saez,francisco.sanchez}@ctag.com

S. Mangosio
Centro Ricerche Fiat, Strada Torino 50, 10043 Orbassano, Italy
e-mail: stefano.mangosio@crf.it

J. Fischer-Wolfarth and G. Meyer (eds.), *Advanced Microsystems for Automotive Applications 2014*, Lecture Notes in Mobility,
DOI: 10.1007/978-3-319-08087-1_8,

Keywords: Active Safety, Automatic Braking, Automotive Radar, Collision Avoidance, Collision Warning, Control Strategies, Steering Recommendation, VRU Protection.

1 Introduction to ARTRAC Project

In 2001, when the European Union set itself the goal of halving the number of road accident victims by 2010, intelligent vehicle safety systems were expected to catch on much faster than they have. In the past the high cost of the available driving assistance systems is the main reason why they were only in use in a few top-of-the-range models, and the EU target is still far from being met. Thus, in the policy orientations on road safety 2011-2020 the European Commission is funding projects that are aligned with the objective of the reduction of fatalities with the help of driving assistance systems and of the more penetration of those driving assistance systems in lower range vehicle models to make them available for more people.

In this framework, the European Commission is funding ARTRAC, whose consortium comprises car manufacturers, research organizations, universities and SMEs with experience in the automotive sector. The ARTRAC project has the goal of providing (develop, test and demonstrate) an active safety system based on a 24 GHz radar to avoid collisions with VRUs that is economically viable in the volume vehicle market. The ARTRAC development promises to deliver a significant technological breakthrough that can contribute to achieve the objectives laid out in the EU and which will open up new possibilities for driving assistance systems.

The ARTRAC safety system will base the perception of the vehicle's surroundings on a single automotive 24 GHz narrow band radar sensor. The system will be tested on two demonstrator vehicles and promoted to relevant bodies and stakeholders, including end-users.

2 System Solution

Based on the environmental detection, the ARTRAC system solution to avoid or mitigate a collision with a VRU in critical situations has been defined with two different approaches which will mainly provide two different reactions in the two vehicle demonstrators: frontal collision warning and automatic braking in the demonstrator from Centro Ricerche Fiat (CRF) and automatic braking and system-initiated steering recommendation in the demonstrator from Volkswagen (VW). However, the global control strategy in both of them has the same basis.

2.1 Global Control Strategies

In the chapters bellow collision warning, deceleration actuation and steering recommendation, as part of the system solution, will be presented separately.

In this chapter the general global control strategy which has a common basis for both vehicle demonstrators is presented.

First of all, the measurement data from the sensor is used within the environment detection module to reconstruct the actual traffic scene and to check whether VRUs are present in the scene or not. Then, an obstacle selection module identifies the main obstacle to be taken into account as possibly provoking a potentially risky situation. A risk assessment unit, based on different strategies in each demonstrator estimates a collision risk that indicates how hazardously the situation is, if an accident is imminent, where is the predicted point of impact (POI) in a vehicle-pedestrian collision, etc.

With the collision risk as input, a decision making which safety strategy in the current traffic constellation is activated takes place. In this sense, the strategies followed in the two demonstrators diverge: the CRF strategy is based on different areas to define the final system reaction: frontal collision warning or automatic braking and in the VW demonstrator, a strategy based on the time to collision (TTC) and the evaluation of the scene selects the actuation between automatic braking and system-initiated steering recommendation.

The global control strategy for each demonstrator will be detailed in the subchapters below.

2.2 *CRF Demonstrator Solution*

If a VRU has been detected by the ARTRAC system, it shall check if the object is in the warning area (defined zone where a warning will be given to the driver in case the pedestrian has a lateral velocity and the predicted trajectory of the pedestrian collides with the trajectory of the vehicle) or in the intervention area (defined zone where an automatic brake intervention will be processed).

If a VRU is in the warning area and the collision risk is over a threshold (a POI is predicted but the VRU is currently outside of the impact zone and all the function operative conditions are verified) a warning will be displayed to the driver and the Pre-Fill will be requested.

Moreover, if the intervention condition is over a threshold (the driver is not braking or the driver is not braking with enough deceleration to avoid the collision) the ARTRAC system will automatically intervene on the brakes in order to avoid the collision or if not possible to at least reduce the impact speed.

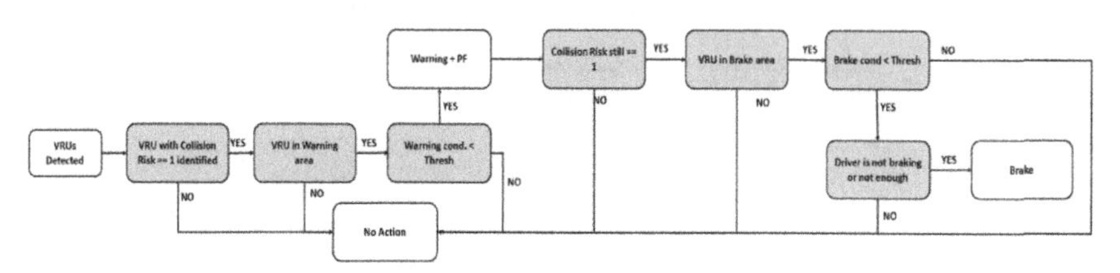

Fig. 1 CRF demonstrator global strategy

The actuation unit in CRF ARTRAC vehicle is implemented into a dSPACE MicroAutoBox 1401/1511 ECU (MABX).

The overall functionality is a composition of three sub-functionalities: Warning: to advise the driver about the presence of a dangerous pedestrian, Pre-Fill: to precondition the vehicle brake system optimizing brake performances in case of driver or autonomous braking, and Autonomous Brake: to brake autonomously (or enhance a driver initiated braking) in order to avoid/mitigate the impact.

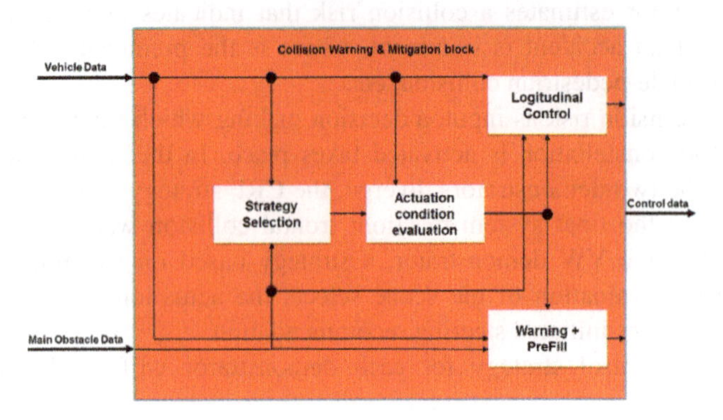

Fig. 2 Block diagram of actuation unit in CRF Demonstrator

2.3 VW Demonstrator Solution

If a VRU has been detected by the ARTRAC system in the vicinity of the host vehicle and the collision risk is over a threshold (corresponds to a probability that indicates how hazardously the situation is and if an accident is imminent) the system deploys the actuators under determined considerations.

If the intervention condition is under a threshold (based mainly on a TTC) and the contact with VRU is not avoidable by steering (the POI is predicted to be in the centre of the bumper or oncoming traffic in the near side lane) the ARTRAC system applies a braking maneuver with increasing brake pressure.

If the intervention condition is below a threshold, the collision is avoidable by steering (the POI is predicted to be at the edge of the front bumper) and there is an

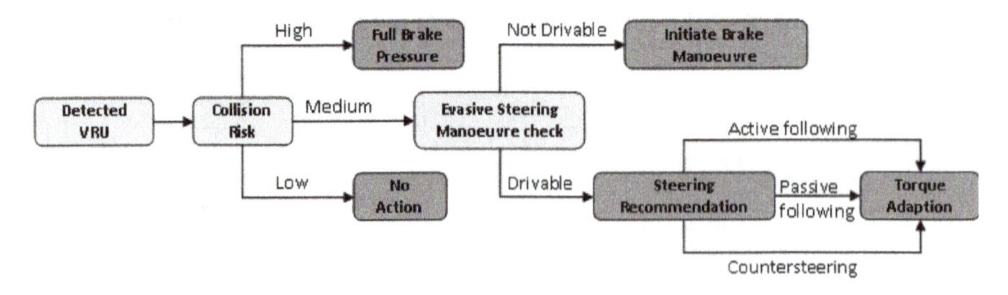

Fig. 3 VW demonstrator global strategy

obstacle in the nearside lane the ARTRAC system applies a braking maneuver with increasing brake pressure. If there is no obstacle in the nearside lane the ARTRAC system applies a steering recommendation to the driver.

If the intervention condition is over a threshold the ARTRAC system applies a braking maneuver with full pressure.

The ARTRAC sensor is feeding, via sensor fusion algorithm and risk assessment algorithms, a vehicle actuation unit. The Volkswagen demonstrator will have a MABX as a main actuation unit, which has the possibility of switching between two different submodels. These submodels are alternative implementations of control strategies developed by the project partners CTAG and VW.

The actuation unit makes a decision on what kind of vehicle safety actuation is activated. This decision is based on information delivered by vehicle data (e. g. host vehicle motion parameters), detected object data (e. g. object position, velocity vector, object class, etc.) as well as the previously calculated collision risk.

Inside of the actuation unit different further calculations are conducted. Firstly the vehicle's path is predicted into several time steps in the future. Depending on the collision risk, the received object list is parsed and an object selection is performed. This selected object (what is the endangered pedestrian) is considered for the intervention decision making concerning longitudinal or lateral intervening. The output of the actuation unit directly controls brake and steering torque of the vehicle.

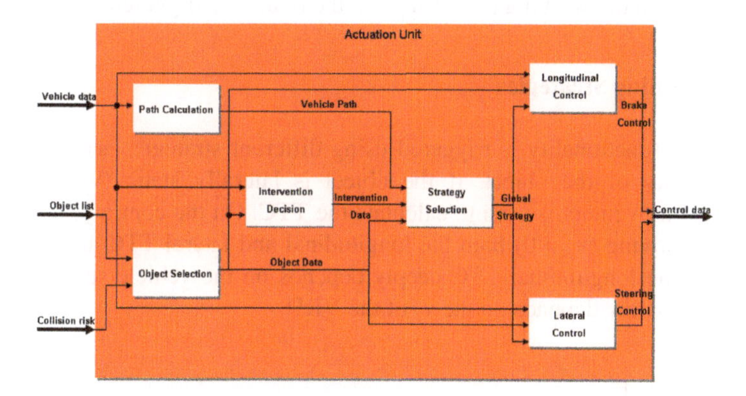

Fig. 4 Block diagram of actuation unit in VW Demonstrator

3 Development of Specific Software for the Warning and Actuation

General description of the software for the on board system is given in the following.

3.1 CRF Demonstrator Solution

As before mentioned, The CRF actuation strategy is based on two defined areas: a warning area and a brake area (brake corridor) that have been defined in order to trigger the brake only if the VRU is inside this corridor and the warning if the VRU is in danger but outside the brake corridor.

The brake corridor width is wider than the vehicle width and is currently defined as Veh_width + Delta.

In the following picture an example is reported:

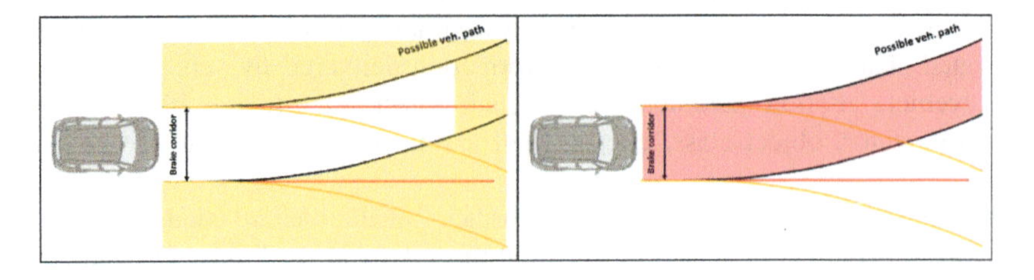

Fig. 5 Example of brake corridor (red) and warning area (yellow)

Among all the obstacles detected by the sensor the one with collision risk equal to 1 is passed to the warning and brake logics. If the warning or brake conditions are verified than the warning (+PreFill) and the brake are triggered.

3.1.1 Warning Strategies

The warning functionality is triggered using different strategies according to the lateral velocity of the object. If the object is laterally static (v_y=0) only the longitudinal movement is used to calculate the TTC. On the contrary, if the object is laterally moving ($v_y \neq 0$) both the longitudinal and lateral TTC are considered. Notice that the longitudinal TTC deeply depends on the vehicle speed while the lateral TTC mainly depends on the v_y of the VRU.

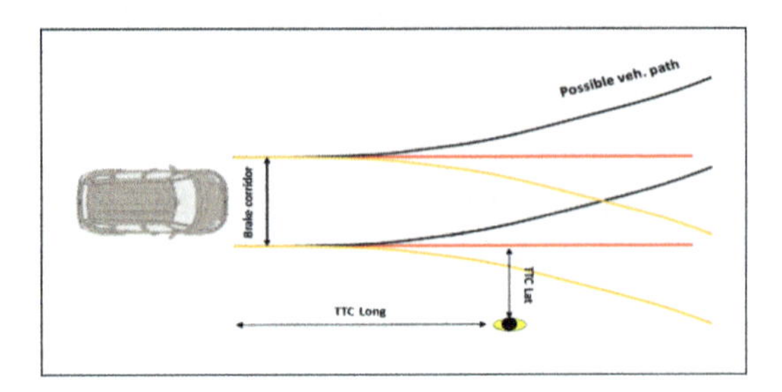

Fig. 6 TTC Lat and TTC Long explanation

To calculate the lateral and longitudinal TTC the intersection between the VRU movement path and the brake corridor is evaluated than the TTC needed to the VRU to reach the brake corridor edge is calculated.

3.1.2 Automatic Braking Strategies

The braking intervention actuation depends on the scenario around the demonstrator, on the dynamics of the vehicle demonstrator and on the dynamics of the obstacle selected as the most dangerous one. The function shall be able to identify dangerous situations and properly provide the brake actuation.

The required final expected result is to reach vehicle speed equal to zero and final distance from pedestrian over a certain threshold in the whole functionality speed range (20-60 km/h). A maximum deceleration profile depending on the vehicle speed has been defined.

The logic represented in Figure 7 is applied in the brake control to reach the expected results: the distance needed to accomplish the maneuver reaching the expected results (final vehicle speed/ final distance vehicle-pedestrian) is calculated and if the actual relative distance between the pedestrian and the vehicle is different from the expected one to accomplish the maneuver than the deceleration level is modulated. Due to functional safety the initial deceleration applied is limited to A_max (< 1g) if the vehicle speed is over a certain threshold.

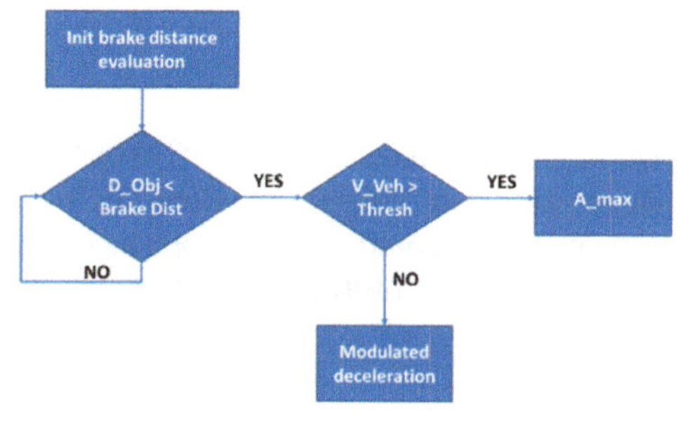

Fig. 7 Deceleration profile definition logic

Legend: D_Obj →Distance of the object seen by the sensor from the frontal bumper of the vehicle
V_Veh → Vehicle Speed
A_max → Maximum allowed deceleration

3.2 VW Demonstrator Solution

In the following paragraphs the function implemented into the VW demonstrator will be described.

The longitudinal control strategy and the lateral control strategy are fed by a collision risk module able to identify between all the VRUs detected by the ARTRAC sensor the most dangerous one. The logic takes as input the coordinates (x,y) of the VRU and its speed component (v_x,v_y).

3.2.1 Longitudinal Control

One very effective design criterion is to apply full brake pressure only in situations where the collision is unavoidable by driver actions, it means that the collision risk is over a collision risk threshold, and the intervention condition is over an intervention condition threshold. If the collision is avoidable, the collision risk is still over a threshold but the intervention condition is under an threshold, the strategy is to increase steadily the brake pressure to smoothly reduce vehicle speed. This increasing pressure strategy allows either driver reaction, given that TTC is still enough, either automatic full brake pressure once the intervention condition gets over a threshold. See Figure 8.

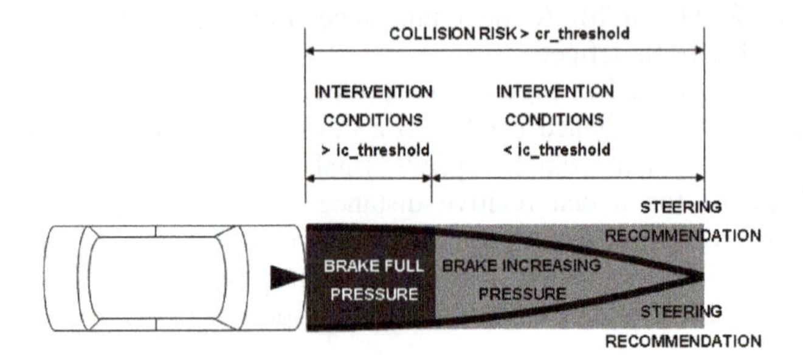

Fig. 8 Longitudinal control strategy

The automatic braking results in a deceleration of the vehicle. When the control system applies a braking maneuver with full pressure, the maximum deceleration value is set to -6.0 m/s^2. When the control system applies a braking maneuver with increasing pressure, the deceleration profile is based on the distance travelled and the velocity as shown in Figure 9.

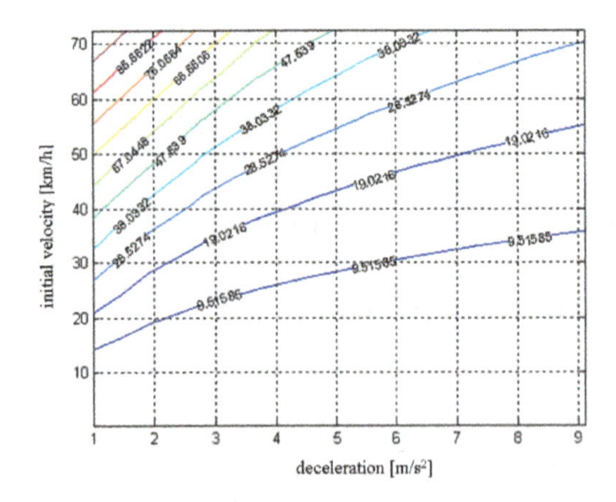

Fig. 9 Deceleration profile

3.2.2 Lateral Control

The goal of the lateral support is to control the vehicle on an evasive maneuver which has a trajectory like an S-curve. The S-curve is required to add some lateral motion without changing the vehicle's driving direction after the intervention. So, the vehicle is parallel translated in lateral-direction, it has the same heading as the inertial head was. That control strategy guarantees that the vehicle will not drive into another direction as it did before the maneuver, minimizing another conflict with other traffic participants.

The Volkswagen approach for lateral control follows a geometric trajectory planning strategy. Idea of this approach is to assemble the trajectory of a set of principle geometric functions. The number of geometric functions is high which provides some degree of freedom in the optimization process. Due to the fact that in many countries roads are laid out based on clothoids this function seems to be a well suited candidate for these calculations. The curvature of a clothoid increases linearly. This enables a linear turning speed of the steering wheel and results in jerk-free driving dynamics.

A sigmoid function is used to define an evasive trajectory instead of a clothoid. While the sigmoid function $y(x) = \frac{B}{2} \cdot \left(1 - \tanh\left(a \cdot \frac{x-c}{2}\right)\right)$ has more parameters, it can be easier adapted to the required shape.

4 Results and Conclusions

In this paper, two different system solutions for VRU protection have been presented. Both of them perfectly fulfill the goal of avoiding, or at least, mitigating a possible collision in a risky situation with a VRU from different approaches, both of them fed by the environmental detection of single 24 GHz radar developed within the ARTRAC project.

The software developed for the demonstrator vehicles will be tested in a number of different scenarios.

The depicted test scenarios in Figure 10 will be varied in different parameter dimensions in order to assess the functional performance. Because of the manifoldness, the total number of test to be conducted has to be limited. Otherwise the number of test explodes to extreme high values. For practical reasons a subset of the possible combinations will be conducted within this research project.

The following parameters are dedicated for being varied in the different test cases: ego vehicle speed, lateral offset of ego vehicle in lane, pedestrian speed, pedestrian walking direction, POI, and road surface condition.

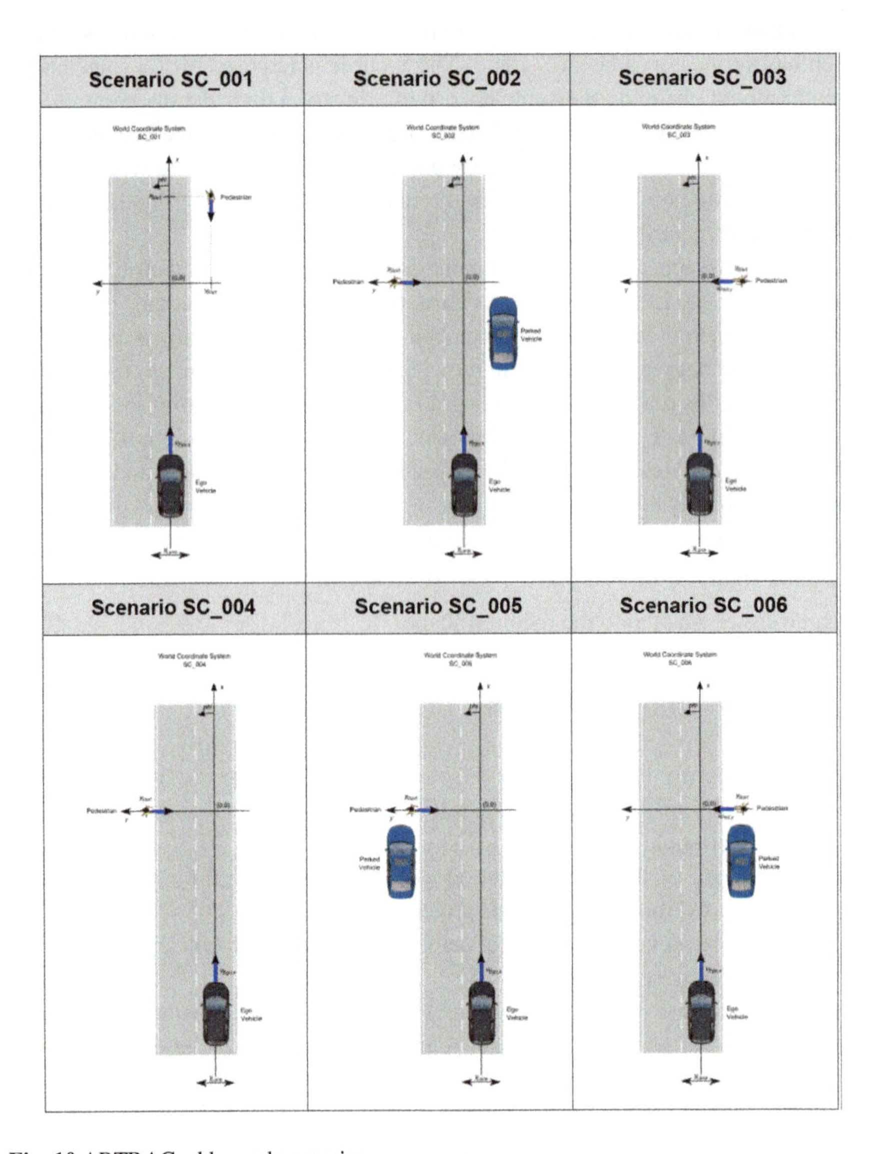

Fig. 10 ARTRAC addressed scenarios

Fig. 11 Pedestrian test facility to conduct functional performance measurements

References

[1] Brandt, T., Sattel, T.: Path planning for automotive collision avoidance based on elastic bands. In: 16th IFAC World Congress, Prague, Czech Republic, July 3-8 (2005)

[2] Heuer, M., Al-Hamadi, A., Meinecke, M.-M.: An Observation Model for High Resolution Radar Data in the Context of an Automotive Pedestrian Safety System. In: International Radar Symposium, Dresden, Germany, June 19-21 (2013)

[3] Lu, M., Wevers, K., van der Heijden, R., Heijer, T.: Adas applications for improving traffic safety. In: Proc. IEEE Int. Systems, Man and Cybernetics Conf., vol. 4, pp. 3995–4002 (2004)

[4] Meinecke, M.-M., Heuer, M., Achmus, S., Olders, S., Rohling, H., Heuel, S., Mende, R., Häkli, J., Wegner, A., Sanchez, D., Mangosio, S.: User needs and Requirements for VRU protecting systems based on multipurpose narrowband radar. ARTRAC Deliverable D2.1 (2012)

[5] Sáez, M., Álvarez, E., Meinecke, M.-M., Heuer, M., Mangosio, S., Cuevas, Á.: Driving Intervention Actuation. ARTRAC Deliverable D5.1 (2014)

References

Power Saving in Automotive Ethernet

Thomas Suermann and Steffen Müller

Abstract. A network sleep and wake-up concept designed to promote in-vehicle energy efficiency is nowhere to be found in the BroadR-Reach automotive Ethernet standard [1]. However, a variety of approaches which would be compatible with the standard are conceivable. Below we set out some objectives in relation to the implementation of a sleep and wake-up process and take a closer look at various possible approaches. On technical grounds we shall come down in favour of an approach which permits the control of the sleep and wake-up processes using the network's own resources. We go on to discuss partial networking as a network management operational state for an Ethernet-based transmission system and show how the associated sleep and wake-up process can be implemented at physical level. The proposed solution employs familiar AUTOSAR control mechanisms which are already in use in today's vehicle networks.

Keywords: Automotive Ethernet, Physical Layer, Switch, Power Saving, Partial Networking, OpenAlliance.

1 Background to Partial Networking

Familiar in-vehicle network technologies for data communication, such as CAN, LIN and FlexRay, provide mechanisms for switching off individual vehicle functions which are not being used in a controlled fashion. This will happen, for instance, when the driver turns off the engine and leaves the vehicle. In this case it is necessary to switch off electrical loads in the parked vehicle because otherwise undesired power consumption would cause the battery to go flat.

It must of course be simultaneously possible to initiate network communication at any moment in order to regain complete system control if needed. This could be done through the vehicle's access control, which would send a message that the

T. Suermann · S. Müller(✉)
NXP Semiconductors Germany GmbH,
Stresemannallee 101, Hamburg, Germany
e-mail: {thomas.suermann,st.mueller}@nxp.com

J. Fischer-Wolfarth and G. Meyer (eds.), *Advanced Microsystems for Automotive Applications 2014*, Lecture Notes in Mobility,
DOI: 10.1007/978-3-319-08087-1_9,

driver has opened the door. Today, control units are woken up via the network itself, either through the detection of activity or the sending of special wake-up signals.

The task of the transceiver (PHY from now on) is to distinguish between network activity or special wake-up tokens on the one hand and interference noise on the other, thus avoiding unintended wake-ups and unnecessary power consumption. Such interference is highly likely in the vehicle environment, and this factor will be considered in detail when assessing the robustness of the network design.

While a vehicle is being driven, not all individual functions will be in use at any one time. To take one example, the rear-view camera used for parking is switched off during normal driving. This feature is used infrequently, and would indeed be a distraction for the driver if it transmitted images continuously. To realise the potential for saving power and, by extension, also fuel, the CAN protocol has been extended to allow partial networking [2]. As a result, control devices can now be configured to only switch on when they receive specific wake-up messages.

Should Ethernet become established as the in-vehicle backbone bus, partial networking capability will become increasingly important. This applies even more so in the case of electric vehicles, where every unit of power that can be saved will assist the vehicle in attaining a greater range.

Below we discuss various approaches to partial networking for an automotive Ethernet network, with special reference to the necessary services to be provided, including at physical (signal) level.

2 Requirements

In developing a wake-up concept targeted at partial networking, the following objectives need to be addressed:

- A node wake-up which takes at most 250 ms (from wake-up to communication)
- Power consumption of a wake-able port must be less than 10µA
- No unintended wake-ups due to electromagnetic interference
- The system must not induce any changes in the MAC layer
- Integration into AUTOSAR network management system possible
- Open standard, able to update to future standards (Gigabit Ethernet [3])
- There must be no need for any additional hardware

3 Potential Solutions

3.1 Selective Wake-Up

In CAN (Controller Area Network), nodes can be configured in such a way that they can be woken up individually via targeted messages. In Ethernet networks with point-to-point connections between nodes and switches, the switch determines the onward wake-up pathway, i.e. which of its own ports are to be wakened. To this end network management messages are evaluated. The process propagates from switch to switch until the whole wake-up pathway has been traversed and all addressed nodes have been reached.

Fig. 1 depicts the selective wake-up process. The outlined network comprises the functional clusters A, B and C. Each node is assigned to at least one cluster (colour coding). The clusters are represented by a VLAN with its own network management. The illustration assumes an initially quiescent network, with the switch and microcontroller receiving no power while all PHY components are in the inactive Sleep state. In this state only the wake-up receiver is active, so that it can wake up the node if it detects either activity or a wake-up message.

The left-hand node in Fig. 1 detects a wake-up event and initiates a wake-up request for cluster A. The network node itself initiates the wake-up process and sends an 'idle' sequence. The adjacent PHY detects the activity and wakes up the connected switch. As soon as the connection has been established, the left-hand node sends network management messages for cluster A so that the following switch 'knows' which ports (PHY) it next has to activate. This concludes phase 1. This is followed in phases 2 and 3 by identical wake-up procedures for the remaining routes, until the entire wake-up pathway has been activated.

This mechanism can in principal be used on most IEEE 802.3 baseband standards, including 100Base-TX, 1000Base-T and BroadR-Reach. The critical factor here is the time taken to establish a link. According to the BroadR-Reach specification this can be up to 200 ms, depending on the channel characteristics. In our example four phases are required; meaning that up to 800 ms might be needed to initiate a cluster, which is unacceptably long for many in-vehicle functions.

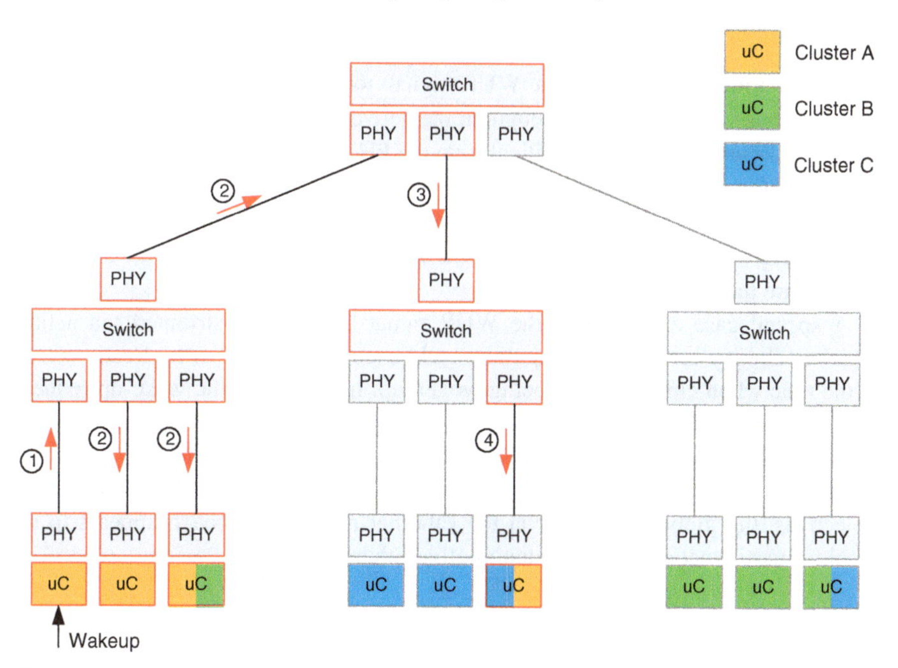

Fig. 1 Mechanism for selective wake-up. Phases 1 – 4 have separate connection structure.

3.2 Global Wake-Up via a Separate Wake-Up Line

This approach employs a wake-up line which is connected with each network node. The network is wakened globally by activating the wake-up line. As a result, all the network nodes are activated at the same time and all links are established simultaneously. Functional clusters which are not needed are subsequently switched off by the network management system. This approach is characterised by its simplicity and the fact that it delivers low wake-up times of less than 250 ms in the case of BroadR-Reach. However, a disadvantage is its additional use of a wake-up line.

3.3 Global Wake-Up via the Ethernet Network

A general feature of this wake-up process is the introduction of a dedicated wake-up phase during which continuous wake-up (WUP) signals are sent. These are distributed throughout the network, leading to a global network wake-up. The challenge here is to propagate the wake-up signal throughout the network rapidly enough for the individual links to be established virtually simultaneously.

Initially the network is switched off and in the quiescent Sleep state. Only the wake-up receivers in the PHY components are active, then the left-hand node in Figure 2 detects a wake-up event and orders the wake-up of cluster A. The node itself initiates the wake-up process by sending a wake-up signal for a specified time to its adjacent PHY. The next switch then detects this signal at one of its ports and immediately sends a WUP signal to its other ports. The wake-up message thus spreads swiftly throughout the network.

The wake-up time depends critically on the PHY's WUP detection time and the delay in transmitting the signal to the other ports. According to our initial estimates, both time parameters are typically in the double-digit µs range, meaning that cluster wake-up times of less than 250 ms can be guaranteed. Unneeded clusters are then switched off again.

A special case arises when the WUP signal has to be distributed via a link which is already active. This situation would occur if cluster B was already active at the time when cluster A was being woken up. In this event the wake-up request would be coded within the idle sequence, similar to the coding of the loc_rcvr_status signal in the BroadR-Reach standard's idle sequence.

In view of the fact that this method achieves all of the above-mentioned objectives, the global wake-up via the Ethernet network approach appears to be ideal for partial networking with automotive Ethernet.

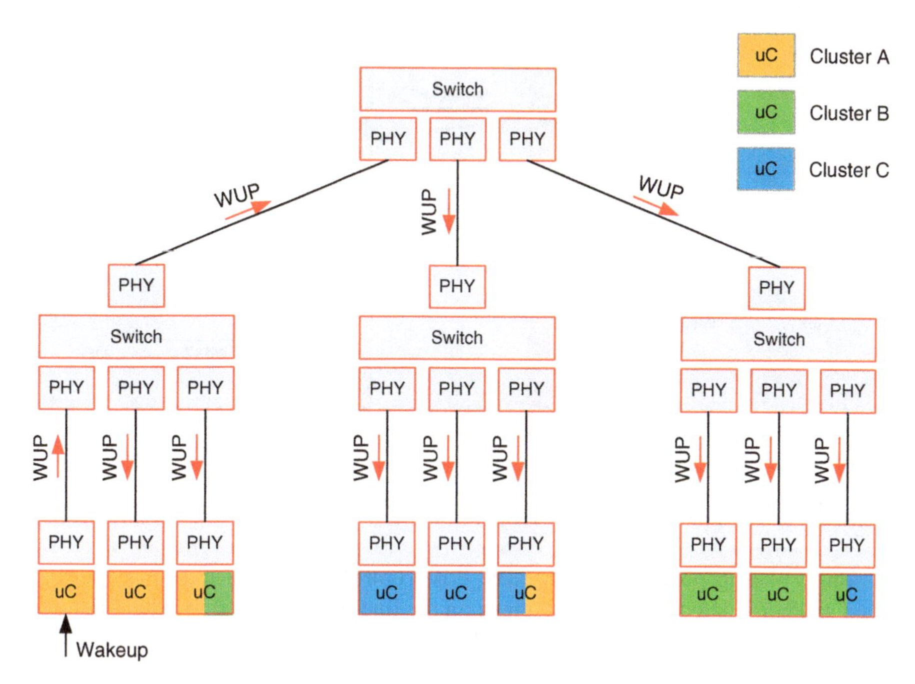

Fig. 2 Mechanism for global wake-up via WUP wake-up tokens

4 Sleep and Wake-Up Process at the Physical Link Level

The above discussion of partial networking relates to the system level. Below we focus on the sleep and wake-up process at the link level for the BroadR-Reach standard, as well as on the physical layer services included here. In Figure 3 we first examine the switching off process, i.e. how both of a link's PHYs assume Sleep status in coordinated fashion.

The switch-off process is initiated by the node's network management [4] by switching PHY1 from Normal state to Sleep Request via an SMI command. Next, this transmits LPS code groups. When PHY2 detects these code groups it switches to Sleep Request, whereupon PHY2 now also sends LPS code groups. PHY1 treats the reception of these LPS code groups as confirmation of the switch-off process. After a set time has elapsed, PHY1 switches into Sleep state and switches the left-hand node off. PHY2 also switches into Sleep state after the same interval and signals this by switching off the INH signal.

The wake-up process at link level is depicted in Fig. 4. At the outset both PHYs are in Sleep state. When an end node (here on the left-hand side) detects a wake-up event (e.g. via a CAN or LIN interface), the management software initiates the wake-up process by first switching PHY1 from Sleep to Normal state, and then instructs PHY1 to send the WUP signal.

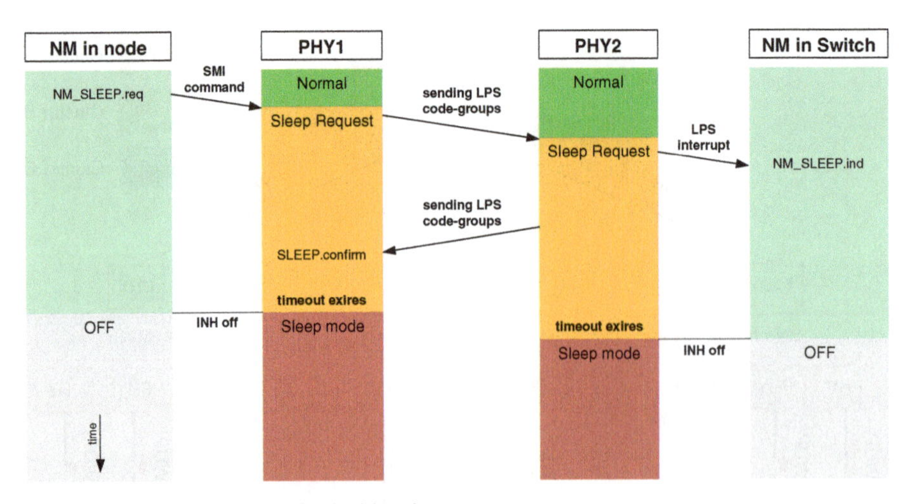

Fig. 3 Switch-off process at physical level

This wake-up phase ends when the link_control signal is set in the PHY. As soon as PHY2 detects the WUP signals on the line, it switches from Sleep through Standby to Normal state, while the PHY device simultaneously activates the control device's voltage supply (if this was still switched off).

A wake-up event is displayed by the management software via a WU Interrupt. The wake-up source is registered in the PHY. After a wake-up event in PHY2 the link_control signal is automatically set to enable, and PHY2 can initiate the link start-up directly through its training phase. Once the link is established, the network management can commence sending network management messages for the cluster.

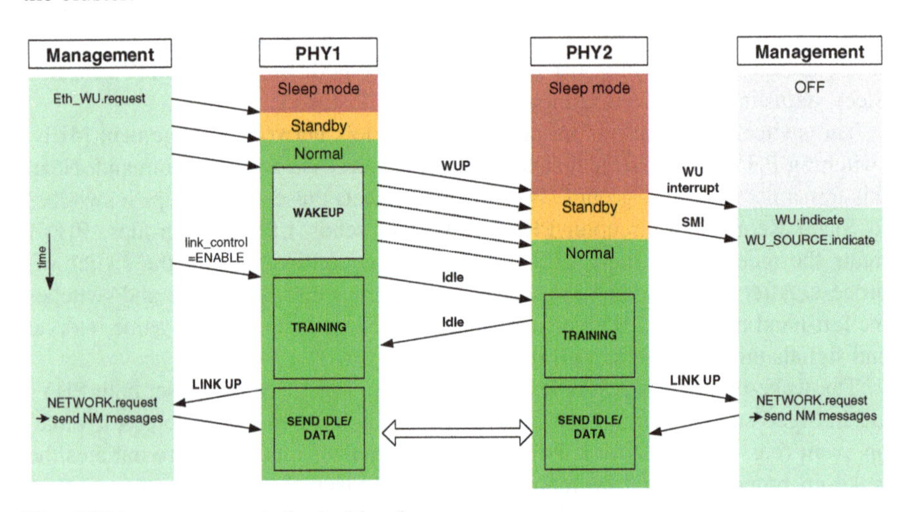

Fig. 4 Wake-up process at physical level

5 Operating Status at Physical Level

For the account of the sleep and wake-up process given in Section 4, the various operating states of a PHY had already been introduced. In this section we take a closer look at their functional characteristics and interrelations by reference to a status diagram (see Fig. 5).

After the voltage supply is switched on the PHY goes into Standby state. The transmitter and receiver remain switched off and are not activated until the PHY enters Normal state. Once in Normal, after the control signal 'link_control' is set, the PHY attempts to establish a link, thereby initiating the training phase in the Master PHY.

After the system switches to Sleep Request, the switch-off process is initiated at link level. The synchronised link between the two communication partners remains in place, with the LPS code groups being sent separately through an idle phase. The time spent in Sleep Request state is limited, and once this time has elapsed, the PHY switches into Sleep state provided LPS [code groups] have previously been detected. Only the wake-up receiver remains active in this state, and as a result power consumption is reduced to a few μA. As soon as WUP signals are detected, the PHY switches from Standby to Normal state in order to establish a new link.

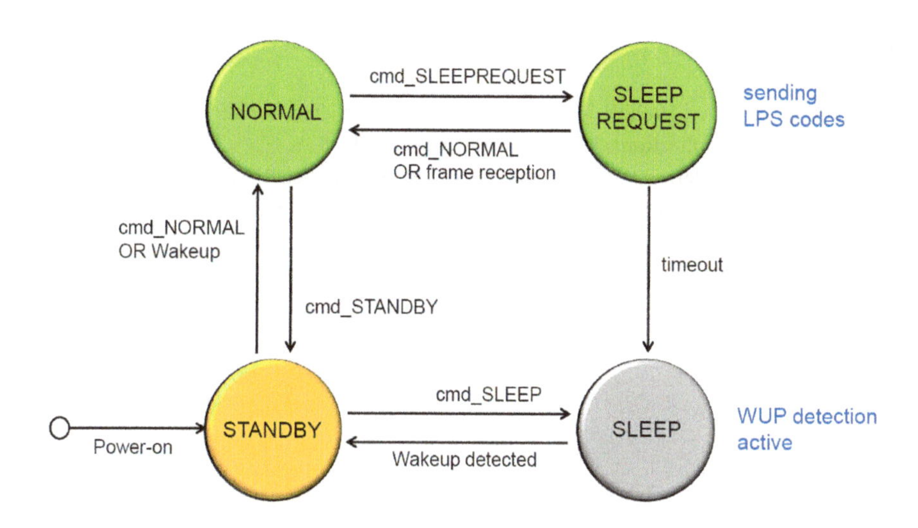

Fig. 5 Status diagram for a typical PHY component

6 Summary

All essential requirements for the successful introduction of partial networking in automotive Ethernet are met by the global wake-up via the Ethernet network approach. It uses the existing network infrastructure, and the network is wakened with the required speed via a WUP signal propagating through the network. The subsequent establishment of the physical links then takes place in parallel.

The described mechanisms can be applied to a variety of IEEE baseband standards and the new required services like LPS and wake-up signalling could be implemented in a straightforward way into existing PHY solutions. The amount of power used by the network in quiescent state can be calculated from the number of PHY components in the network and their quiescent state power consumption. The assumption is that, along with the PHY device, the remaining functionality of the nodes (control devices) is completely switched off when they are in Sleep.

In this article we have not considered the issue of electromagnetic compatibility. However, the principles behind the proposed mechanisms have been used in other network technologies and represent the state of the art, and the same applies to their interaction with AUTOSAR.

The introduction of partial networking into automotive Ethernet is vital for the success of automotive Ethernet. Over the past decade the importance of a vehicle's overall power consumption has increased dramatically, and despite the fact that demand for transmission bandwidth has risen exponentially, the vehicle network must make its contribution in this sphere. NXP is sharing responsibility here both in terms of concept work and the provision of products.

References

[1] OPEN Alliance, BroadR-ReachTM Physical Layer Transceiver Specification for Automotive Applications, V2.0 (September 18, 2013)

[2] Elend, B., Müller, S.: NXP Semiconductors, Teilnetzbetrieb für mehr Reichweite (partial networking for greater range), Hanser Automotive (November 2011)

[3] http://www.ieee802.org/3/RTPGE/

[4] AUTOSAR, Specification UDP Netw. Management, V3.0.0, R4.1 Rev1

Analysis of Cluster Ring Controller/Area Networks for Enhanced Transmission and Fault-Tolerance in Vehicle Networks

Po-Cheich Chiu, Yar-Sun Hsu and Ching-Te Chiu

Abstract. The Controller Area Network (CAN) [1][2] is widely adopted in vehicle networks due to the simple communication protocol. However, with the increasing node number in vehicle network, insufficient bandwidth and faulty nodes or links, become two important problems in a single CAN bus. We propose a cluster ring topology [3][4] for CAN bus to tackle both the bandwidth and fault tolerance problems. By applying the cluster ring topology, the extra bandwidth can also be used to fault tolerance for link or node fails. In addition, we estimate the injection rate versus schedulable messages in the three cluster ring topologies. The throughput models under different link or node faults for the three cluster ring topologies are also analyzed. Then we provide simulation results to verify the developed theoretical models.

Keywords: CAN bus, In-vehicle network, real-time system.

1 Introduction

Due to the security and durability issues, car industries use the controller area network (CAN) to communication between nodes in a car. This bus based communication and control is low cost and efficient. However, the bandwidth of the CAN bus is not sufficient to connect automotive components which increase rapidly. The relative researches of the CAN for timing analysis usually adopt the worst case response time [5][6][7] method to calculate the worst case of the message to be waited in the queue. However, the worst case response time method fails to find out the number of messages to be injected into the topology such that some bandwidth is wasted. The fault tolerance scheduling method provided in [8] can't estimate the injection rate

P.-C. Chiu(✉) · Y.-S. Hsu
Department of Electrical Engineering National Tsing Hua University,
Guangfu Rd. Sec. 101, Hsinchu, Taiwan 30013, R.O.C.
e-mail: tennis4558581@gmail.com

C.-T. Chiu
Department of Computer Science Institute of Communications Engineering
National Tsing Hua University, Guangfu Rd. Sec. 101, Hsinchu, Taiwan 30013, R.O.C.

J. Fischer-Wolfarth and G. Meyer (eds.), *Advanced Microsystems for Automotive Applications 2014*, Lecture Notes in Mobility,
DOI: 10.1007/978-3-319-08087-1_10, © Springer International Publishing Switzerland 2014

for different topologies. The previous research on the cluster CAN topology [3] dramatically enhances the bandwidth than the traditional CAN. In this paper, we analyze the random traffic in several different cluster CAN topologies, single ring, single ring 2 phase, dual ring 2 phase, and build up the network model to estimate the injection rate and the schedulable message percentage. We also analyze the link fault scenario to provide the model to describe the injection rate drop of the dual ring topology. To verify the results from the estimation, we build up the simulation to verify the estimation with the proposed model. The throughput of the proposed topology is three time better than the single ring topology, and it also provide the fault tolerance when link fault occurs.

The paper organization is shown below. Section 2 is the cluster CAN model construction. Section 3 is the simulation. Section 4 is the conclusion.

2 Cluster CAN Model Construction

2.1 Cluster CAN Topology

With the increasing node number in vehicle network, insufficient bandwidth and faulty nodes or links, become two important problems in a single CAN bus. We propose a cluster ring topology [3][4] for CAN bus to tackle both the bandwidth and fault tolerance problems. By applying the cluster ring topology, the extra bandwidth can also be used to fault tolerance for link or node fails. There are three cluster ring topologies including the single ring, single ring 2 phase, dual ring 2 phase and they are described below.

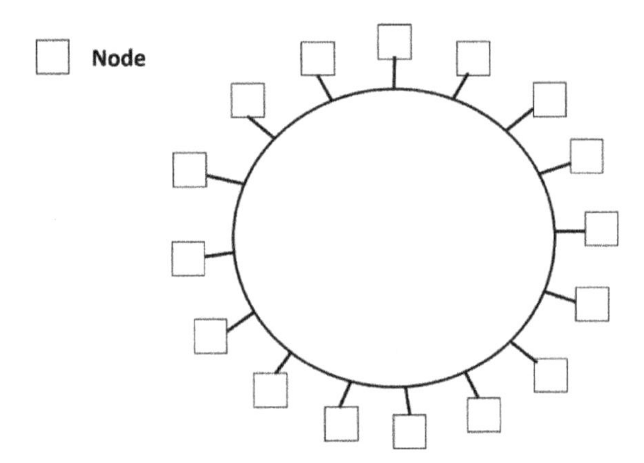

Fig. 1 Single ring topology

Fig.1 shows the single ring topology on a shared bus that is connected as a ring. All the nodes are connected to the shared bus. The nodes on the ring can send the messages on the shared bus according to their priority. In the CAN protocol, every node has its unique priority and the scheduler uses the priority to do the arbitration.

Fig.2(a) is the single ring 2 phase topology. In this example, there are 16 nodes shared the ring bus. The topology also has two sets of switches to control the

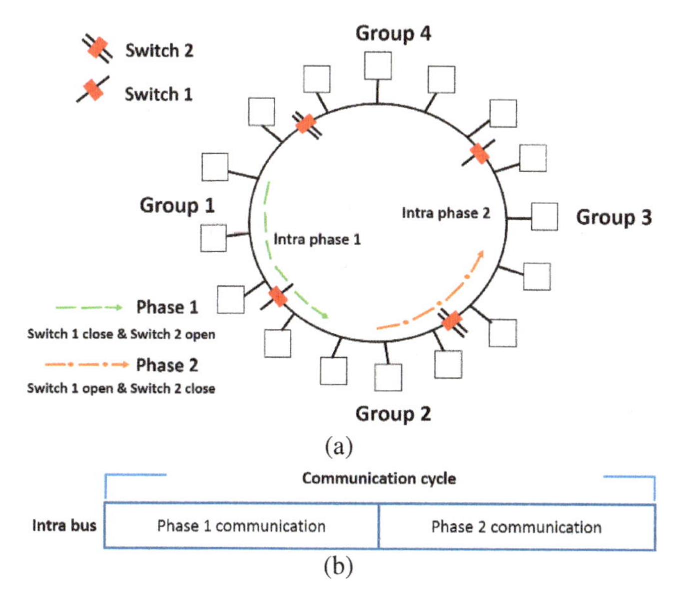

Fig. 2 (a) Single ring 2 phase topology. We use switches to control the communication between groups. (b) In this example, group 1, 2, and group 3, 4, can do the communication at phase 1; group 1, 4, and group 2, 3, can do the communication at phase 2.

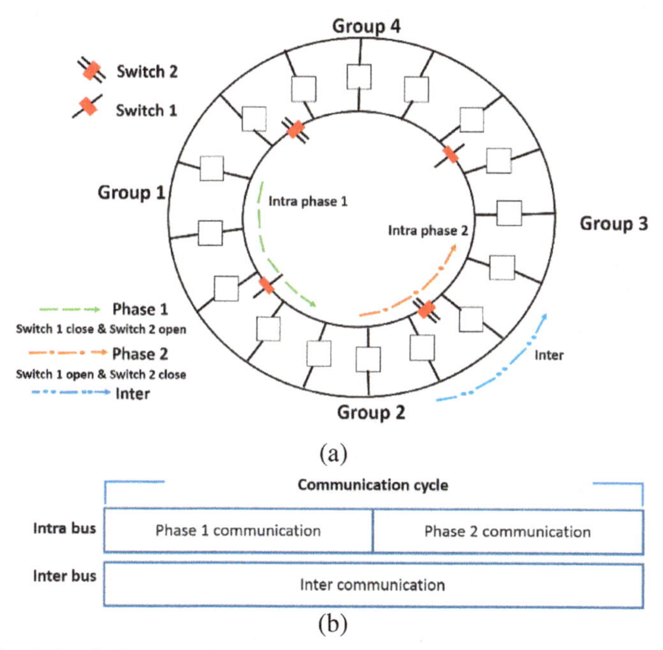

Fig. 3 (a) Dual ring 2 phase topology. We add the inter bus to let group 1, 3, and group 2, 4, nodes can communicate. (b) Dual ring 2 phase topology communication cycle.

communication phase to increase the transmission bandwidth [9]. The bus with switch mechanism is called intra bus. The nodes are divided by the switches into different group. Fig.2(b) is the communication cycle of the single ring 2 phase topology that include two phases. During the communication phase one, when the set of the switch 1 is closed and the set of switch 2 is open, the nodes in the group 1 and group 2 can transmit messages to each other, and so do the nodes in the group 3 and group 4. During the communication phase two, when the set of the switch 2 is closed and the set of switch 1 is open, the nodes in the group 1 and group 4 can transmit messages to each other, and so do the nodes in the group 2 and group 3. By switching the phase, we can gain extra bandwidth from this mechanism.

Fig.3(a) is the dual ring 2 phase topology. We add another bus outside the previous topology and all the nodes are also connected into this outer ring. This extra bus is called inter bus. In the single ring 2 phase topology, the nodes in group 1 and group 3 nodes or nodes in group 2 and group 4 cannot transmit messages between these groups under the two phase communication scheme. The main function of this inter bus is to communicate between the group 1 and group 3 nodes, and so do the nodes in group 2 and group 4. Fig.3(b) is the communication cycle of the dual ring 2 phase topology.

2.2 Injection Rate Model Construction

To find out the maximum number of messages that can be sent into a network, we construct the injection model for the three cluster ring topologies. We start from the single ring topology. The following table is the notation of the symbol and definition used in the derivation.

Table 1 Notation of the symbol used in the following expression

Symbol	Definition
C	Maximum transmission time of a message and the unit is millisecond (ms).
T	Period of time slice and the unit is millisecond (ms).
S	Total time slices number of a simulation.
R	The maximum injection rate. Under this injection rate, every node can let all their messages to be schedulable.
R_f	The injection rate with fault link. Under this injection rate, every node can schedule all their messages.
M_t	The maximum number of messages that a topology can carry.
M_{all}	The total message numbers in a simulation.
M_i	The total message number generated by all the nodes
M_c	The message number created by nodes per time slice
L_f	The link fault number of a topology. The link fault is the fault for transmission from the node to the ring bus.
N	The number of nodes in a topology.
W	The total weight of the topology
W_f	The sum of the link fault weight of the topology
U	The parallel percentage of the cluster CAN bus.
U_{inv}	The non-parallel percentage of the cluster CAN bus

In the single ring topology, T represents the time period of the time slice. C is the time that a message is transmitted to the destination. Since the transmission time to different destinations is different, we use the maximum value in the derivation. In every time slice, the maximum number of transmitted messages per time slice is $\frac{T}{C}$. S is the total time slice in a simulation. Therefore, the total maximum number of transmitted messages M_t is shown below.

$$M_t = S \times T \div C \tag{1}$$

M_i is the total message number generated by all the nodes. Let R be the injection rate of all the nodes and assume that the injection rate is the same for all nodes. N is the number of nodes in a topology. M_c is the message number created by a node per time slice and we assume that one message is generated by a node in every time slice. The total message number generated by all the nodes is expressed in the following equation.

$$M_i = R \times N \times S \times M_c \tag{2}$$

Let the $M_t = M_i$ and $M_c = 1$, we can get the injection rate for the single ring topology below.

$$R = T \div C \div N \tag{3}$$

In the cluster CAN topology, we separate the bus to transmit messages in parallel and gain more bandwidth. In the single ring 2 phase topology, the bus is separated by the switches into two buses to transmit messages in parallel. Therefore, the injection rate gains double in the single ring 2 phase topology, and the injection rate gains triple in the dual ring 2 phase topology in ideal case. The ideal injection rate for these three cluster ring topologies are shown below.

$$R = \begin{cases} T \div C \div N & \text{single ring topology} \\ 2 \times T \div C \div N & \text{single ring 2 phase topology} \\ 3 \times T \div C \div N & \text{dual ring 2 phase topology} \end{cases} \tag{4}$$

2.3 *Link Fault Model Construction*

We assume the fault of the topology only occur at the link from the nodes to the ring bus. There are no faults occurring at nodes and the ring bus. To fit the fault link scenario, we analyze the contribution of the intra and inter bus. Under the dual ring 2 phase topology, we gain more bandwidth due to the intra bus clustering. When the link fault occurs, the injection rate drops. The more the link fault occurs, the more the injection rate drops. We separate the injection rate in the dual ring 2 phase case into two parts as follows.

$$R = \frac{T}{C \times N} + 2 \times \frac{T}{C \times N} \tag{5}$$

The first part is the contribution of the inter bus, and the second part is the contribution of the intra bus. While the link fault occurs, the contribution of the intra bus starts to fall. Once all the link faults occur, the contribution of the intra bus drops to zero. So we define the injection rate of link fault as R_f that every node can schedule their message under this injection rate. Then we update the equation as follows.

$$R_f = \frac{T}{C \times N} + 2 \times \frac{T}{C \times N} \times \frac{W - W_f}{W} \tag{6}$$

We give every node a weight. The weight is based on the priority. For example, the priority 1 node get the weight 16, the priority 2 node get the weight 15 and so on when there are 16 nodes in the topology. W_f is the sum of the weight of those nodes with link fault. W is the sum of the weight of all the nodes in the topology. The new term represents the percentage of the link fault effecting the injection rate in the topology.

We obtain from the simulation results and find out that the intra bus cannot contribute twice of the bandwidth in the dual ring 2 phase topology. It is because the setting of the communication cycle in the intra bus forces some messages to wait till its phase for transmission. In particular, when the traffic is high, this effect becomes more significant. Therefore, we define the U as the percentage of bandwidth that the extra bandwidth contributes in the cluster CAN bus. The range of U is between zero and one. Let $U_{inv} = 1 - U$. The U value that we obtain from the simulation is 0.8, and the U_{inv} under this case is 0.2. We modify the expression as follows.

$$R_f = \frac{T}{C \times N} + 2 \times \frac{T}{C \times N} \times \frac{W - W_f}{W} \times \left(U + U_{inv} \times \frac{L_f}{N}\right) \tag{7}$$

L_f is the number of link fault in the topology. The link fault means that the link is broken and the node cannot transmit messages to the ring bus. The new term represents the extra bandwidth percentage which is gained from intra bus. When there is no link fault, the intra bus has relative low extra bandwidth because the traffic is crowded. However, once the link fault occurs, the traffic is not as crowded as no link fault occurs. So the intra bus has relative high extra bandwidth. We modify the formula as follows. For the single ring topology and the single ring two phase topology, when one of the link is broken, the link fault injection rate becomes zero. It is the node with broken link cannot transmit any messages.

$$R_f = \begin{cases} 0, \ if \ L_f > 0, & single\ ring\ topology \\ 0, \ if \ L_f > 0, & single\ ring\ 2\ phase\ topology \\ \frac{T}{C \times N} + \frac{2 \times T}{C \times N} \times \frac{W - W_f}{W} \times \left(U + U_{inv} \times \frac{L_f}{N}\right), & dual\ ring\ 2\ phase\ topology \end{cases} \tag{8}$$

3 Simulation

3.1 Simulation Construction

The simulation environment used to verify the proposed method is described below. First, we construct the message class data type. This class contains following properties: priority, wait time, deadline, dead flag, send flag, and destination. Priority represents the message priority in the CAN bus, and also represents the source node of this message. Wait time represents the time period that a message waits in the queue. Deadline represents the maximum time of this message can wait in the queue. Send flag represents that the message can be transmitted from the queue. Dead flag represents that the wait time is larger than the deadline time, and this message must be canceled from the queue. When a message is generated, the three parameters (priority, deadline and destination) must be initialized.

Second, a random traffic generator is created. Three parameters (node ID, inject threshold, and time slice) are used to control the random traffic generation. The node ID indicates the source of the random traffic. The inject threshold is the probability of the message generated per time slice. The time slice is the number of total time slice in a simulation.

Then we define the parameters for a simulation that includes the injection rate, time slice, node number, the maximum time of transmission, and deadline. Since the CAN bus uses the priority for arbitration, we build up a queue array to store the messages to be sent. The queue array sends a messages according to the message's priority.

In every time slice, every node takes the message generated from the random traffic, and puts the message into the queue. Then the messages are sorted according to the message's priority. After the sorting process is done, the queue starts to send the message from the first element of the queue. When a message is sent, the property and send flag of this message are set to 1. We also record this message's destination. In this way, we can add up the total number of messages of this node which can be scheduled successfully. The queue keeps sending the messages from the queue until the remained time becomes zero.

After the sending process, the queue removes the object messages in the queue which has the property send flag set to one. The queue also checks the dead flag. If the dead flag is true, then the queue also removes this message object and records it as well. These processes are repeated until the simulation time achieves. Finally, the simulation adds up the number of scheduled messages, failed messages, and total messages.

3.2 Simulation Results

In this section we show the simulation results of different topologies. In Fig.4, the y-axis is the scheduled percentage. The x-axis is the injection rate. Fig.4 shows that single ring topology reaches the maximum traffic capacity when the injection rate increase to 10%. The single ring 2 phase topology reaches the maximum traffic capacity when the injection rate increases to 35%, and so do the dual ring 2 phase topology. From the figure we find out that the dual ring 2 phase topology can achieve more bandwidth than others. Also, the throughput can gain more than the single ring topology. We also see the single ring 2 phase topology extends the bandwidth. Since the single ring 2 phase topology cannot transmit the inter messages, so the scheduled percentage cannot achieve 100%.

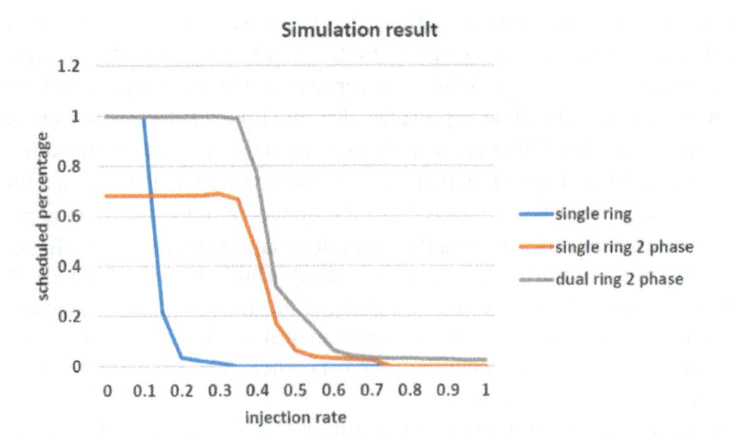

Fig. 4 The scheduled percentage versus injection rate simulation result

Fig.5 is the simulation result of throughput versus different topologies under the no link fault case and the eight link fault case. In Fig.5, the y-axis is the total scheduled message number and the x-axis is the injection rate. From Fig.5, we see that the topology with no link fault achieves the bandwidth saturation earlier. The dual ring 2 phase topology can gain about 3 time injection rate than the single ring topology in both cases.

Fig.6 plots the theoretical injection rate drop versus the fault link. Once the fault link occurs, the single ring topology and the single ring 2 phase topology cannot let the fault link node transmit it message at all, so the injection rate drops to zero. The dual ring 2 phase topology can still transmit the message when the link fault occurs due to the connection to the inter ring bus. From Fig.6, we see that in the dual ring 2 phase case, injection rate drop is very close to the theoretical injection rate drop.

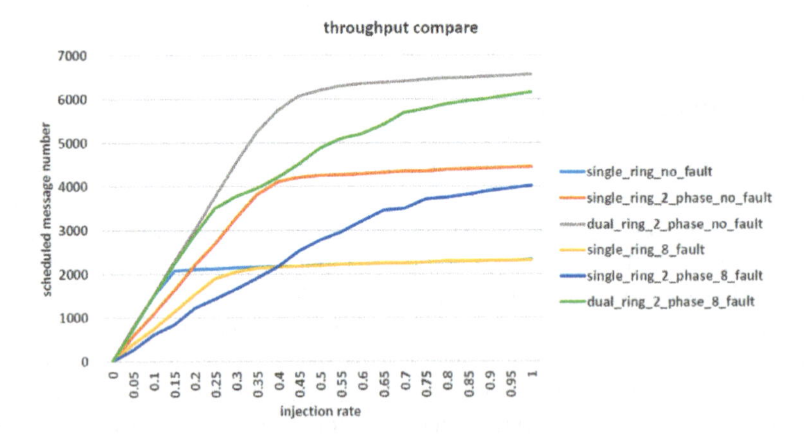

Fig. 5 The throughput versus injection rate simulation result

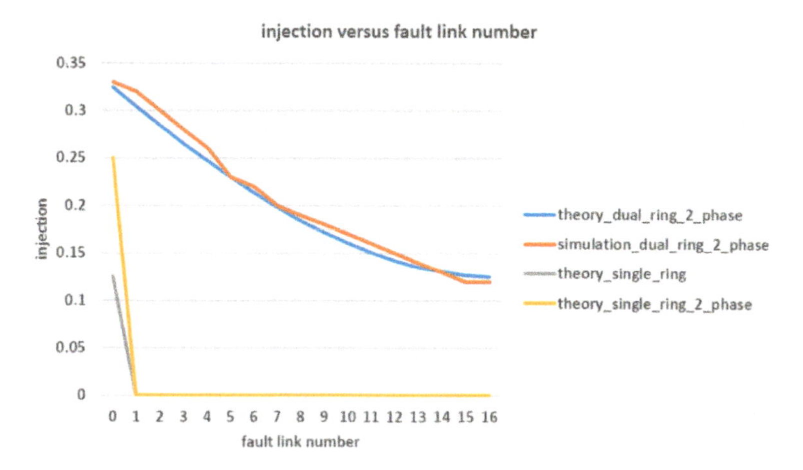

Fig. 6 The theoretical injection rate drop and the simulation injection rate drop result

Fig.7 is the result of the throughput change versus the link fault and under 45% injection rate. We see that the single ring topology maintains saturated when the link fault occurs. It is because the injected messages is far more than the topology can afford under 45% injection rate. However, the single ring 2 phase topology and dual ring 2 phase topology throughput do not encounter this problem.

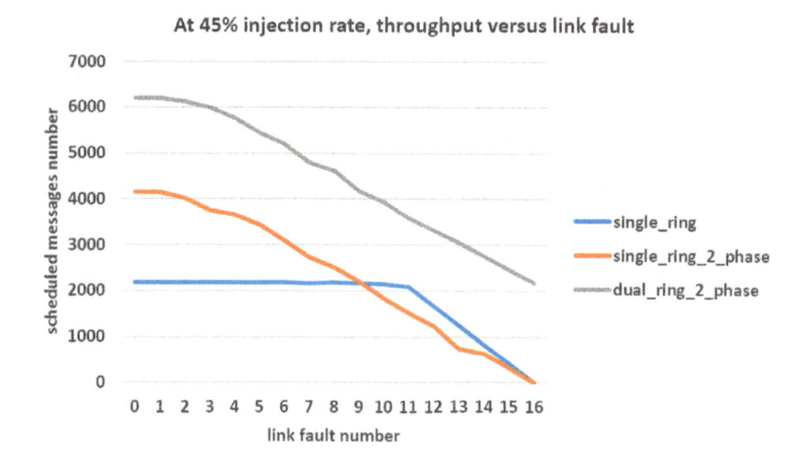

Fig. 7 At 45% injection rate, throughput versus link fault simulation result

4 Conclusion

In this paper, we provide a theoretical model to analyze the relation between the injection rate and the scheduled messages for three topologies including single ring, single ring 2 phase, dual ring 2 phase. We provide a step by step simulation construction and use it to verify the model. Under this model, we can estimate the injection rate of every node correctly when the link fault occurs. It helps to figure out the maximum injection message can be sent through the bus under the fix number nodes condition for building up the CAN protocol system.

References

[1] Bosch, Can specification version 2.0. Robert Bosch GmbH, Postfach30 02 40 D-70442, Stuttgart (1991)
[2] Natale, M.D.: Understanding and using the controller area network. Handout of a lecture at U.C. (October 2008)
[3] Liu, W.-C.: Cluster-Based CAN with Enhanced Transmission Capability for Vehicle Networks. In: International Conference on Connected Vehicles and Expo, pp. 43–48 (2012)
[4] Chen, J., Jone, W., Wang, J., Lu, H.-I., Chen, T.: Segmented bus design for low-power systems. IEEE Transactions on Very Large Scale Integration (VLSI) Systems 7(1), 25–29 (1999)
[5] Tindell, K., Hansson, H., Wellings, A.: Analysing real-time communications: controller area network (can). In: Proceedings of the Real-Time Systems Symposium 1994, pp. 259–263 (December 1994)
[6] Davis, R., Burns, A., Bril, R., Lukkien, J.: Controller area network (can) schedulability analysis: Refuted, revisited and revised. Real-Time Systems 35, 239–272 (2007), http://dx.doi.org/10.1007/s11241-007-9012-7, doi:10.1007/s11241-007-9012-7
[7] Bril, R.J., et al.: Message response time analysis for ideal Controller Area Network (CAN) refuted. In: Proceedings of the 5th International Workshop on Real-Time Networks (RTN (2006)
[8] Aysan, H., Dobrin, R., Punnekkat, S.: Fault Tolerant Scheduling on Controller Area Network (CAN). In: 13th IEEE International Symposium on Real-Time Distributed Computing, pp. 226–232 (2010)
[9] Seceleanu, T., Stancescu, S.: Arbitration for the segmented bus architecture. In: Proceedings of the 2004 International Semiconductor Conference, CAS 2004, vol. 2, pp. 487–490 (October 2004)

Context-Based Service Fusion for Personalized On-Board Information Support

Alexander Smirnov, Nikolay Shilov, Aziz Makklya and Oleg Gusikhin

Abstract. Current in-vehicle information systems make it possible to benefit from integration of new ideas to provide richer driving experience. The paper presents a concept, main supporting technologies and an illustrative case study for improved on-board information system. The key idea of the proposed approach is to implement context-based service fusion supported by a negotiation model. This would provide a new, previously unavailable level of personalized on-board information support via finding compromise decisions taking into account proposals of various services and driver preferences.

Keywords: personalized on-board information support, context, service fusion.

1 Introduction

Modern navigation systems incorporate such ideas as average traffic speed on roads, generation of different routes (e.g., fastest, "green", easiest, etc.) and indicate various points of interests (POI) along the route. However, one cannot create a route from point A to point B e.g., "with a feature to see the most interesting POIs, crossing the country border where and when it the least crowded, and be in time for the ferry (all at the same time)". Besides, the system has to propose such routes based on the driver's explicit and tacit preferences even though he/she has never been in this area before.

A. Smirnov(✉) · N. Shilov
St. Petersburg Institute for Informatics and Automation of the Russian Academy of Sciences (SPIIRAS), 39, 14 Line, St. Petersburg, 199178, Russia
e-mail: {smir,nick}@iias.spb.su

A. Makklya · O. Gusikhin
Ford Motor Company, P.O. Box 6248, Dearborn, MI 48126, USA
e-mail: {amakkiya,ogusikhi}@ford.com

A. Smirnov
University ITMO, 49, Kronverkskiy pr., St. Petersburg, 197101, Russia

J. Fischer-Wolfarth and G. Meyer (eds.), *Advanced Microsystems for Automotive Applications 2014*, Lecture Notes in Mobility,
DOI: 10.1007/978-3-319-08087-1_11,

However, current developments of on-board information systems (i.e., Ford's AppLink, Chrysler's UConnect, Honda's HomeLink, etc.) make it possible to benefit from their integration with other information and decision support systems to provide a richer driving experience and seamless integration of information from various sources.

The proposed approach is a step to "infomobility" infrastructure, e.g. towards operation and service provision schemes whereby the use and distribution of dynamic and selected multi-modal information to the users (drivers), both pre-trip and, more importantly, on-trip, play a fundamental role in attaining higher traffic and transport efficiency as well as higher quality levels in travel experience by the users [1].

Correct decisions can only be made in the right context related to the current situation. Context is any information that can be used to characterize the situation of an entity where an entity is a person, place, or object that is considered relevant to the interaction between a user and an application, including the user and applications themselves [2]. The context is updated depending on the information from the environment. The context updates the service's knowledge, which in turn defines its behaviour. The ability of a system to describe, use, and adapt its behaviour to its context is referred to as self-contextualization [3]. The presented approach exploits the idea of self-contextualization to adapt autonomously behaviours of multiple services to the context of the current situation in order to provide their services according to this context and to propose context-based decisions.

The idea of service fusion originates from the concept of knowledge fusion, which implies a synergistic use of knowledge from different sources in order to obtain new information [4]. Thus, service fusion in this work can be defined as synergistic use of different services to have new driver support possibilities not achievable via usage of the services separately.

Context-based service fusion can provide a new, previously unavailable level of personalised on-board information support via finding compromise decisions taking into account proposals of various services and driver preferences. For each particular situation, there can be a large amount of feasible solutions for the drivers to choose from (i.e., the fastest route, the most spectacular route, a trip scheduled in accordance ferry schedule, etc.). The proposed technology could help to choose the most acceptable one for the driver in the current situation. Such a system could be capable to:

- Propose most feasible trip route and schedule taking into account not only preferences of the driver but also taking into account preferences of the previously served drivers with similar interest
- Take into account the context model of the current situation and its predicted development
- Automatically apply for required services (such as ferry fees)
- Continuously learn from all the users

For example, such system could support the following scenarios:

- You want to re-fuel the car and have a dinner in a decent restaurant. Instead of finding a cheapest gas station, the system finds a gas station located near a restaurant, which has good feedback from its customers or belongs to the brand preferred by you.
- You are driving to a meeting. For some reason the meeting is postponed and you have some free time. The system will recommend visiting a POI, which is on your way.
- You are driving home after work. You need to buy some groceries and have a couple of shops to do this. The system recommends visiting a particular one of these, because your old friend Jack will be there at the same time.

The paper is structured as follows. Sec. 2 describes the service fusion processes. Sec. 3 explains the mechanism of self-configuration supporting service fusion. A case study is presented in sec. 4. Major results are summarized in sec. 5.

2 Context-Based Service Fusion

In the presented approach we aim to achieve service fusion on the basis of the idea of self-organizing systems. The long-term vision of the presented research is to facilitate the development of "truly" self-organizing service networks systems providing driver infomobility support suggesting different solutions based on different information sources and taking into account different factors. The paper concentrates on the network architecture and its self-organization and proposes some ideas in the areas of context-based self-organization and adaptive behaviour for providing new capabilities.

Self-organizing systems are characterised by their capacity to spontaneously (without external control) produce a new organization in case of environmental changes. These systems are particularly robust, because they adapt to these changes, and are able to ensure their own survivability. Knowledge fusion from different sensors, services and components is decisive for efficient context-based self-organization [5].

A self-organizing network is an interconnected network of multiple entities (or self-interested agents) that exhibit adaptive action in response to changes in both the environment and the system of entities itself [6]. The key mechanisms supporting self-organizing networks are self-organization mechanisms and negotiation models. The following self-organization mechanisms are usually selected [7]: intelligent relaying, adaptive cell sizes, situational awareness, dynamic pricing, intelligent handover.

We argue that in addition to the traditional and well-understood centralized solutions for pre-defined tasks, emerging technologies of semantic service-oriented architectures - agents, ontology, and Web services - are equally significant for self-organization and adaptation to continuously changing environments. Instead of responding to requests for information, they permit, intelligently anticipate, adapt, and automate driver support tasks. The following negotiation models can be mentioned [8]:

- Different forms of spontaneous *self-aggregation*, to enable both multiple distributed services / agents to collectively and adaptively provide a distributed service, e.g. a holonic (self-similar) aggregation.
- *Self-management* as a way to enforce control in the ecology of services / agents if needed (e.g. assignment of "manager rights" to a service / agent.
- *Situation awareness* – organization of situational information and their access by services / agent, promoting more informed adaptation choices by them and advanced forms of stigmergic (indirect) interactions.

The fusion of services can only be achieved if the following major principles of collaboration are used as the basis for self-organization:

1. *Contribution*: the services have to cooperate with each other to make the best contribution into the overall system's benefit – not into the agents' (services') own benefits.
2. *Task performance*: the main goal is to complete the task performance – not to get profit out of it.
3. *Non-mediated interaction*: the services operate in a decentralized community and in most of the negotiation processes there are no services managing the negotiation process and making a final decision.
4. *Common terms*: since the services work in the same system they use common terms for communication. This is achieved via usage of the common shared ontology.
5. *Trust*: since the services work in the same system they can completely trust each other (the services do not have to verify information received from other services).

As was mentioned above, the service fusion problem refers to integration of functionalities of different services to obtain new functionality, which is not possible when the services are used separately. The main feature of the service fusion lies in creation of synergetic effect from the integration of services. Basically, such effect can be achieved through integration of both tacit and explicit knowledge as well as through their combination [4]. Based on the analysis of knowledge fusion studies, five service fusion processes eventuated in different results can be distinguished (Table 1).

The importance of knowledge sharing in self-organizing service networks, both at technical and semantic levels, has been shown by [16, 17]. Ontologies are widely used for problem domain description in modern information systems to support semantic interoperability. Ontology is an explicit specification of a structure of a certain domain. It includes a vocabulary for referring to notions of the subject area, and a set of logical statements expressing the constraints existing in the domain and restricting the interpretation of the vocabulary. Ontologies support integration of resources that were developed using different vocabularies and different perspectives of the data. They are usually built in a semi-automatic way based on existing databases and electronic documents. To achieve semantic interoperability, systems must be able to exchange data so that the precise meaning of the data is readily accessible and the data itself can be translated by any system into the form that it understands [18].

An ontological model is used in the approach to solve the problem of service heterogeneity. This model makes it possible to enable interoperability between heterogeneous information services due to provision of their common semantics and terminology. Application of the context model makes it possible to reduce the amount of information to be processed.

“Self-contextualization” is one of the key enablers for defining the current context of the "car-driver" system. It is the ability of the system to describe, use and adapt its behaviour to its context [3]. The context updates the parametric knowledge of services, which in turn defines their behaviour (capability to perform certain actions in order to change the own state and the state of the environment from the current to the preferred ones). The proposed approach exploits the idea of self-contextualization to autonomously adapt behaviours of multiple services to the context of the current situation in order to provide their services according to this context and to propose context-based decisions. For this reason, the proposed conceptual model enables context-awareness and context-adaptability of the planner and driver and car context services.

Table 1 Knowledge fusion processes and their possible results

#	*Service fusion process*	*Service fusion result*
1	Intelligent fusion of several heterogeneous services into a view that may be used by systems and humans as the basis for problem solving and decision making [9, 10]. Intelligent fusion assumes taking into account the semantic content of the services being fused.	New “complex” service solving a new problem
2	Inference of explicit knowledge from information / knowledge hidden in services fused [11]	New knowledge about knowledge object
3	Combining different autonomous services in different ways in different scenarios, which results in discovery of new relations between different services [12-13]	New relations between services
4	Re-configuration of the service network to achieve a new configuration with new capabilities or competencies [14]	New capabilities / competencies of a service network
5	Involvement of knowledge from various services in problem solving that results in a new service [15]	New “complex” service solving a problem

The advantage of a such system is that the system can adapt to the current situation. For example, if the driver changes the route for some reason, the system will adapt the solution based on the new information (e.g., new GPS coordinates) and provide new directions in the car navigation system.

3 Service Network Self-configuration via Agent-Based Negotiation

As was mentioned, self-organizing networks provide a better coordination of independent services. Each service of the network has its own knowledge stored in its knowledge base. This knowledge is described by a portion of the common shared ontology related to the current service's tasks and capabilities. Capabilities, preferences and other information about the service are stored in its profile that is available for viewing by other services of the system. This facilitates communication, which is performed via the communication module responsible for meeting protocols and standards that are used within the system.

In order to make services capable not only to respond to requests for information, but also to intelligently anticipate, adapt, and automate their behaviour, they are assigned agent functionality (such service will be referred as "agent-based service"). Intelligent agent is an autonomous software entity that can navigate a heterogeneous computing environment and can, either alone or working with other agents, achieve some goals [19].

The services communicate with other services for two main purposes: (1) they establish links and exchange information for better situation awareness; and (2) they negotiate and make agreements for coordination of their activities for a proposed solution. The services may also get information from various information sources (e.g. provided by Ford's AppLink).

Most multiagent systems require using agent negotiation models to operate. The negotiation models are based on the negotiation protocols defining basic rules so that when agents follow them, the system behaves as it is supposed to. The main protocols include voting, bargaining, auctions, general equilibrium market mechanisms, coalition games, and constraint networks.

However, any usage protocol requires an objective function to operate. We propose to use "utility" of the solution as the objective function taking into account driver preferences. This utility characterizes the "usefulness" of the solution for the driver. This utility can be calculated as a weighted sum of utilities of various activities including in the solution. An example of the utility usage is presented in the next section.

4 Case Study

The approach demonstration is based on the following scenario: *You need to re-fuel the car (based on the automatic gas level identification) and have some rest and a dinner in a decent restaurant (based on the automatic fatigue level identification depending on how long you have been driving). Instead of finding a cheapest gas station, the system finds a gas station located near a restaurant, which has good feedback from its customers or belongs to the brand preferred by you.*

This solution consists of two actions (visiting a restaurant and refuelling the car) and involves three negotiating services: restaurant advisor, gas station advisor [20] and planner (this service, responsible for time keeping, is involved in almost any scenario in order to avoid solutions which would suggest driving too far away). Each of the three services are assigned certain functions calculating degree of usefulness of their suggestions for the driver (e.g., visiting a café with average customer ratings has a lowest utility, visiting a nice restaurant with high customer ratings is estimated has a higher utility, and visiting the favourite driver's restaurant has the highest utility). The utility scale of the planner service might depend on usual distances driven by the driver, his/her preferences and current schedule. The total utility of the solution depends on the contributions of each participating service. The appropriate mathematical models are yet to be developed.

In order for such a mechanism to operate efficiently, it requires a continuous adjustment of the services' utilities. This can be done through collecting information and knowledge from different sources. A taxonomy of knowledge sources can be found in [21], p. 369. Among those, mentioned in the taxonomy, the following ones can be mentioned:

1. User feedback (the driver can increase or reduce the utility of a certain service). This is a reliable information source; however, in real life it is very unlikely, that the driver will provide such feedback.
2. Initial driver profile (the driver can fill out the initial preferences in his/her profile). This is also a reliable information source but such information will be outdated after some time.
3. Analysis of driver decisions (the system can analyse if the driver followed the proposed solution, or which solution is preferred if several alternative solutions are presented to the driver). This is a less reliable information source, but such information will never be outdated and development of learning algorithms can significantly improve such feedback.
4. Analysis of decisions of drivers with similar interests/habits. This source originates from the method of collaborative filtering used in group recommendation systems.

The interaction between services is presented in fig. 1. It is based on usage of AppLink for interaction with the vehicle. In addition to the information already stored in the services (associated databases, user settings, revealed preferences, etc.), they acquire the following information from other services, namely:

- Gas station advisor obtains current car location, gas level, and predefined driver preferences.
- Restaurant advisor obtains current car location and predefined driver preferences.
- Planner obtains driver's schedule from his/her smartphone and predefined driver preferences to estimate current time restrictions.

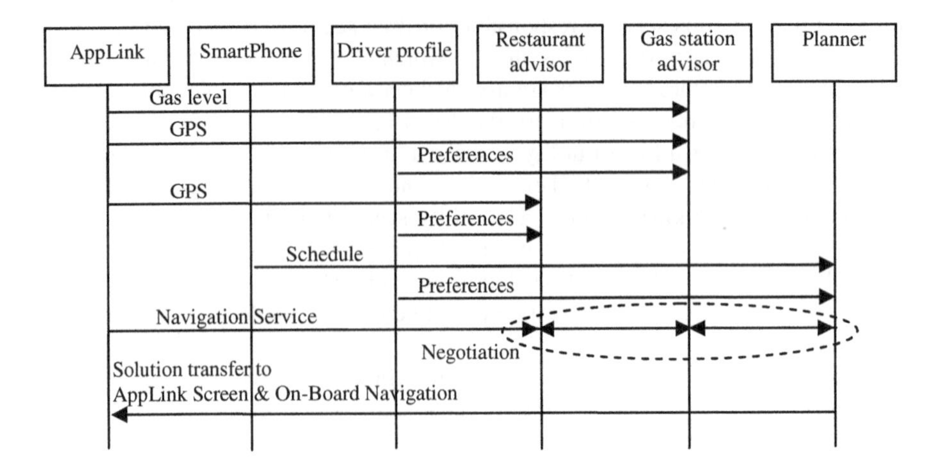

Fig. 1 Service interaction example

After that, the services negotiate in order to generate one or several alternative solutions based on the driver requirements. During this negotiation, they can query available navigation system to estimate the driving time between different locations. Finally, the generated solutions are transferred to the AppLink screen so that the driver could choose the most appropriate one, and to the in-car navigation system.

5 Conclusion

The paper presents the concept, main supporting technologies and an illustrative case study for improved on-board information system. It is proposed to extend existing information systems with additional services, providing useful information. The key idea is to implement context-based service fusion supported by a negotiation model. The context-based service fusion can provide a new, previously unavailable level of personalised on-board information support via finding a compromise decisions taking into account proposals of various services and driver preferences.

Acknowledgements. The research was supported partly by projects funded by grants # 13-07-13159, # 13-07-12095, # 14-07-00345, # 12-07-00298, and # 12-07-00302 of the Russian Foundation for Basic Research, project 213 (program 15) of the Presidium of the Russian Academy of Sciences, and project #2.2 of the basic research program "Intelligent information technologies, system analysis and automation" of the Nanotechnology and Information technology Department of the Russian Academy of Sciences. This work was also partially financially supported by Government of Russian Federation, Grant 074-U01.

References

[1] Ambrosino, G., Boero, M., Nelson, J.D., Romanazzo, M. (eds.): Infomobility Systems and Sustainable Transport Services, 336 p. ENEA Italian National Agency For New Technologies, Energy And Sustainable Economic Development (2012)

[2] Dey, A.K., Salber, D., Abowd, G.D.: A Conceptual Framework and a Toolkit for Supporting the Rapid Prototyping of Context-Aware Applications. Context-Aware Computing, A Special Triple Issue of Human-Computer Interaction 16, 229–241 (2001)

[3] Raz, D., Juhola, A.T., Serrat-Fernandez, J., Galis, A.: Fast and Efficient Context-Aware Services. John Willey & Sons, Ltd. (2006)

[4] Smirnov, A., Levashova, T., Shilov, N.: Patterns for Context-Based Knowledge Fusion in Decision Support. Information Fusion (in press), http://dx.doi.org/10.1016/j.inffus.2013.10.010

[5] Smirnov, A., Sandkuhl, K., Shilov, N.: Multilevel Self-Organisation of Cyber-Physical Networks: Synergic Approach. International Journal of Integrated Supply Management 8(1/2/3), 90–106 (2013)

[6] Choi, Y.T., Dooley, K.J., Rungtusanatham, M.: Supply Networks and Complex Adaptive Systems: Control versus Emergence. Journal of Operations Management 19, 351–366 (2001)

[7] Telenor R&D, Report, Project No TFPFAN, Program Peer-To-Peer Computing (2003), http://www.telenor.com/rd/pub/rep03/R_17_2003.pdf

[8] De Mola, F., Quitadamo, R.: Towards an Agent Model for Future Autonomic Communications. In: Proceedings of the 7th WOA 2006 Workshop From Objects to Agents (2006), http://sunsite.informatik.rwth-aachen.de/Publications/CEUR-WS/Vol-204/P07.pdf

[9] Scherl, R., Ulery, D.L.: Technologies for Army Knowledge Fusion. Final report, Monmouth University, Computer Science Department, West Long Branch; Report No. ARL-TR-3279 (2004)

[10] Alun, P., Hui, K., Gray, A., Marti, P., Bench-Capon, T., Cui, Z., Jones, D.: Kraft: an Agent Architecture for Knowledge Fusion. International Journal for Cooperative Information Systems 10(1-2), 171–195 (2001)

[11] Roemer, M.J., Kacprzynski, G.J., Orsagh, R.F.: Assessment of Data and Knowledge Fusion Strategies for Prognostics and Health Management. In: Proceedings of 2001 IEEE Aerospace Conference, vol. 6, pp. 2979–2988 (2001)

[12] Laskey, K.B., Costa, P., Janssen, T.: Probabilistic Ontologies for Knowledge Fusion. In: Proceedings of 2008 IEEE 11th International Conference on Information Fusion (2008), http://ieeexplore.ieee.org/xpl/freeabs_all.jsp?arnumber=4632375

[13] Jonquet, C., LePendu, P., Falconer, S., Coulet, A., Noy, N.F., Musen, M.A., Shah, N.H.: NCBO Resource Index: Ontology-Based Search and Mining of Biomedical Resources. Journal of Web Semantics 9(3), 316–324 (2011)

[14] Lin, L.Y., Lo, Y.J.: Knowledge creation and Cooperation between Cross-Nation R&D Institutes. International Journal of Electronic Business Management 8(1), 9–19 (2010)

[15] Smirnov, A., Pashkin, M., Chilov, N., Levashova, T., Haritatos, F.: Knowledge Source Network Configuration Approach to Knowledge Logistics. International Journal of General Systems 32(3), 251–269 (2003)

[16] Sandkuhl, K., Smirnov, A., Shilov, N.: Configuration of Automotive Collaborative Engineering and Flexible Supply Networks. In: Cunningham, Cunningham (eds.) Expanding the Knowledge Economy – Issues, Applications, Case Studies, pp. 929–936. IOS Press, Amsterdam (2007)

[17] Smirnov, A., Shilov, N., Kashevnik, A.: Developing a Knowledge Management Platform for Automotive Build-To-Order Production Network. Human Systems Management 27(31), 15–30 (2008)

[18] Heflin, J., Hendler, J.: Semantic Interoperability on the Web. In: Proceedings of Extreme Markup Languages, pp. 111–120. Graphic Communications Association (2000)

[19] Franklin, S.: Is It an Agent, or Just a Program?: A Taxonomy for Autonomous Agents. In: Jennings, N.R., Wooldridge, M.J., Müller, J.P. (eds.) ECAI-WS 1996 and ATAL 1996. LNCS, vol. 1193, pp. 21–35. Springer, Heidelberg (1997)

[20] Klampfl, E., Gusikhin, O., Theisen, K., Liu, Y., Giuli, T.J.: Intelligent Refueling Advisory System. In: Proceedings of 2nd Workshop on Intelligent Vehicle Control Systems, Madeira, Portugal, pp. 60–72 (2008)

[21] Burke, R., Ramezani, M.: Matching Recommendation Technologies and Domains. In: Ricci, F., Rokach, L., Shapira, B., Kantor, P.B. (eds.) Recommender Systems Handbook. Springer (2011)

Prediction of Switching Times of Traffic Actuated Signal Controls Using Support Vector Machines

Toni Weisheit and Robert Hoyer

Abstract. At signalized intersections there is a significant saving potential of emissions by an energy-efficient and fuel-optimized approach to the stop line. For this purpose, various assistance systems have already been developed. Among other things these systems provide the driver with speed recommendations to cross the next traffic light without stopping. However, accurate information about forthcoming traffic signal switching times is required. Modern traffic signal systems adapt their switching times depending on the current traffic flow. So a predicted phase transition will only occur with a smaller probability than 100%. The paper identifies specific challenges by developing an algorithm for a prediction of traffic actuated signal controls and it presents its mathematical foundations and the results of the prediction.

Keywords: Prediction, Switching Times, Traffic actuated Signal Controller, Support Vector Machines.

1 Background and Motivation

Stops at signalized intersections and resulting acceleration processes on restarting have a significant impact on fuel consumption and emissions of motorized traffic. The cooperation between infrastructure and vehicles via an exchange of data and information is a promising way in order to improve traffic efficiency in urban areas with simultaneous reduction of emissions. For several years the number of driver assistance and information systems in mass-production vehicles has been increasing more and more. Because these systems were previously largely autarkical working, future systems to be developed should be interconnected with

T. Weisheit(✉) · R. Hoyer
University of Kassel, Department of Traffic Engineering and Transport Logistics,
Mönchebergstr. 7, 34125 Kassel, Germany
e-mail: {toni.weisheit,robert.hoyer}@uni-kassel.de

J. Fischer-Wolfarth and G. Meyer (eds.), *Advanced Microsystems for Automotive Applications 2014*, Lecture Notes in Mobility,
DOI: 10.1007/978-3-319-08087-1_12,

the infrastructure for a preferable energy-efficient driving. The potentials concerning the reduction of emissions by an exchange of data and information between infrastructure and vehicles have already been shown in [1]. Especially with the information about forthcoming switching times of fixed timed traffic signals, the driver is able to drive through the signalized road network in an energy- and emission-optimal way. The supply of future switching times of fixed timed traffic signals is trivial. However, modern traffic signal systems adapt their phases and phase transitions to the current traffic situation. So their switching times may vary cycle by cycle. Consequently, a prediction of the switching times is required. This is an indispensable prerequisite for the realization of vehicle functions which shall support a fuel-efficient and a low-pollution driving in urban areas. However, a predicted phase transition, respectively a switching time, will only occur with a probability smaller than 100%. Several approaches for a prediction of traffic actuated switching times already exist. The mathematical approach of Markow chains was used in [2] to calculate the probability of occurrence of different states which are represented by different phases and phase transitions. In [3], the past circuit information of the traffic lights are used to generate a prediction. By additional information, such as predicted arrival times of public transport vehicles and traffic demand prognosticated in the traffic management center, an enhancement of the prediction is possible. Furthermore, an approach based on the theory of a finite state machine is presented in [4]. Here, all detector data like time headways or degree of occupation were used to model the correspondent traffic actuated signal control. For the development of driver assistance systems, which shall enable an energy-efficient driving in urban areas, the supply of the switching times is not only required for individual traffic signal systems but also for their extensive availability in the road network. Therefore, an easy transferability of the prediction algorithms to other traffic signal systems has to be noted while being developed. Regardless of the methods utilized for the prediction, some important constraints have to be considered in the development process, which have a direct impact on the quality of the prediction.

2 Boundary Conditions for a Prediction of Switching Times

The modality of traffic dependency has the greatest influence on the quality of the prediction. In Germany, it can be divided into signal program adaption and signal plan generation as generic terms. With a signal program adaption, modifications of green periods may be carried out depending on the fulfilment of certain criteria. Furthermore, a request of a demand phase is possible. Here, irregular phase insertions for temporary traffic streams like public transport vehicles are conducted. With a signal program adaption several temporal core areas for red and green still exist in a cycle. With a signal plan generation, the phase sequence and cycle time vary in addition to the number of phases and green periods. In these cases, a recurring pattern of the phases is hard to discern. In addition, the demand on a comprehensive supply of switching times complicates the development of algorithms in this context.

The latency period given by the technical system is another important factor. The latency is defined as a time lag between the times of data acquisition by the devices in the field and the availability in the traffic management center. The mathematical approach of Markow chains mentioned above requires short latencies in the low single-digit range of seconds. The implementation of an online prediction is more and more impeded with increasing latency because essential system correlations are occasionally not identified in time by the algorithms.

The application-specific horizon of the prediction is a third condition to be considered. An efficient routing, which depends on the mode of drive, requires a prediction of switching times of all traffic lights on possible links with a forecast of several minutes. However, to reduce emissions by switching off the engine while stopping at a red signal head, a horizon in the range of a phase length is sufficient. The larger the horizon is, the worse the quality of the prediction will be, since the switching times depend on the future traffic situation which cannot be predicted exactly. Furthermore, a once calculated prediction can immensely vary during the access to a traffic light system by using updated data. So the driver´s acceptance would be affected regarding the resulting policy proposal.

3 Approach for a Prediction

3.1 Preliminary Remark

The simplest method to predict future switching times is a calculation of their relative frequencies using switching times of passed cycles. With a large population of data, those frequencies correspond to an occurrence probability (law of large numbers). Figure 1 shows the signal layout plan of an intersection in the German City of Duesseldorf. As an example, a prediction of the end of green period of signal head named DR is used.

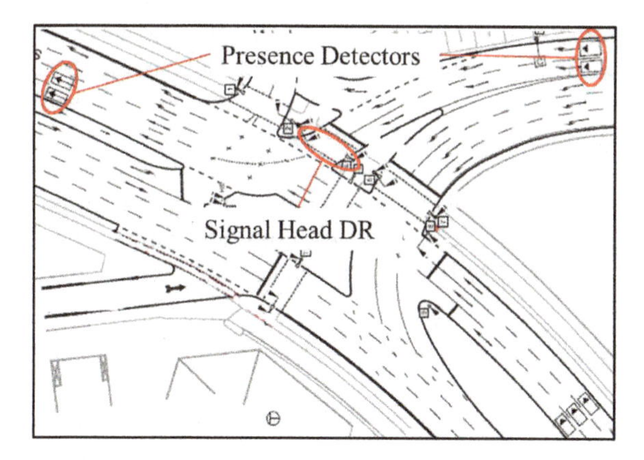

Fig. 1 Signal layout plan of an intersection in the city of Duesseldorf

The cycle time is constant at 70 seconds. The green period of the signal head will be extended for a certain time slice which depends on two occupancy criteria of the marked detectors. Traffic signal systems occasionally provide different signal programs for different times of day. So the observed switching times are analysed by their corresponding signal programs in order to get a preferably good result. The signal control of the chosen example consists of three different signal programs. As population for a calculation of occurrence probabilities of possible ends of green period, the switching times of about 1,700 cycles were evaluated (approximately 33 h). Due to the relation of the observed switching times to the total number of cycles for the appropriate signal programs, following occurrence probabilities arose.

Table 1 Occurrence probabilities of end of green period with separated signal programs

End of Green Period [Cycle Second]	17	20	23	24
Occurrence Probability [%] Signal Program 1	0	15	76	9
Occurrence Probability [%] Signal Program 2	33	21	0	46
Occurrence Probability [%] Signal Program 3	46	18	27	9

This approach is not a satisfactory solution due to partly significant uniformly distributed values. For this reason, the inclusion of traffic data, in particular the detector data, is the obvious next step. For this purpose, the so called Support Vector Machines (SVM) are used as a mathematical approach. This algorithm divides a set of objects into classes. Thereby, a widest possible area around the class limits remains free of objects. The class limit is called hyperplane in this context. Thus, SVM can be considered as a classifier. In this particular case, the objects are represented by different detector data. The basis for the classification is an appropriate training data set used by the algorithm to learn the formation of the class limits. Afterwards, the accuracy of the prediction can be verified by a test data set. This approach has already been used in [5] to predict switching times of traffic actuated signal controls in Singapore. Thereby, phase lengths and traffic volumes of the last five cycles were used as training data. The general functionality of SVM will be briefly outlined below.

3.2 Support Vector Machines

A set of feature vectors $x_1, x_2, \ldots, x_n \in X$ and the respective class labels $y_1, y_2, \ldots, y_n$ with $y_i \in \{+1, -1\}$ constitutes the basis for this machine learning algorithm. The dimension of the feature vectors corresponds to the number of input variables respectively detector data, used for a prediction. To find the best possible parting plane for the feature vectors, an edge is inserted on both sides of the plane. This edge is widened until it contacts feature vectors of the two classes which are called support vectors. The resulting plane is called hyperplane. Figure 2 shows the principle of separation of the feature vectors by a hyperplane.

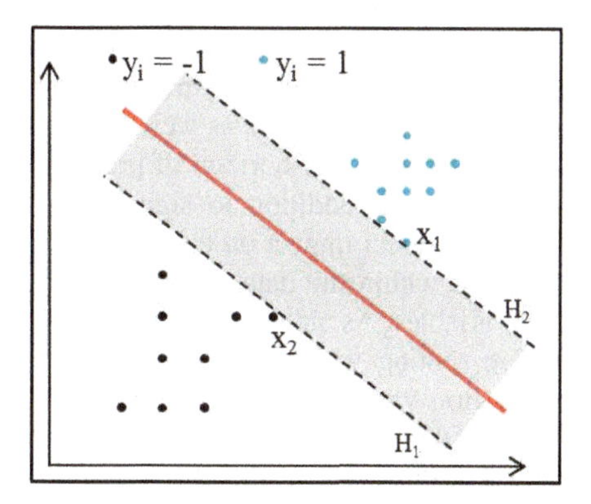

Fig. 2 Separation of feature vectors by a hyperplane

To find the best hyperplane, SVM maximize the distance between the planes H_1 and H_2 (Figure 2). Finally, the classification of a new feature vector is calculated by a decision function that only depends on the support vectors. Due to shortage of space, [6] is referred for a more detailed description of the derivation of this decision function at this point. In reality, however, the training data are often not linearly separable. Accordingly, a suitable non-linear hyperplane has to be found. For this purpose SVM use the so called kernel trick. Thereby, the data are mapped into a space of higher dimension which is called feature space. In such a space, a linear hyperplane can be calculated by scalar products. The hyperplane becomes non-linear by transforming the data back into the lower dimensional space. Figure 3 illustrates this principle with the input space on the left and the feature space on the right.

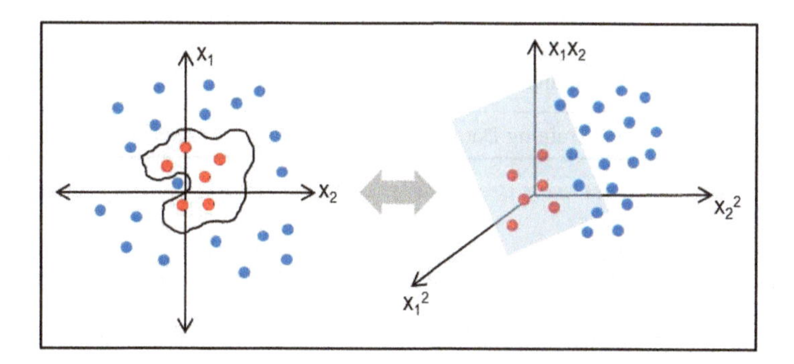

Fig. 3 Principle of generating a complex hyperplane by transforming data into a feature space

At this point [6] and [7] are referred for a more detailed description of the kernel trick.

3.3 Generation of Data Models

The underlying data model is decisive for the classification and consequently for the prediction accuracy. With its help the algorithm learns the relationships between the various data. For instance, these data could be time gaps, traffic volumes or even logons and logoffs of public transport vehicles as well. The data have to be prepared in various ways according to the required horizon of prediction. For the example in Figure 1, detector data are used in addition to signal specific states to train the classifier, because these have a direct impact on the signal control. The detector data are handled as input variables within the data model, while the resulting switching times are treated as target variables. As already described, a horizon of several cycles is required for an effective routing which depends on the mode of drive. So the switching times in combination with the traffic volumes (q) of three cycles before (q_U(i-3)) are used for a first prediction. Its accuracy can be improved by using updated data subsequently. For this purpose, the traffic volumes of the previous cycle (q_U(i-1)) are used as input variables. The traffic volume may change cycle by cycle whereby different switching times may occur too. However, the traffic volume which depends on the time of day, can be approximately classified by using detector data of cycles U(i-3) and U(i-1). So a prediction can be improved. An advantage of the approach shown here is that the data from all detectors of the considered intersection are included into the data model. Thus, an assignment of detectors to the individual signal heads is not necessary in advance. So the prediction approach can be easily transferred on other signalized intersections. Seven detectors exist for the example in Figure 1. It means that seven different input variables will arise for each horizon of the prediction. Table 2 summarizes the variables for a prediction of the end of green period (t_EG) of signal head (SH) DR for one, respectively three cycles in the future. By using detector data of the current cycle, a further correction of the calculated prediction is possible. For this purpose, traffic volumes are not used but detection events are transferred into a data model. For the example shown, the end of green period depends on two certain occupancy criteria of the marked detectors in Figure 1. In this way, the particular cycle second for the last detection event before the

Table 2 Data model for a prediction with different horizons

	Training Data		Prediction data	
Horizon of Prediction	U(i+1)	U(i+3)	U(i+1)	U(i+3)
Target Variable	t_EG (SH DR)_U(i)	t_EG (SH DR)_U(i)	t_EG (SH DR)_U(i+1)	t_EG (SH DR)_U(i+3)
Input Variables	q (Det 1)_U(i-1) ... q (Det 7)_U(i-1)	q (Det 1)_U(i-3) ... q (Det 7)_U(i-3)	q (Det 1)_U(i) ... q (Det 7)_U(i)	q (Det 1)_U(i) ... q (Det 7)_U(i)

Table 3 Data model for a short-term correction of prediction

	Training Data	Prediction Data	
Horizon of Prediction	U(i)	U(i)	with i, k ∈ ℕ
Target Variable	t_EG (SH DR)_U(i-k)	t_EG (SH DR)_U(i)	
Input Variables	t_LDG (Det 1)_U(i-k)	t_LDG (Det 1)_U(i)	
	...	...	
	t_LDG (Det 7)_U(i-k)	t_LDG (Det 7)_U(i)	

occurred end of green period of each cycle (t_LDG) is used as input variable. However, because of the very short horizon in this case, a preliminary calculated policy proposal of a driver assistance system cannot be corrected in time. Table 3 summarizes the variables for this data model.

3.4 Results of Prediction

In advance, it has to be mentioned that the results described below, using a data model according to Table 2, exclusively refer on a horizon of U(i+1), because the results of U(i+3) are very similar. For examining the prediction accuracy, real data of the intersection shown in Figure 1 were used with a basic population of about 5,600 cycles. An automatic classification of the input variables according to the principle of SVM was implemented using the statistical program R. The algorithm uses 70% of data to generate a hyperplane and 30% to test the classification. The results shown below refer to the test data. Table 4 shows a comparison of predicted switching times and actual occurred ones for the end of green period (t_EG) of signal head DR and a horizon of one cycle for signal program 1 (compare table 1). It must be noted, that the switching times of DR have the greatest variance for that intersection. So the results for the remaining signal heads are considerably better.

Table 4 Comparison of predicted and occurred wwitching times for one signal program

	t_EG	predicted		
	[Cycle Second]	20	23	24
occured	20	19	99	0
	23	19	933	0
	24	9	32	1

Altogether, there are 1,112 predictions for this example. In this case the algorithm classifies 19 (cycle second 20) + 933 (23) + 1 (24) = 953 times the end of green period correctly. This corresponds to a prediction accuracy of about 86%. This is an improvement compared to the best value of table 1 which has been calculated on the same database. The prediction values of the remaining signal programs are similar to the presented, so their depiction is disclaimed at this point. For a further improvement of the prediction in the current cycle, the algorithm uses training data according to the data model of table 3. However, the use of this data model is only possible if very short latencies are available. Furthermore, an additional benefit for a fuel-efficient access is hard to achieve due to the very low horizon of the prediction. Table 5 shows appropriate results for the same signal program.

Table 5 Comparison of predicted and occurred switching times for low latencies

	t_EG	predicted		
	[Cycle Second]	20	23	24
occured	20	105	5	1
	23	8	918	4
	24	15	2	35

Here, an accuracy of about 97% has been achieved by referring the matches of predicted and occurred switching times on the total number of cycles. This is a significant increase compared with the values of table 4. The prediction of switching times of the remaining signal heads is carried out analogous to the example presented here. The results were even better in these cases, since the switching times of signal head DR have the greatest variance. For each data model, only the target variable had to be modified. It then corresponded to the switching times of the signal head to be predicted.

Because the algorithm does not generate a correct prediction in each case, a calculation of a probability of green for each cycle second is necessary for a corresponding policy proposal. This probability is obtained by the distributions of the occurred switching times for each column according to the tables shown here. So for each switching point predicted by the algorithm, there is an appropriate probability of green for each cycle second.

4 Conclusions

The knowledge of forthcoming switching times is an important prerequisite for the development of applications for an energy-efficient driving in urban areas. However, since modern traffic lights adjust their phases and phase transitions to the current traffic volume, a prediction of switching times is required. The accuracy of a prediction depends on various boundary conditions, at which the modality of traffic dependency is of basic importance. However, the application-specific horizon of prediction and the latency period make different demands on the algorithm and have direct impact on the prediction quality as well. The algorithmic approach of Support Vector Machines (SVM) was investigated regarding its suitability for a prediction of switching times. The data model which is used by the algorithm to learn relationships between detector data and switching times is crucial for the success of the prediction. Assuming different latencies and horizons, data models have been generated using different traffic parameters as input variables. Subsequently, the achieved prediction quality was evaluated. An advantage of the examined approach is its easy transferability to other signalized intersections, since the use of input variables which have no direct influence on the target variable, is largely harmless. The influence of time headway criteria and public transport vehicles can also be learned by the algorithm using specific data models. For reasons of space, this aspect has not been discussed in this paper. For a provision of application-specific information, a preparation of prediction values into a probability distribution of green-occurrence is still required.

Acknowledgements. This research is financially supported by the German Federal Ministry of Economics and Technology (BMWi) under grant number 19 P 11007 R (UR:BAN). The authors would also like to thank the Department of Transport Management of the City of Düsseldorf for their cooperation in UR:BAN. Furthermore, we would like to thank the project partners, especially Gevas software for technical suggestions to the research.

References

[1] Otto, T.: Kooperative Verkehrsbeeinflussung und Verkehrssteuerung an signalisierten Knotenpunkten, Schriftenreihe Verkehr der Universität Kassel Heft 21, Kassel (2011)

[2] Menig, C., et al.: Der informierte Fahrer - Optimierung des Verkehrsablaufs durch LSA-Fahrzeug-Kommunikation, Heureka 2008, Stuttgart (2008)

[3] Krumnow, M.: Schaltzeitprognose verkehrsadaptiver Lichtsignalanlagenim Rahmen des Forschungsprojektes EFA 2014/2, 8. Tagung verkehrliche Informationspattform für Management- und Optimierungssysteme, Dresden (2012)

[4] Hoyer, R., et al.: Modelling of vehicle actuated traffic lights. In: Schulze, T., et al. (eds.) Simulation und Visualisierung 2004, Society for Modeling and Simulation International SCS, pp. 359–369. European Publishing House, Erlangen (2004)

[5] Koukoumidis, E., et al.: Leveraging Smartphone Cameras for Collaborative Road Advisories. IEEE Transactions on Mobile Computing 11(5), 707–723 (2012)

[6] Schölkopf, B.: Support Vector Machines – a practical consequence of learning theory. IEEE Intelligent Systems 13(4), 18–21 (1998)

[7] Meyer, D., et al.: Support Vector Machines in R. Journal of Statistical Software 15(9) (2006)

Part III
Vehicle Efficiency and Green Power Trains

Predictive Optimization of the Operating Strategy in Future Volkswagen Vehicles

Jan Bellin, Norbert Weiss, Matthias Breuel, Michael Kurrat and Christoph Stamprath

Abstract. This publication introduces a method for optimization of plug-in hybrid vehicle (PHEV) operating strategies. Due to its structure, the algorithm presented can be applied to various powertrain dimensions and concepts. Besides reducing CO_2 emission it also increases the overall distance travelled using the electric drive, thereby increasing the driver's electric driving experience. This is illustrated using simulation as well as real world measurement data.

A special feature of the system currently being developed is its driver interface. During optimization by the operating strategy it can assist the driver in adapting his driving behavior, thus increasing potential for reduced CO_2 emission and cost of operation.

Keywords: PHEV, hybrid vehicles, predictive operating strategy, energy management, hybrid HMI, low emission, low resource.

J. Bellin(✉)
Volkswagen AG, *elenia* TU-Braunschweig, Rosenweg 11,
30627 Hannover, Germany
e-mail: jan.bellin@volkswagen.de

N. Weiss
Volkswagen AG, EAEF/2, Wolfsburg, Germany
e-mail: Norbert.weiss@volkswagen.de

M. Breuel
Volkswagen AG, EEF, Wolfsburg, Germany
e-mail: Matthias.breuel@volkswagen.de

M. Kurrat
elenia, TU-Braunschweig, Schleinitzstraße 23, 38106 Braunschweig, Germany
e-mail: m.kurrat@tu-bs.de

C. Stamprath
elenia, TU-Braunschweig, Am Flughafen 13, 38110 Braunschweig, Germany
e-mail: c.stamprath@tu-bs.de

J. Fischer-Wolfarth and G. Meyer (eds.), *Advanced Microsystems for Automotive Applications 2014*, Lecture Notes in Mobility,
DOI: 10.1007/978-3-319-08087-1_13,

1 Introduction

As a bridge technology, closing the gap between conventional and electro mobility, PHEV technology will gain more and more relevance on the vehicle market. To fully use its potential regarding ecological sustainability and customer acceptance, intelligent systems for automatic optimized drivetrain control are required. Figure 1 shows a comparison of the CO_2 emission of different drivetrain technologies in the NEDC[1]. It illustrates the advances in CO_2 reduction made possible by the increasing electrification of vehicles. As suggested by [3], however, the potential for reducing CO_2 emission in PHEVs strongly depends on the operating strategy and the route that it is applied to (illustrated by shaded area above the NEDC emission of PHEV in figure 1). Especially, if an operating strategy with no predictive functionality[2] is well suited for the respective route, the margin for optimization with predictive functionality is significantly reduced.

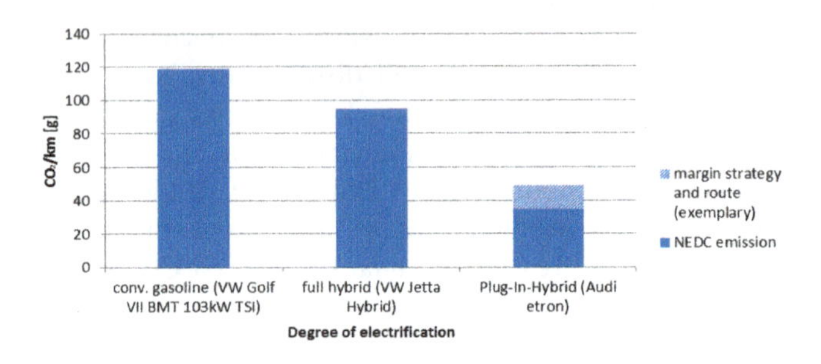

Fig. 1 NEDC CO_2 emission of electrified vehicles [1][2][3]

Various solutions, aiming at PHEV drivetrain optimization using predictive route data, have been published. Many recent publications, for example [4], represent research projects that rely on highly accurate route data and/or extensive driver input to minimize CO_2 emission using CPU-intensive algorithms. Currently, however, highly detailed route and traffic data as well as extensive computational power is not available in series production vehicles. In addition to requirements regarding serviceability and testability, this requires software capable of providing quality energy management whilst relying on low resolution input data.

In 2015, Volkswagen will launch the *Passat* PHEV. This will be its first PHEV using knowledge of the entire route ahead for optimization of the operating strategy. The system will enable Volkswagen customers using route guidance reduce CO_2 emission and increase phases of purely electric drive.

[1] "New European Driving Cycle": Reference driving cycle used to determine standard CO_2 emission of vehicles registered in the European Union.

[2] This is considered the reference strategy to determine any operating strategy's potential. It starts out in electric drive until the traction battery is fully depleted.

2 Optimization Method

The method outlined in this chapter features the following optimization targets:

- predictively allocating discrete operation modes[3] on a known route ahead to reduce CO_2 emission of a PHEV
- increasing the distance travelled electrically without compromising fuel consumption, thereby increasing customer value
- reserving electric energy to travel the last segment in electric drive[4]

Distinguishing between discrete operation modes increases plausibility before customer and, at the same time, reduces algorithmic complexity as well as resource requirements. Actual distribution of drive torque in any operation mode is performed by a basis operating strategy module, similar to that used in conventional hybrid vehicles, factoring in the vehicle's current drive state. This guarantees low impact on the vehicle's drivability.

2.1 Overview and Architecture

The predictive operating strategy is hierarchically structured, as illustrated in figure 2. At the highest level, one function module, the *Long Range Prediction* produces a preliminary operation mode based on data containing information on the entire route ahead. At the same time, by the *Short Range Prediction*, a preliminary operation mode is produced based on the current driving condition, the distance until the next external charging process as well as information on a short segment of the route ahead[5]. Depending on navigation data quality, one of these two modes is transferred to the next hierarchy level as demanded operation mode.

On a second level, this operation mode is revised using data on the current drivetrain state to overrule any operation mode that, although seemingly ideal with respect to predictive data, will cause an increase in CO_2 emission due to an aspect of the vehicle's current driving condition that was not included in the predictive data (for example minimum engine runtimes, engine diagnoses ...). Based on this, a target state of charge (SOC) window and a threshold torque for maximum electric driving is generated, which is used by the basis operating strategy for actual distribution of torque between the internal combustion engine (ICE) and the electric motor.

If, for any reason, the *Long Range Prediction* cannot produce a plausible operation mode (for example absence of route information), the operation mode of the *Short Range Prediction* provides a fall back operation mode still factoring in

---

[3] *Sustain battery charge, maximum electric driving, increase battery charge;* Few exceptions are possible to implement special customer features (see chapters 2.2 and 2.3).

[4] The current application will reserve enough energy for the last 1000 m to be travelled electrically.

[5] Since only a short road segment (< 1000 m) has to be covered, resolution of this road data is substantially higher than that used for the *Long Range Prediction*.

all remaining information about the route ahead. As mentioned above, the drivability of a vehicle mostly depends on the actual distribution of torque done by the *basis operating strategy*. Therefore a short failure of a hierarchically higher level of the predictive operating strategy can only result in a change in the demanded operation mode but never in an implausible step in the torque demanded from either the ICE or electric motor.

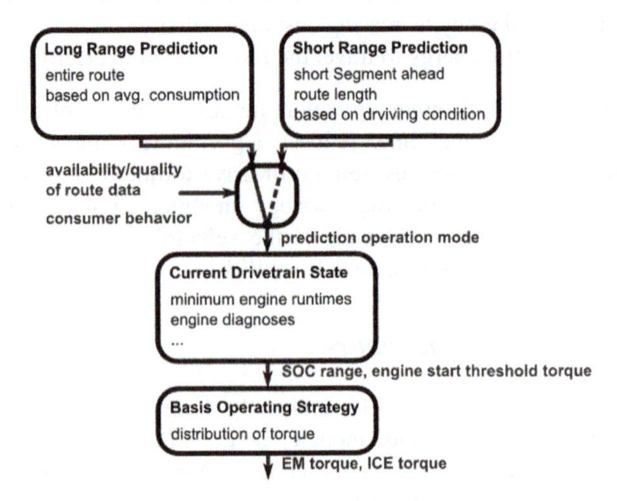

Fig. 2 Hierarchically structured predictive operating strategy

2.2 *Short Range Prediction*

The *Short Range Prediction* function module generates a drive state based on the current vehicle driving condition and a short road segment ahead. Furthermore, if the overall distance until the next external charging process is entered by the driver or determined by the system itself, the short range prediction module is able to ensure an empty battery upon reaching the next charging station.

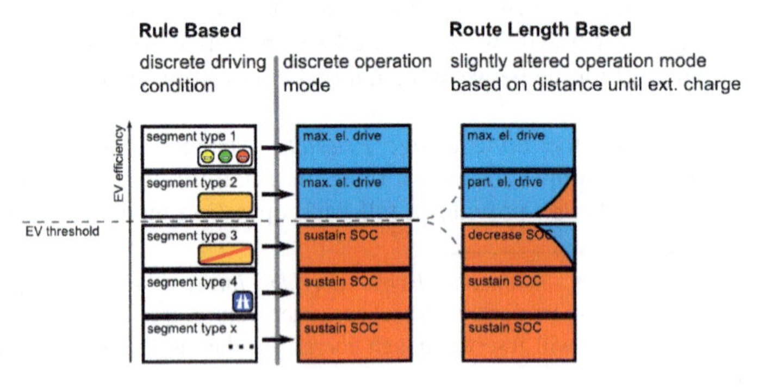

Fig. 3 Short range optimization

The functionality of the *Short Range Prediction* module is shown in figure 3. A preliminary operation mode is generated, rule based, from the current driving condition[6]. It is based on the assumption that some vehicle conditions have to result in mainly electric and some in mainly ICE drive in order to minimize fuel consumption on a standard route. This demands a threshold driving condition that will result in a discrete changeover from mainly electric to mainly ICE drive and vice versa. Subsequently, based on the distance until the next external charging process, the threshold is widened. By introducing more electric driving into mainly ICE drive phases and more ICE driving into mainly electric drive phases, the battery is emptied exactly upon reaching the next external charging stop whilst preserving the feel of discrete drive states before customer.

2.3 Long Range Prediction

The long range function module takes into account every road segment that will be passed until the destination entered into the navigation system by the driver is reached. On the route ahead it allocates segments to be travelled in one of the three discrete drive states (see beginning of chapter 2) to reach the destination with a predefined SOC[7], minimizing CO_2 emission while maximizing distance travelled on purely electric drive. Its low complexity allows continuous computation of the best current operation mode which results in an immediate adaption to disturbances along the route.

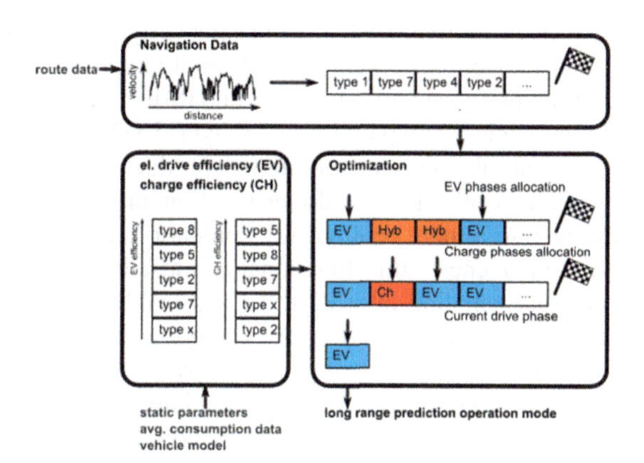

Fig. 4 Long range optimization

[6] This can either be determined from the vehicle's velocity profile and/or from the navigation system.

[7] The predefined SOC can usually be assumed as the minimum SOC tolerated by the battery management system. See also Chapter 4.

The algorithm is based on different parallelizable processes as illustrated in figure 4. Upon receiving route data from the navigation system, the route is interpreted as an array whose elements each represent one of nine segment types[8].

In a parallel process, each of the nine segment types is assigned an attribute representing the efficiency for electric driving and an attribute representing the efficiency for charging the battery using the ICE on a specific segment type. This can be done by using an online simplified vehicle model, fixed values or by interpreting segment specific drivetrain consumption data collected on previous trips.

With this information the actual optimizing algorithm module generates an array with allocated drive states, similar to the array containing the segment types, in the following manner:

- allocating the segments with the highest efficiency for electric driving to be travelled in maximum electric drive until the energy currently stored in the vehicle's battery is completely distributed on the route ahead
- allocating charging phases on the remaining road segments to enable allocating additional maximum electric phases if this increases the overall efficiency
- finding the vehicle's current segment and demand the operation mode that is allocated for this segment

3 Driver Interface and Human Machine Interaction (HMI)

As stated in [5], the main motivation for purchasing a hybrid vehicle today is the belief in ecological sustainability as well as the desire to express affiliation with the ecologically aware fraction of society. Therefore an HMI control concept displaying where and how the ecological compatibility of a hybrid vehicle is increased by an intelligent drivetrain control represents additional product value before customer. Furthermore, immediate feedback on a driver's action in terms of CO_2 emission along the route can lead to a more ecologically compatible driving behavior. While increasing vehicle attractiveness, this functionality must be seamlessly integrated into the driver's habitual handling in order to be applicable for everyday use. This chapter presents a concept of HMI functionality satisfying those requirements.

Figure 5 shows an example of how the simple driver's action of switching on the long range optimization results in an immediate change in the vehicle's standard navigation display.[9] Figure 5b illustrates how the optimization will use the electric energy of the traction battery to decrease local emission in inner city areas as well as in the proximity of the driver's home address (to support an ecological appearance in her or his social environment) rather than depleting it inefficiently on the motorway.[10] In case of a disturbance along the route, such as minimum engine runtimes, implausible driver action or a traffic jam (figure 5c),

[8] Segmentation criteria: road class, slope.

[9] Figure 5 a): manual maximum electric driving; b): long range optimization.

[10] Assumption: The driver charges the vehicle's battery at home.

the navigation display responds immediately, allowing the driver to add a resting period, detour or a charging stop to his route planning in order to ensure maximum electric drive in the city. This last aspect can be supported additionally by a display of available charging stations, from which the driver can choose, to be considered in the long range optimization process.

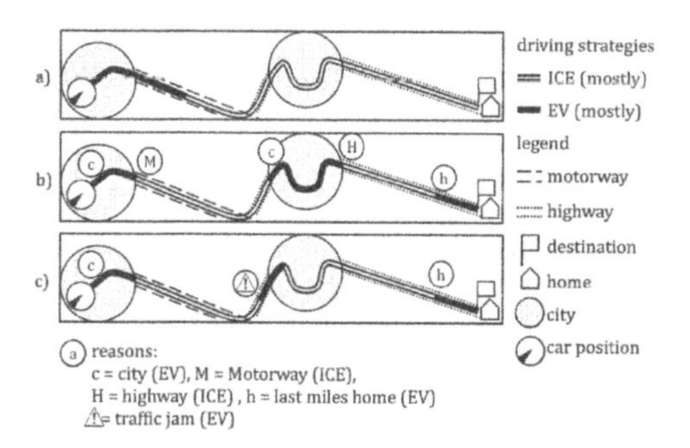

Fig. 5 Operating strategy – HMI concept

4 Potential of Predictive PHEV Optimization

To qualitatively evaluate the potential of the predictive PHEV operating strategy, two criteria were applied:

- *to evaluate CO_2 reduction*: reduction of fuel consumption/CO_2 emission by efficiently using the energy from the vehicle's traction battery
- *to evaluate electric driving experience*: increase of distance travelled electrically

4.1 Test Vehicle and Simulation Model

The test vehicle as well as the simulation model used for all real world measurements is a prototype front wheel drive PHEV of the *Golf* segment, see table 1.

Table 1 Technical specifications of test vehicle[11]

feature	*Test vehicle*	*simulation*
drive train architecture	parallel PHEV	parallel PHEV
peak electric power	80 kW	80 kW
peak ICE power	110 kW	74 kW
maximum system performance	150 kW	154 kW
battery capacity used (20%-90% SOC)	5 kWh	6 kWh

[11] Due to the algorithm's portability, test vehicle and simulation model as well as driving cycles are not required to be identical in a qualitative evaluation.

4.2 Simulation Study

To develop and test the method for optimizing PHEV drivetrain operation, a drivetrain model with properties, comparable to those of the test vehicle, was used as reference. It implements a scalable optimizing algorithm that was downsized to correlate with the functionalities currently implemented on the engine control unit of the test vehicle. The model was then stimulated using representative driving cycles. To keep results comparable, it was assumed that the ICE is always at its optimum operating temperature, the environment is at 20°C and no auxiliary consumers are used.

As exemplary simulation results using predictive functionality, figure 6 shows a comparison of three scenarios, differing in complexity of the operating strategy. The driving cycle features 29.7 km of motorway, 27.8 km of highway, 12.2 km of inner city driving and an altitude difference of 259.4 m. It was chosen since it consists of a variety of different road classes, a complex elevation profile and a length greater than the all-electric range of the vehicle model implemented in the simulation environment.

While all strategies ensure an empty battery upon completion of the driving cycle, the full scale long range optimization is able to save 0.33 kg (28%) of fuel compared to the reference strategy. Had a navigation error occurred, the fall back strategy, only using limited information on the route, would still have saved 0.17 kg (15%) of fuel. Besides reducing fuel consumption, the strategies using route information were also able to increase the overall distance travelled electrically. This is due to the fact that segments suitable for electric drive usually require less energy when passed electrically than others.

Table 2 Scenario comparison with exemplary simulation results

Scenario	*kg fuel used*	*Wh used*	*% el. drive*
reference strategy	1.17	6216.05	68.5
full scale optimization[12]	0.84	6173.70	79.2
known route length[13]	1.00	6100.26	72.2

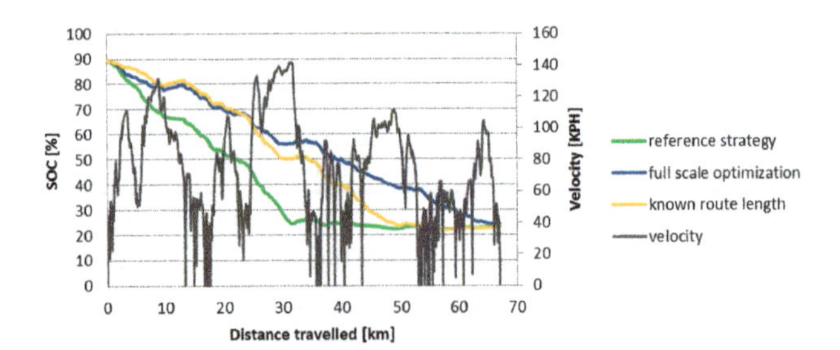

Fig. 6 Comparison of three scenarios via simulation

[12] *Long Range Prediction* module active, see chapter 2.3.
[13] *Short Range Prediction* module active, see chapter 2.2.

4.3 Real World Test Drives

To validate the results generated from the simulation study and further test the functionality of the optimization method, real world test drives were conducted on a driving cycle representing a typical day trip to and from work with the vehicle described in chapter 4.1. The overall distance of 66.2 km is slightly greater than the average daily distance travelled by German commuters age 30-49[14][6] as well as the all-electric range of the test vehicle. Furthermore it is assumed that when driving to work, the motorway and from work a highway and inner city dominated route is chosen. This results in a route with 34.7 km of motorway, 16.9 km of highway and 14.7 km of inner city driving. Few deviations were accepted and inevitable due to traffic and weather conditions. Except for slight alterations in parameterization to fit the test vehicle's technical specifications and electronic systems, the algorithm was transferred from the simulation model.

The results from the test drives (see figure 7) mostly correlate with those from the simulation study as shown in table 3. The full scale optimization reduced fuel consumption by 0.36 kg (15 %), the fall back strategy still saved 0.32 kg (14 %). Besides, the distance travelled electrically was increased by 7.3 %, and 4.4 %, respectively. This illustrates the continuity of the algorithm's saving potential throughout different environment, route and vehicle driving conditions. The fact that, compared to the strategy with known route length, there is only a small improvement in fuel consumption when applying the full scale optimization shows, however, how a different route influences the efficiency of a PHEV operating strategy. In this case, due to an unexpected disturbance, the full scale optimization was not able to use the entire energy stored in the vehicle's battery, resulting in increased fuel consumption.

Table 3 Scenario comparison with exemplary test drive results

Scenario	*kg fuel used*	*Wh used*	*% el. drive*
reference strategy	2.33	5173.0	54.3
full scale optimization	1.97	5078.9	61.6
known route length	2.01	5173.8	58.7

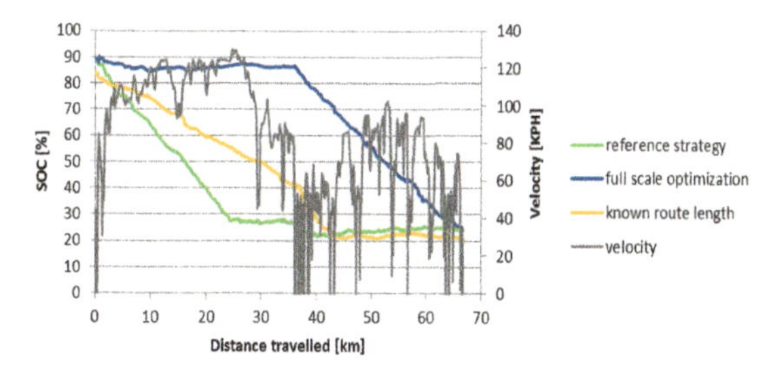

Fig. 7 Comparison of three scenarios in real world test drives[15]

[14] As suggested by [5] this represents a typical Plug-In-Hybrid customer.

[15] Velocity profile is taken exemplary from one of the test runs and shows slight deviations between measurements due to traffic conditions.

5 Conclusion and Outlook

The results presented in this paper have shown how a simple optimization approach, only distinguishing between discrete operation modes, can significantly decrease CO_2 emission of PHEVs. Besides maintaining its functionality plausible before customer as well as simplifying software maintenance, the method's modular structure ensures portability to various vehicle and drivetrain concepts. This is especially relevant since it also enables use in small, low-budget vehicle concepts, thereby contributing to the current downsizing trend in the automotive industry.

However, the potential for optimization strongly depends on the route that it is applied to, disturbances from traffic and driving behavior as well as the customer's use of external charging stations. This implies that in the future, further including the driver into a PHEV's optimizing strategy through intelligent HMI still bears great potential for increasing its overall efficiency.

References

[1] price list Golf/Jetta, http://www.volkswagen.de/ (accessed December 2013)
[2] http://www.audi.de/de/brand/de/neuwagen/tron/a3-sportback-e-tron.html (accessed December 2013)
[3] Roussea, A., Pagerit, S., Gao, D.: Plug-in Hybrid Electric Vehicle Control Strategy Parameter Optimization. Journal of Asian Electric Vehicles 6(2), 1131 (2008)
[4] Ozatay, E., Onori, S., Ozguner, U., Rizzoni, G.: On the use of geographical, traffic management, and weather information in the cloud to optimize the energy use of vehicles. In: 13th Stuttgart International Symposium: Automotive and Engine Technology, pp. 41–56 (2013)
[5] Schäfers, T., Esch, F.: Der Hybrid-Fahrer, Das unbekannte Wesen? ATZagenda, 90–95 (January 2012)
[6] Bundesministerium für Verkehr, Bau und Stadtentwicklung. Ergebnisbericht Mobilität in Deutschland, Fig. 3.11 (2008)

(Cost)-Efficient System Solutions e.g. Integrated Battery Management, Communication and Module Supply for the 48V Power Supply in Passenger Cars

Harald Gall, Manfred Brandl, Martin Jaiser, Johann Winter, Wolfgang Reinprecht and Josef Zehetner

Abstract. Although the implementation of the 48V voltage supply in next generation passenger vehicles is certain to go ahead, economic and technical challenges remain to be overcome. The cost of lithium based batteries is falling, but costs for energy and power per liter/kg still remain an impediment to the application of the new voltage domain in high volumes. Engine cold-start, aging, efficient use of available energy and effective battery management systems are all facing problems that require technical solutions. ams AG provides (cost)-efficient IC solutions for lithium based battery systems, in-vehicle communication and power supplies for electronic modules for 48V systems. The Virtual Vehicle Competence Center provides the co-simulation environment for system integration and simulation-based evaluation of these systems.

Keywords: passenger vehicles, 14V supply, 48V supply, dual-voltage supply, (cost)-efficient, Battery Management System, in-vehicle communication, CAN, FlexRay, module supply, high-voltage CMOS, scalable process technology, CO_2 reduction, electrification, ams AG, Virtual Vehicle Kompetenzcenter.

H. Gall(✉) · M. Brandl · W. Reinprecht
ams AG, Tobelbaderstrasse 30, Unterpremstaetten, 8141, Austria
e-mail: {harald.gall,manfred.brandl,wolfgang.reinprecht}@ams.com

M. Jaiser · J. Winter
ams Germany GmbH, Erdinger Strasse 14, Aschheim b. München, 85609, Germany
e-mail: {martin.jaiser,johann.winter}@ams.com

J. Zehetner
Virtual Vehicle Research Center, Inffeldgasse 21/A, Graz, 8010, Austria
e-mail: josef.zehetner@v2c2.at

J. Fischer-Wolfarth and G. Meyer (eds.), *Advanced Microsystems for Automotive Applications 2014*, Lecture Notes in Mobility,
DOI: 10.1007/978-3-319-08087-1_14,

1 Introduction and Drivers for the 48V Power Supply

Electrification of individual mobility gained huge momentum in recent years. Different systems, from hybrid electric vehicles (m)HEV to plug-in electric vehicles (PEV), have been introduced to the market. Those vehicle types follow the same trend as conventional internal combustion engine (ICE) vehicles: more electrical and electronic functions are being implemented to enhance comfort and safety, improve efficiency and comply with CO_2 regulations laid down by the authorities.

Compared to the conventional components which are mechanically coupled to the combustion engine, electrically actuated components can be used in a more controlled fashion and thus more efficiently.

The conventional 14V power supply has reached the limit of its ability to support further electrification. Due to the low voltage level and power capacity of the 14V power supply, the high power demand of the new electrical systems [1] may lead to unacceptable line losses. In particular, high transient loads in the vehicle power supply can make the supply voltage for ECUs drop below a critical level, which may lead to system reboot in the worst case.

Studies published by major automotive Tier Ones like Continental Automotive [2] and Valeo [3] have shown that the introduction of the dual-voltage power supply with the extension of the 48V domain can reduce fuel consumption and CO_2 emissions by up to 13%. This is mainly because high power components in passenger cars can be more efficiently supplied within the 48V domain, wiring and component weight is reduced and the components are more efficiently operated. Energy recovery systems providing up to 10kW peak may be implemented with the 48V power supply and provide extra energy for the drive cycle. True, the first vehicles equipped with the 48V power supply might not provide a 48V supply to all electro-mechanical and electrical components with more than 1 kW peak power consumption. However, systems like Electric Power Steering (EPS), Electronic Braking Systems (EBS), heating, ventilation and air-conditioning (HVAC) systems, water, oil and fuel pumps and charging and recuperation systems are the high power systems in conventional passenger vehicles and will be the first systems to be connected to the 48V power supply.

Comfort systems (e.g. HVAC and infotainment systems) are common high-energy loads, which have a direct impact on the customer's appreciation of the vehicle.

Safety relevant systems (e.g. vehicle dynamics control systems) are often high-power loads, and they require a robust power supply in order to meet regulatory requirements.

While the integration of systems with moderate power demand is still possible to a limited extent for the conventional 14V power supply, the energy demand of new high-energy systems cannot be met by it any more. The introduction of the complementary 48V power supply is meeting that challenge due to the increased energy and power density that it offers.

2 System Approach and Challenges for the Dual Voltage System in Passenger Cars

By introducing a higher voltage level into the vehicle's power supply, it becomes possible to integrate further electrical systems with high power demand into the vehicle. In particular, with the 48V power supply proposed by [4], the integration of high-energy loads (e.g. hybrid modules for propulsion [5], HVAC systems etc.) for long operation is possible. The dual-voltage power supply of 14V and 48V has a huge potential with regard to system capability, energy efficiency (e.g. measured by CO_2 emission) and new functions. However, the additional scope to the power supply design should be considered carefully, since the best result can only be achieved with a multidisciplinary system approach and appropriate tool support.

During operation, several vehicle electrical systems are activated, including electrical power steering (EPS), vehicle dynamics control (VDC) and active suspension system. These systems are safety-critical and require a robust power supply in all operating conditions, which must be provided by the power supply. It's quite obvious that the impact of the vehicle dynamics' load on the power supply can only be properly investigated, if both the vehicle dynamics and the electrical and electronic aspects of the systems are taken into account.

Of course, the main goal is to keep system costs for the dual-voltage power system as low as possible, therefore the preferred solution is to have the 14V and 48V ground electrically connected and to totally avoid the need for galvanic isolation to decouple the voltage levels in control units (CU) or integrated circuits (IC). Figure 1 shows a proposed system layout with a bi-directional DC/DC converter connecting both low voltage 14V and 48V domains.

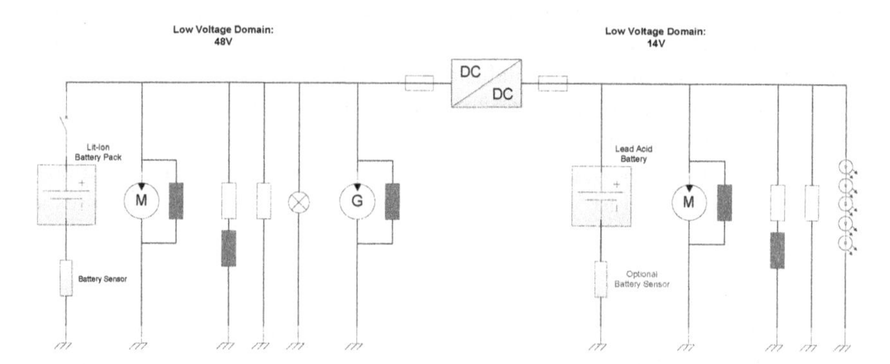

Fig. 1 System layout of the dual-voltage power supply without galvanic decoupling

In supplying electrical and electronic components, the bi-directional DC to DC converter can handle failure modes such as loss of ground at the 48V side and can ensure that a high voltage above the specified standard levels at the 14V domain cannot occur. For communication devices, this is becoming a challenge as the communication network interconnects components of both domains as a matter of course without isolation.

A severe failure case is identified during loss of ground at one electronic control unit (ECU) which is connected to an automotive bus system such as CAN or FlexRay™ [6].

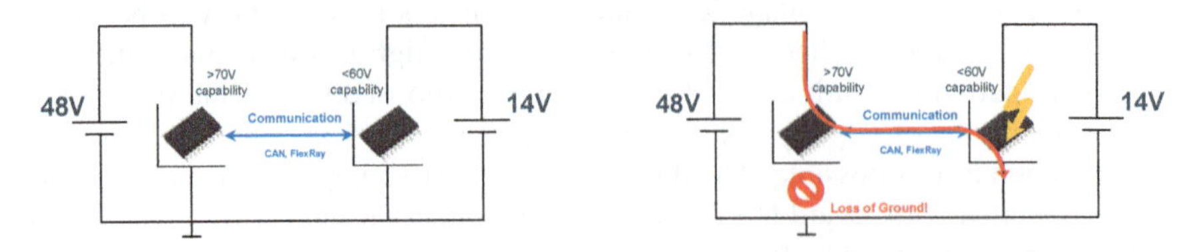

Fig. 2 Failure mode - loss of ground (left: no failure / right: loss of Ground at the 48V side)

In case of loss of ground at one ECU in the 48V domain, the system must ensure that the voltage from the high side of any communication transceiver is not forwarded on to the bus lines and influence the 14V domain. In the worst case, the high voltage can destroy devices that cannot withstand voltage levels higher than 48V. Communication transceivers commonly used today are specified to withstand voltage levels of less than 48V. Although the nominal voltage for 48V and the operating range is up to 54V [7], during specific operating conditions the minimum voltage level must be assumed to be capable of reaching much higher levels.

3 Requirements for Semiconductor Devices

For the semiconductor industry a significant and costly investment is needed to be ready for the technical specifications of the 48V power supply. Every semiconductor fabrication process is designed to support one narrowly-defined voltage domain. The automotive industry's 48V domain, with its demand for at least 70V breakthrough voltage, introduces a new process requirement.

Most semiconductor fabrication processes used for automotive applications are qualified for temperature ranges (AEC-Q100 grade 2 to grade 0) and the lifetime requirements are limited to 50V or 60V. For industrial applications, processes are qualified to above 100V.

The gap between 50V/60V and >100V must be closed in order for cost effective solutions to be provided.

Increasing the breakthrough voltage level has an exponential impact on the size of transistors. Therefore it is very important to specify the requirements for the 48V power supply thoroughly in order that the overall costs are kept low.

Significant area in any chip design is devoted to fulfilling the automotive electro-static discharge (ESD) requirements; this area as well increases exponentially as the maximum voltage level increases.

For global pins such as supply (battery connection and ground) and communication pins which are connected to the vehicle harness, the standard requirements of ±6kV according to the AEC-Q100-002 Human Body Model apply[1]. Generally for communication interfaces with differential bus signaling such as CAN and FlexRay™ interfaces, protection against reverse polarity has to be guaranteed.

ESD protection devices are needed to protect the IC against ESD which might occur during fabrication, assembly, test and board mounting. Further, exposed pins on the connector of a board can get ESD strikes at any point in the life of the component, due to handling, repairing or exchanging of components. An ESD protection circuit does not help the IC perform its intended function; its sole purpose is to prevent damage to the IC. It is important that the ESD circuit does not itself impose any parasitic load on the IC.

Maintaining two voltage domains in the passenger car alters the specification parameters for electronic and electrical components. It's not just that components for the new voltage domain are affected: traditional control units, some operating for decades on the lower voltage, must be able to communicate with control units supplied by the 48V power supply. The communication link needs to provide the high levels of reliability normally expected of automotive components in order to provide for fault-free operation.

System simulations in a co-simulation environment performed by Virtual Vehicle showed that voltage peaks over 80V lasting for a few milliseconds are reached in the event of a load dump, which could occur due to battery failure or malfunction of the battery management system. With the same co-simulation system model, load cases for a more detailed component-level simulation can be derived, and further simulation shows that voltage peaks of over 90V are reached, in lasting for a few microseconds.

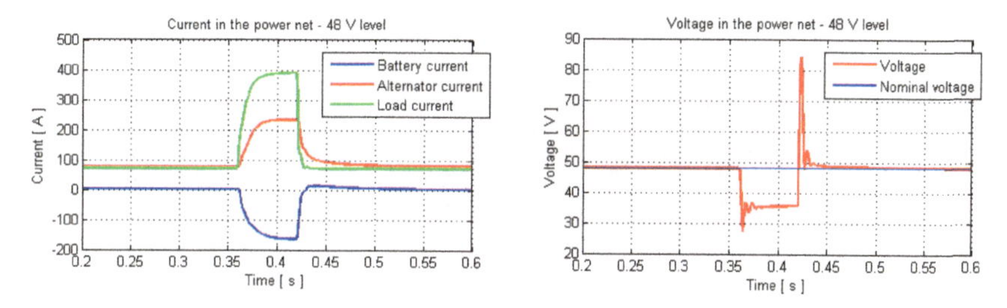

Fig. 3 Simulation showing voltage peak due to load dump in the 48V power supply

The first results from simulations show that transients much higher than 70V can easily occur in the power supply especially during a load dump. With reasonable effort the maximum voltage peaks can be limited to 90V.

Hence, for supply and communication ICs a breakthrough voltage of over 90V appears to be a very reasonable starting point for the semiconductor suppliers to align their process developments.

[1] German car makers have specified an ESD test pulse based on IEC-61000-4-2 with ±6kV.

4 (Cost)-Efficient Solutions for Integrated Battery Management of 48V Li-Ion Batteries, In-Vehicle Communication and Supply for Electronic Modules

ams provides a scalable process node on its 0.35μm CMOS technology targeting the application for 48V systems. This process technology provides transistors for any voltage capability between 20V and 120V. It gives flexibility, so that, no matter what are the requirements of any one IC, no additional process needs to be characterized or qualified: the entire voltage operating range is supported up front. Thus the ams chip designers are choosing N- and P-MOS devices for the voltage capability required according to the specification, and can immediately start the development of the IC. A combination of different voltage requirements is possible.

4.1 Module Supply and In-Vehicle Communication in Dual-Voltage Environment

ams power management ICs for the low-power onboard supply and communication devices (e.g. CAN, FlexRay™) are designed for maximum voltage ratings up to 90V. Communication devices for 48V are fully compatible with 14V devices available today (with a maximum voltage capability lower than 40V at the bus lines). During possible loss of ground (Fig. 2) in a 48V control unit, the ams 48V communication devices will prevent any voltage transients exceeding 40V on the bus lines, and therefore no damage to the "older" 14V communication devices can happen.

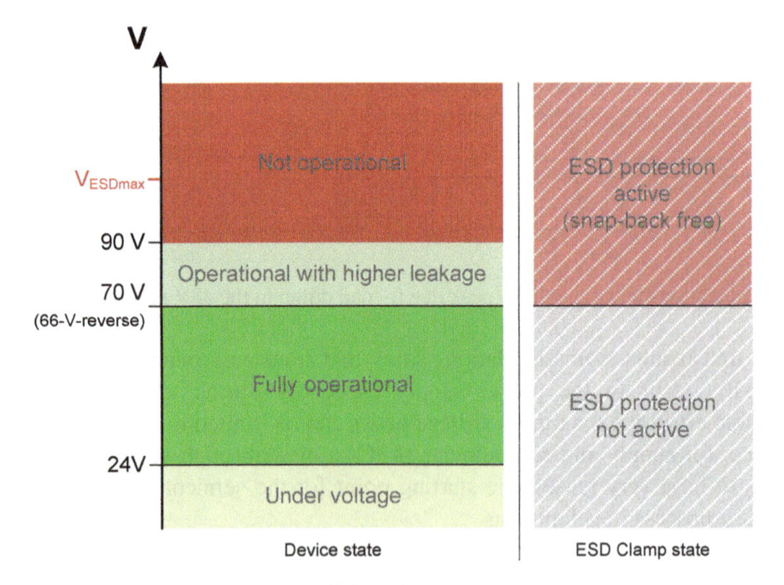

Fig. 4 Voltage operating areas of ams 48V devices

Furthermore, communication will be maintained without limitation if the voltage at the bus pins in the forward direction is lower than 70V and 66V in the reverse direction, which is some way above the standard operating area of the 48V domain (see Fig. 4 for the ams operating areas). There is a margin for the safe communication of at least 12V (66V max. reverse voltage – 54V operational area according to [7]).

Above that level a leakage current[2] may be introduced on the bus lines: this could impair communication to some extent[3], but it will still be maintained.

In any case, the devices are capable of withstanding transient voltage levels up to 90V without damage, guaranteed over lifetime and temperature ranges.

4.2 Battery Management for 48V Lithium Based Batteries

Highlights for the choice of energy storage for the 48V dual-voltage supply are high efficiency in energy storage, fast charge acceptance, high discharge currents, excellent cycle robustness, battery life and energy to weight ratio. Lithium based batteries fulfill those requirements and are the ultimate choice for 48V energy storage in highly electrified ICE cars as well as mild hybrid vehicles.

Lithium batteries are, to a certain extent, intrinsically unsafe and therefore require cell supervising circuitries (CSC) for cell monitoring and balancing, a current sensor, temperature sensors, battery disconnect function and a battery controller combined in a battery management system (BMS) to safely operate the battery and to ensure an acceptable battery lifetime.

In contrast with traction batteries in full electric vehicles (FEV), specific issues need to be considered when developing the battery management system for the 48V application:

- State of charge (SOC) control to keep the battery on average well below full charge. This ensures that energy from the alternator or from the electric motor/generator can be absorbed and stored in the battery at any time. The amount of energy to be recovered from the vehicle's kinetic forces (e.g. braking) is unpredictable. For reasons of energy efficiency and CO_2 reduction, that energy should be stored whenever it is generated.
- The battery is operated rather in a micro cycle mode and there might be rare idle phases at low or virtually zero currents. During drive cycles, loads might be as high as 10kW or even more depending on the degree of electrification of the car's systems. In idle mode, currents from energy exchange between the 48V and 14V domains as well as from small polling loads will most likely remain.
- In a dual-voltage architecture with a mixture of 14V and 48V loads and sources, the BMS system needs to ensure failure-tolerant and safe interoperability for in-vehicle communication and the power grid.

---

[2] Sum of currents from operating conditions to ground.

[3] E.g. the common mode range for CAN or FlexRay™ Transceivers is up to the specified ±12V respectively ±15V.

State of charge (SOC) has to be as good as possible controlled in this application, so accurate determination is required. For a battery in which current is flowing virtually all the time, the primary method of state of charge measurement is Coulomb counting: Charge flowing in and out of the battery is continuously tracked, integrated and the state of charge is calculated in relation to the capacity of the lowest-capacity cell in the stack.

Accurate and offset-free current measurement in an accurate time grid is a precondition of accurate SOC assessment. However, real world charge integration suffers from inaccuracies in current capture, and cells lose capacity due to ageing. Thus open circuit voltage (OCV) based recalibration of SOC from time to time is essential as a secondary SOC measurement method. The OCV recalibration needs to take place when the battery is idle or the load is very low, the battery has recovered from charging / discharging history and state of charge to be as high as possible.

Keeping the SOC of the individual cells equal helps to avoid loss of battery capacity. The usual method to compensate slowly accumulating mismatches due to differences in self-discharge or differences in current drawn by cell supervisor circuits is to run cell balancing at set time intervals.

Balancing the cells is much easier when the battery is almost fully charged because in this state the cell voltage correlates well with the state of charge. This is more difficult at intermediate SOC.

ams has therefore developed a chipset for battery management systems (BMS) which offers high accuracy and low system costs:

- AS8510 data acquisition IC for shunt-based current sensing (using for example a 100μOhm Manganin alloy resistor with a low temperature coefficient). Currents can be measured from 3mA up to 1600A with accuracy better than 1% in standard configuration or below 0.5% with additional temperature calibration, taking into account automotive operating conditions and product lifetime.
- AS8506 cell supervising IC. One device can monitor and balance up to 7 cells; multiple ICs can be chained to support batteries with virtually any number of cells. Ideally for 48V power batteries, 2 x 7 cells of 3.6V nominal voltage where two AS8506 ICs are used or 3 x 5 cells of 3.2V nominal (e.g. LFP cells) with three AS8506 supervising ICs are used.

The device can run the majority of the monitoring and balancing tasks autonomously, hence software tasks are reduced to a minimum, which safes costs in computing power. Safe cell voltage monitoring and balancing are implemented in hardware through simultaneous cell comparison with programmable internal DAC references or analog external references. Absolute voltage and temperature measurement of the cells is provided for the SOC and SOH calculation and can be used for diagnosis purposes.

- AS8801 is a high precision, low drift resistor based attenuator, essential for the improvement of the battery pack voltage acquisition. The device facilitates the fast SOC determination and fast validation of the cell supervision.

- AS8605 48V high-speed CAN Transceiver. This is the first CAN Transceiver from ams with the voltage levels for the 48V voltage supply referred to above (Fig. 4).

The device is directly supplied by the 48V battery voltage and protects the CAN bus lines in case of loss of ground. Therefore the 14V and 48V CAN network can be directly connected.

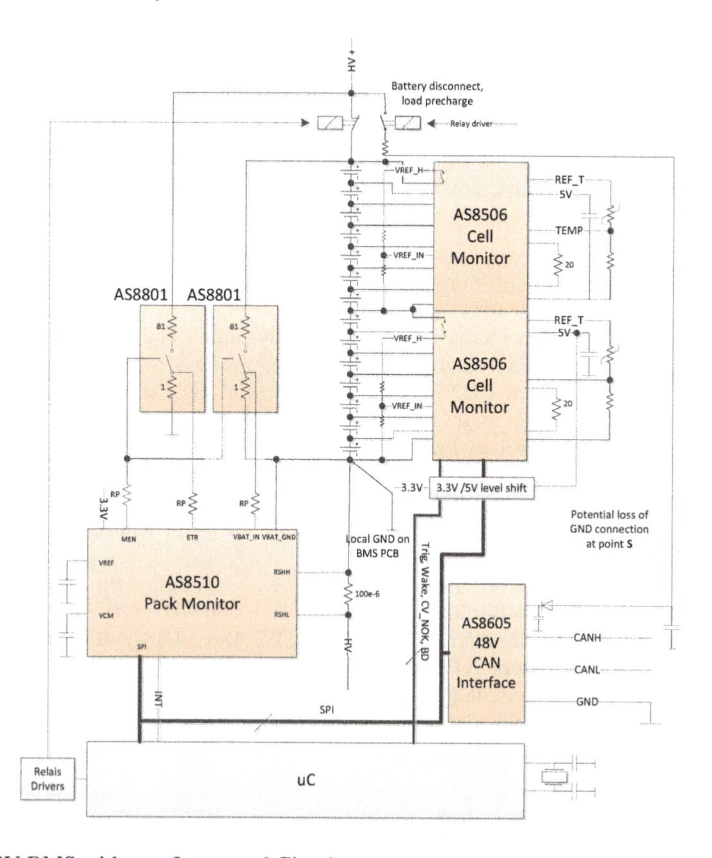

Fig. 5 48V BMS with ams Integrated Circuits

5 Conclusion

ams offers a unique set of semiconductor devices addressing the requirements of the 14V/48V dual-voltage power supply. The AS8510 pack monitor IC together with the AS8506 cell monitoring IC provides a very (cost)-efficient solution for 48V lithium based batteries. Conventional BMS systems require failure-intolerant high-speed communication for cell monitoring and balancing decisions in the host controller. The proposed system solution from ams on the one hand offers a fully synchronized voltage and current measurement, so that a highly accurate and fast calculation of SOC and SOH is possible, and on the other hand system component requirements (e.g. cell monitoring micro-controller, communication links and passive components) are reduced while providing better performance and offering the strongest safety features. The majority of cell supervising functions are integrated in hardware, which increases safety, reliability and performance of the battery management system.

CAN and FlexRay™ transceivers for 48V systems from ams support an in-vehicle communication set-up with no need for galvanic isolation between the voltage domains. Furthermore, 14V transceivers can be electrically connected with ams 48V devices. Even during loss of ground failure condition, the ams transceivers prevent damage to 14V devices. Voltage levels up to 90V are guaranteed; therefore system designers can avoid expensive protection measures for 48V-connected modules.

References

[1] Hesse, B.: Wechselwirkung von Fahrzeugdynamik und Kfz-Bordnetz unter Berücksichtigung der Fahrzeugbeherrschbarkeit, Univ. Dissertation (2011)
[2] Knorr, T.: Reduction of CO2 in a low-Voltage Mild Hybrid Vehicle – Conditions, Challenges and Realization. In: EEHE 2013, Bamberg (2013)
[3] Frossier, M.: Hybrid4All: A low voltage, low cost, mass-market hybrid solution. In: EEHE 2013, Bamberg (2013)
[4] Dörsam, T., Kehl, S., Klinkig, A., Radon, A., Sirch, O.: Energieeffiziente Antriebstechnologien: Die neue Spannungsebene 48 V im Kraftfahrzeug, pp. 184–189. Springer Fachmedien, Wiesbaden (2012)
[5] Picron, V., Fournigault, D., Baudesson, P., Armiroli, P.: Kostengünstiges Hybridsystem mit 48-V-Bordnetz. ATZelektronik 10, 802–803 (2012)
[6] Gall, H., Knaipp, M., Roehrer, G., Reinprecht, W.: Das 14/48-V-Boardnetz aus Sicht eines Halbleiterherstellers. ATZelektronik 06, 458–464 (2013)
[7] LV148:2011-08: Elektrische und elektronische Komponenten im Kraftfahrzeug 48V-Bordnetz, Anforderungen und Prüfungen, aktuelle Version, Audi AG, BMW AG, Daimler AG, Porsche AG, Volkswagen AG

International Research Projects

- ESTRELIA – Safe Batteries with power. ams AG is coordinating company of the European funded research project ESTRELIA. http://www.estrelia.eu
- BattMan – Solar Powered Efficiency. ams AG is partner company and work package leader of the European funded research project BattMan. http://www.eniac-battman.eu

Safety Simulation in the Concept Phase: Advanced Co-simulation Toolchain for Conventional, Hybrid and Fully Electric Vehicles

Stephen Jones, Eric Armengaud, Hannes Böhm, Caizhen Cheng, Gerhard Griessnig, Arno Huss, Emre Kural and Mihai Nica

Abstract. Modern vehicle powertrains include electronically controlled mechanical, electrical and hydraulic systems, such as double clutch transmissions (DCT), powerful regenerative braking systems and distributed e-Machines (EM), which leads to new safety challenges. Functional failure analysis of events such as the sudden failure of a DCT or EM, and the development and the validation of suitable controllers and networks, can now be evaluated using co-simulation techniques, from the early stages of product development. A co-simulation toolchain with a 3D vehicle and road model, coupled with a 1D powertrain model, is used to enable the definition of hardware and software functions, and also to support the rating of the Automotive Safety Integrity Level (ASIL) during hazard analysis and risk assessment in the context of ISO 26262. This innovative approach may be applied to a wide range of powertrain topologies, including conventional, hybrid electric and fully electric, for cars, motorcycles, light or heavy duty truck or bus applications.

Keywords: functional safety, safety, hazard, ISO 26262, ASIL, HRA, control, function, controllability, severity, exposure, car, bus, motorcycle, truck, trailer, hybrid, electric, powertrain, DCT, e-machine, co-simulation, vehicle dynamics, driver.

1 Introduction

In the development of vehicles with greater energy efficiency and performance targets, ever more sophisticated powertrain sub-systems and electronic controllers

S. Jones(✉) · E. Armengaud · H. Böhm · C. Cheng · G. Griessnig ·
A. Huss · E. Kural · M. Nica
AVL List GmbH, Hans-List-Platz 1, 8020 Graz, Austria
e-mail: {stephen.jones,eric.armengaud,hannes.boehm,caizhen.cheng,
gerhard.griessnig,arno.huss,emre.kural,nica.mihai}@avl.com

J. Fischer-Wolfarth and G. Meyer (eds.), *Advanced Microsystems for Automotive Applications 2014*, Lecture Notes in Mobility,
DOI: 10.1007/978-3-319-08087-1_15, © Springer International Publishing Switzerland 2014

are being implemented, in conventional, hybrid electric and fully electric vehicles. In order to control and reduce the costs and risks associated with the development of such novel technologies, vehicle manufacturers and suppliers seek to test new functionalities in the very early stage of the development, especially with respect to vehicle safety and energy efficiency.

To systematically minimize the risk of possible hazards caused by the malfunctioning behavior of electrical and/or electronic systems in passenger cars, the ISO 26262 Functional Safety Standard [1] was introduced in November 2011. A revision of the ISO 26262 is planned for the year 2016 which shall also include commercial vehicles and motorcycles. The ISO 26262 standard requires the conduction of a Hazard Analysis and Risk Assessment (HRA) in the concept phase of the development. In this HRA hazard events with corresponding Automotive Safety Integrity Levels (ASIL) are determined. As these directly impact the product development costs and time, the confidence of such determination should be supported by appropriate techniques e.g. holistic system simulation.

To determine the ASIL for a hazard event, three factors must be considered: *exposure, severity* and *controllability* of the hazardous situation. Whilst guidelines for the evaluation of exposure and severity are available [1], the controllability is difficult to evaluate as most of the controllability assumptions (C0 – Controllable in general, C1 – Simply controllable, C2 –Normally controllable, C3 – Difficult to control or uncontrollable) can in many cases only be reliably evaluated at the end of the product development cycle, on the test track in an expensive and perhaps dangerous vehicle test program. Evidently this imposes a major inconvenience and high development risk, especially if the late vehicle tests conclude that the controllability was incorrectly determined in the project and hence the ASIL must be reconsidered, i.e. part of the development must be redone.

To address this problem, activities [2, 3] have been conducted by AVL based upon the implementation of an advanced 1D / 3D co-simulation toolchain able to enable the efficient frontloaded development of electrified and conventional vehicles, whether they are cars, buses, or light and heavy duty trucks. This toolchain includes the simulation software IPG CarMaker/TruckMaker, AVL CRUISE and MATLAB/Simulink. CarMaker/TruckMaker is a vehicle simulation tool with the ability to represent lateral and longitudinal vehicle dynamics, 3D driving routes, complex multi-vehicle traffic scenarios, and highly dynamic driver maneuvers and thus test cases. The integration of a CRUISE powertrain model with CarMaker/TruckMaker significantly extends powertrain modeling capability and flexibility. CarMaker/TruckMaker and CRUISE both have MATLAB /Simulink interfaces that allow for the ready inclusion of xCU control software e.g. VCU, EMS, HCU, TCU and ABS/ESP/TCS [4], enabling fully virtual functional safety testing from the concept phase and onwards in the project development cycle.

This realistic, yet computationally efficient toolchain enables model based development of control strategies and functional safety analysis based on simulation. Advantageously, once vehicles or powertrain hardware are available, equivalent and reproducible safety analysis testing with real failure injection can be conducted on a real testbed or Hardware-in-the-Loop system, using the corresponding real-time toolchain of AVL InMotion and AVL CRUISE RT. Reusing essentially the same simulation models and critically also the same simulated test cases, as in the preceding office co-simulation phase.

2 Functional Safety – An Overview

The main target of functional safety standards such as ISO 26262 is to ensure the residual risk of a product to fail due to a malfunction of its electric / electronic control system(s) is at or below an acceptable level. Functional safety standards usually rely on appropriate quality management during the development process (to avoid systematic failures during product development), and on appropriate risk identification and risk management over the entire product lifecycle, to mitigate the effects of random failures such as component breakdown.

A cornerstone of this approach is the systematic identification and classification of the hazards for a given system. For the ISO 26262 standard, this is performed for the first time early in the concept phase by means of hazard analysis and risk assessment in order to identify the different potential hazards, and classify these according to their severity, i.e. the risk to persons, exposure, i.e. the probability of a situation to occur, and controllability, i.e. the capability for the human driver to identify and appropriately react to the given situation. This analysis is further consolidated by means of safety analysis (e.g., FMEA, FTA) to understand the effects of component failures and the resulting impact at vehicle level.

A major challenge in the concept phase is the correct analysis of the possible impacts of such hazards and the capability for the human driver to appropriately control the complex systems considered, which include multiple electronic controllers, powertrain subsystems, vehicle steering, brakes and tires, as well as the road surface and driving environment. The challenges are composed of (a) the correct understanding of the realized functions and their possible effects in event of a failure within the system, and (b) the capability to judge the controllability of the vehicle behavior by the different human drivers (e.g. beginner, normal, expert) in the different driving situations (e.g. vehicle speed, road friction, environment).

The advantages of system simulation support in the HRA lie in the reproducibility and the higher coverage of different situations which are composed from countless variations of the parameters illustrated in the preceding paragraphs.

3 Co-simulation Model Description

In order to support assessment of ASIL in very early phase of development, virtual evaluation of vehicle controllability with mechanical, hydraulic, electrical, electronic or software failures, an integrated co-simulation toolchain has been developed in AVL.

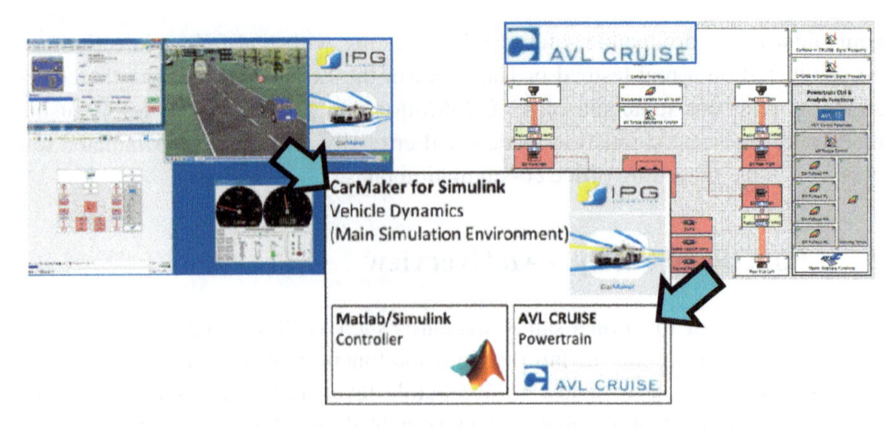

Fig. 1 Simulation environment used for simulation based determination of ASIL. The toolchain includes CarMaker® or TruckMaker®, CRUISE, and MATLAB/Simulink®.

This toolchain connects CarMaker or, if a commercial heavy duty vehicle or bus shall be simulated TruckMaker, with CRUISE, and MATLAB/Simulink (Figure 1). The 3D vehicle dynamics model including wheel suspension and tire model, a realistic human driver and road models are simulated in CarMaker or TruckMaker, whilst the complex 1D powertrain model is developed in CRUISE, and control software is developed and / or integrated via MATLAB/Simulink. This allows the utilization of the capabilities of specialized simulation and authoring tools for maximum overall ease, speed and accuracy.

4 Applications of Concept Safety Simulation for Various Vehicle Configurations

With the described co-simulation toolchain, vehicle dynamics, including controllability, can be readily simulated for many test cases, i.e. a range of different driving maneuvers, with various failure modes injected (e.g. an e-machine failure in an electric vehicle with multiple EMs). Based on the simulation results, the vehicle controllability is evaluated for various selected hazardous situations.

In this paper several simulation scenarios are presented to demonstrate the virtual assessment of ASIL. The first example is a conventional car with a DCT, the second is a heavy duty truck. Thirdly a 4WD HEV in P1 configuration, with a second EM driving the rear wheels is presented. Finally, an EV with the 4 wheels individually driven by 4 separate EMs is shown.

4.1 Vehicles with Conventional Powertrains

4.1.1 Passenger Car with DCT

The first simulation scenario shown involves a conventional front wheel drive passenger car with a DCT. The simulated system malfunction consists of the simultaneous engagement of both the odd and even gear clutches, for a brief period of time (70 ms), which will introduce a sudden brake torque on the driven wheels, and might result in a loss of vehicle control, particularly if failure occurs in a corner on a wet or icy road surface. In this test scenario, the vehicle drives in a circle with radius of 110 m, with a tire-road friction coefficient (μ) of 0.4, at a constant speed of 70 km/h. Figure 2 shows the DCT clutch actuators, clutch torques, and the driver and vehicle responses (the latter two over a much longer time scale). It can be seen that the vehicle laterally diverts more than 2 m, and therefore could run into the path of an oncoming vehicle, or a road side obstacle.

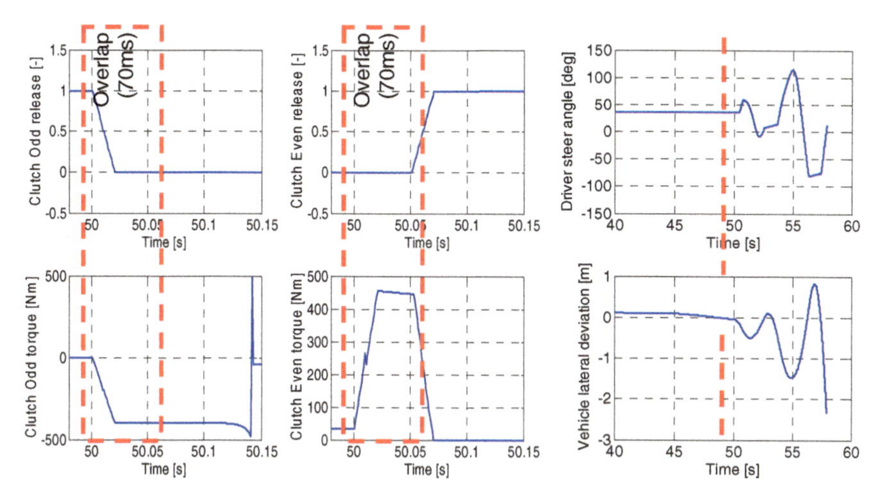

Fig. 2 Simulated signals of DCT clutch actuators and torques, with driver and vehicle response shown over much longer time scales, during a 70 ms double clutch engagement overlap

4.1.2 Heavy Duty Vehicles

The described methodology has also been applied on heavy duty vehicles, including buses and trucks, with semi-trailers and full trailers. A matrix of test cases has been established and simulated involving a tractor or cab with a mass of 7 tons, and a semi-trailer (11 tons when empty or 19 tons when fully loaded) driving round a corner of a radius of 100 m, at a velocity of 65 km/h. Several types of failures were virtually injected, including positive and negative (i.e. braking) torque on inner and outer rear driven wheels, as well as an unintended steering wheel angle reduction (gradually applied over 1.5 s). The human driver reaction time was assumed to be 1.5 s. Simulations were performed with wet (μ=0.7) and dry (μ=1.0) roads. Here selected simulation results are presented of the case involving unintended brake torque (5000 Nm) on the driven rear axle, while driving on a wet road. Figure 3 summarizes the simulated signals. After failure activation (t=200 s), the human driver model seeks to control the vehicle, i.e. minimize the vehicles lateral deviation from nominal track position; here the maximum deviation resulting is ~0.7 m, suggesting the vehicle is likely to just stay in its lane. Aside of perilous lateral vehicular excursions required to control the vehicle after such a possible powertrain failure, the relatively high center of gravity of commercial vehicles, increases the risk of the vehicle tipping over during driver reactions intended to re-gain vehicle control.

Figure 4 shows a modified version of the previously described case in which the vehicle velocity was increased to 71 km/h, and a positive drive torque is introduced and a dry road assumed. The vehicle accelerates for 1.5 s until driver reactions become visible: while the steering wheel angle is increased, the vehicle rolling slightly along its longitudinal axis, the driven inner rear wheel lifting from the ground.

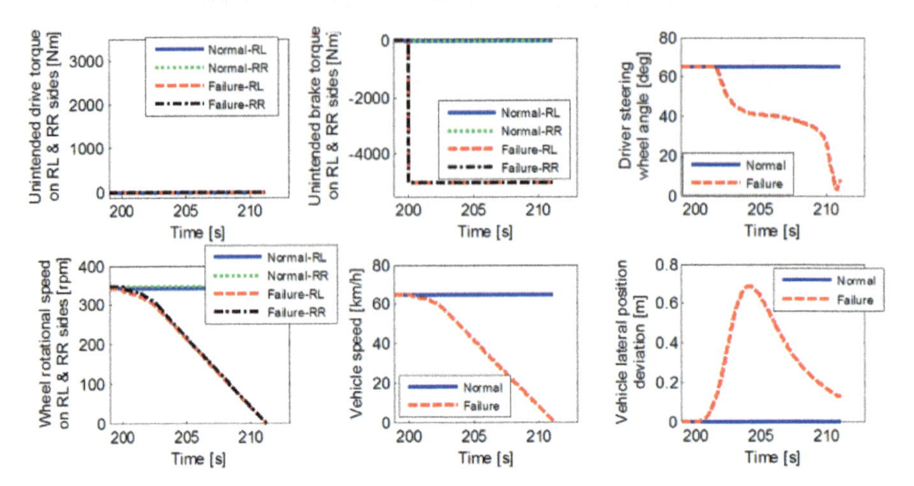

Fig. 3 Simulated signals of tractor with fully loaded semi-trailer at constant vehicle speed of 65 km/h, wet road and unintended brake torque on driven axle. Max. lateral excursion of 0.7 m.

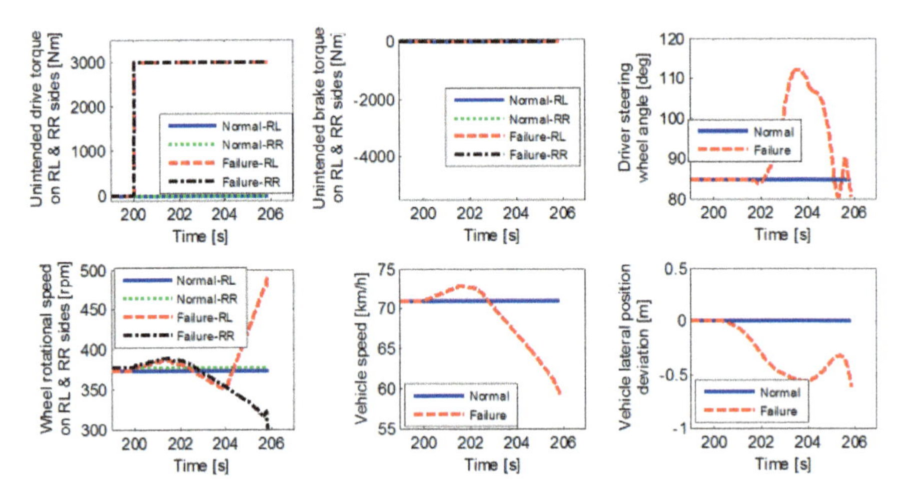

Fig. 4 Simulated signals of tractor with fully loaded semi-trailer at constant speed 71 km/h, dry road, and unintended positive drive torque. At t=204 s the sharp speed increase of the rear left wheel ("Failure-RL", left lower diagram) shows the loss of road contact of this wheel followed by the vehicle rolling over.

As the lifted wheel is still subject to a positive drive torque, its speed sharply increases (see red dashed line in the first graph in the lower row of Figure 4). As the steering wheel angle is reduced too late (around t=204 s), the tipping over of the vehicle is unavoidable (Figure 5).

Fig. 5 Tractor with fully loaded semi-trailer at constant vehicle speed 71km/h, dry road, and unintended positive drive torque (green vehicle; the grey vehicle is the reference vehicle without powertrain failure). Vehicle tipping over cannot be avoided with the unintended positive drive torque active. Snapshots are taken at time ~ 202, 204, 206 s.

4.2 Hybrid Electric Vehicles (HEV)

To demonstrate the applicability of the concept safety simulation approach also for the example of highly complex hybrid electric vehicles, exemplary results are presented here, which were obtained from simulations of a 4WD diesel hybrid electric passenger car (1800 kg) equipped with two EMs, the first of which, is a small integrated starter generator (ISG) of 9 kW power, coupled to the 120 kW diesel driving the front wheels, while a second EM with higher power at 27 kW is directly connected to the rear axle alone.

In the following section, simulation results are shown for selected driving maneuvers including unintended full positive EM torque on rear axle while driving on dry road, and, secondly, while driving on μ-split road conditions with $\mu = 1.0$ (dry) and 0.7 (wet), on inside and outside road strips respectively. Thirdly, a simulation case is presented with a failure consisting of sudden full negative EM torque on the rear axle, whilst driving on wet road. In all cases the hybrid vehicle is driven initially at constant speed of 100 km/h, and in a circle with constant radius of 100 m, and approaching the tire-road friction limits.

Until the virtual failures are triggered, a fixed traction power split of 50/50, between the front and the rear axles is assumed. Figure 6 summarizes the main signals along the maneuver of unintended full positive torque on dry ground. After sudden wheel torque increase the driver gently activates the brakes (not shown), and corrects the steering angle to minimize lateral deviation, which remains below 0.4 m. With μ-split road conditions, the maneuver is quite similar, except that the max. lateral deviation towards center of corner is larger, due to yaw torque introduced by different brake torques on left and right side wheels (data not shown).

Finally, the results of simulation with unintended full negative torque are shown in Figure 7. Initially, the brake torque on the rear wheels causes the vehicle to slow down, and despite driver attempts to regain vehicle stability and control, the lateral excursion reaches 2 m from the center line. Furthermore, once the vehicle velocity has dropped below 70 km/h, the rear EM reaches its max. torque of 200 Nm, which causes the vehicle to quickly yaw towards the corner center point and nearly spins it out of its lane. The maximum lateral deviation during this maneuver is 2 m. At around t = 55 s into the maneuver, the negative torque on the left rear wheel is sufficient to drive the wheel with large negative speed, i.e. in the reverse direction.

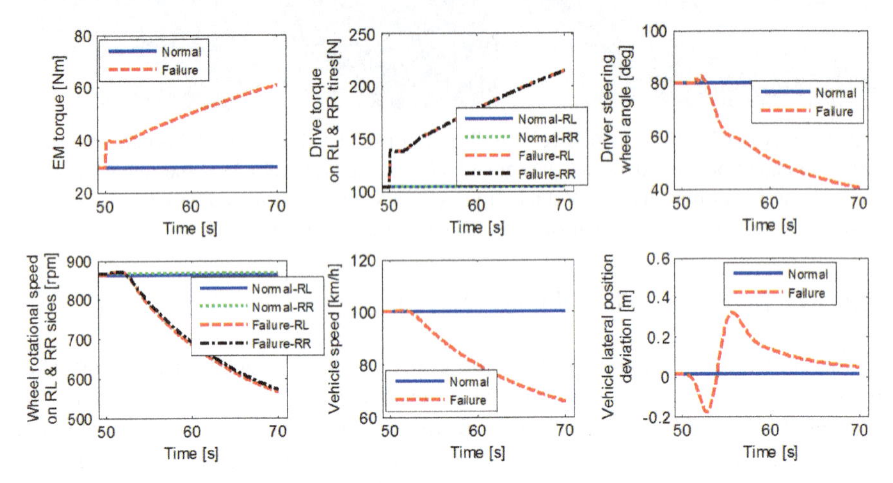

Fig. 6 Simulated signals of hybrid electric vehicle & driver exposed to unintended full positive e-motor torque on rear axle, dry road

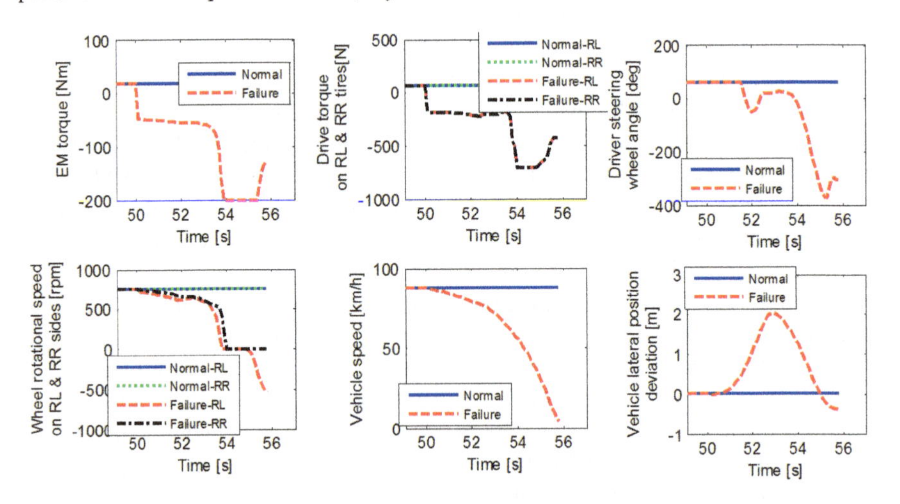

Fig. 7 Simulated signals of hybrid electric vehicle & driver exposed to unintended full negative EM torque on rear axle, wet road; at t = 54 s, the left hand side rear wheel (RL, red curve) speed assumes negative values i.e. it rotates backwards (see lower left most graph)

4.3 Fully Electric Vehicles with Multiple E-Machines

Various failure scenarios should be considered for electric vehicles (EV) with four EMs. For example, one of the four EMs could suddenly generate full positive, full negative or zero torque (contrary to the torque demanded by the driver), due to a sudden failure of EM hardware, related power electronics, control software, or even due to a corrupt controller application calibration.

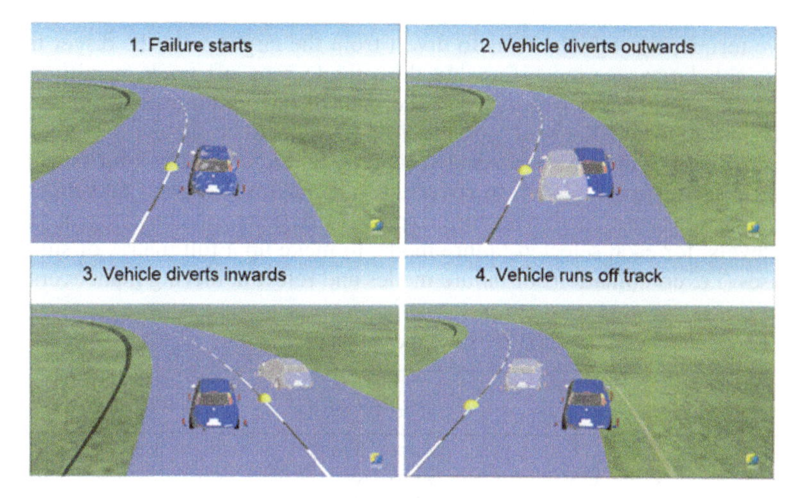

Fig. 8 Snapshots of the vehicle movement when one EM fails with full positive torque on an electric vehicle with four EMs; grey vehicle is reference without failure

Simulation results of one example maneuver are presented here in which the simulated EV runs at a constant speed of 70 km/h, in a circle with radius of 110 m, with tire-road friction coefficient (μ) of 0.4 (corresponding to a wet and slippery

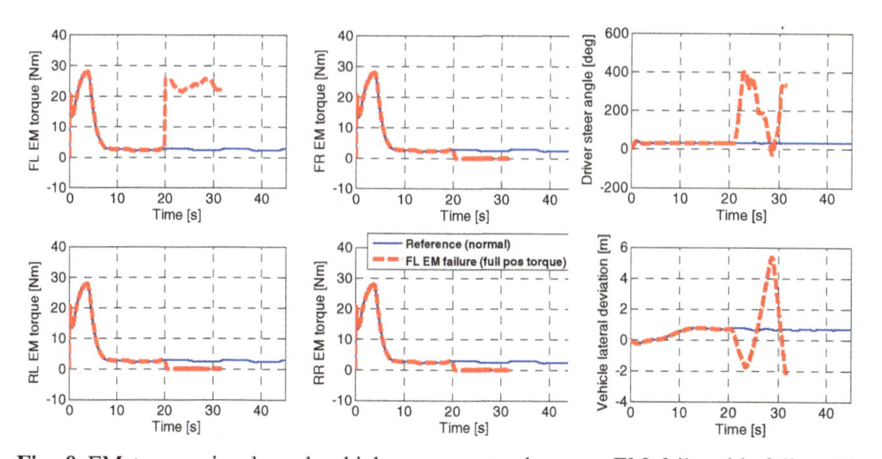

Fig. 9 EM torque signals and vehicle movement when one EM fails with full positive torque on an electric vehicle with four EMs

road). The simulated EV is equipped with four EMs, each with 13 kW mechanical power. In the first failure scenario, the front left EM suddenly provides full positive torque.

Snapshots of the simulated vehicle motion are shown in Figure 8. Two vehicles are shown of which the dark (blue) one has the failing EM, whilst the light grey one is the reference vehicle without failure.

Figure 9 shows the EM torque curves, driver steering and vehicle lateral deviation. The solid (blue) lines represent the reference vehicle, whilst the dashed (red) lines represent the vehicle with a failed EM. When the front left EM fails with full positive torque, the vehicle starts to divert from the target path; the driver tries to correct with large steering effort, but the vehicle runs off track, resulting in an unintended lane change, possibly into the path of oncoming traffic, and later the vehicle partially leave the road, possibly into a road side obstacle.

Analogously to the shown driving scenarios, other potentially critical scenarios can and need to be simulated, to cover the full range of relevant maneuvers e.g. driving on a straight road with differing road frictions, or cornering at various speed levels, for each selected failure mode, to make up a full matrix of simulated test cases (i.e. combinations of failure modes and relevant driving maneuvers).

Table 1 EV controllability matrix for specified driving scenarios

<table>
<tr><th>Driving Scenario</th><th>Road Friction</th><th>EM-Torque Failure</th><th>Controllability</th></tr>
<tr><td rowspan="8">Straight 100km/h</td><td rowspan="4">Dry (μ=0.8)</td><td>Full pos. Trq.</td><td rowspan="8">Controllable</td></tr>
<tr><td>Zero Trq.</td></tr>
<tr><td>Full neg. Trq.</td></tr>
<tr><td>2 Rear EM Full neg. Trq.</td></tr>
<tr><td rowspan="4">Wet (μ=0.4)</td><td>Full pos. Trq.</td></tr>
<tr><td>Zero Trq.</td></tr>
<tr><td>Full neg. Trq.</td></tr>
<tr><td>2 Rear EM Full neg. Trq.</td></tr>
<tr><td rowspan="4">Circling R=110m 100km/h</td><td rowspan="4">Dry (μ=0.8)</td><td>Full pos. Trq.</td><td>Uncontrollable</td></tr>
<tr><td>Zero Trq.</td><td>Controllable</td></tr>
<tr><td>Full neg. Trq.</td><td rowspan="2">Uncontrollable</td></tr>
<tr><td>2 Rear EM Full neg. Trq.</td></tr>
<tr><td rowspan="8">Circling R=110m 70km/h</td><td rowspan="4">Wet (μ=0.4)</td><td>Full pos. Trq.</td><td>Uncontrollable</td></tr>
<tr><td>Zero Trq.</td><td>Controllable</td></tr>
<tr><td>Full neg. Trq.</td><td rowspan="2">Uncontrollable</td></tr>
<tr><td>2 Rear EM Full neg. Trq.</td></tr>
<tr><td rowspan="4">Dry (μ=0.8)</td><td>Full pos. Trq.</td><td rowspan="4">Controllable</td></tr>
<tr><td>Zero Trq.</td></tr>
<tr><td>Full neg. Trq.</td></tr>
<tr><td>2 Rear EM Full neg. Trq.</td></tr>
<tr><td rowspan="4">Circling R=110m 70km/h; driver reaction time 2s</td><td rowspan="4">Dry (μ=0.8)</td><td>Full pos. Trq.</td><td>Uncontrollable</td></tr>
<tr><td>Zero Trq.</td><td>Controllable</td></tr>
<tr><td>Full neg. Trq.</td><td>Uncontrollable</td></tr>
<tr><td>2 Rear EM Full neg. Trq.</td><td>Controllable</td></tr>
</table>

Controllability (here primarily evaluated on the basis of acceptable lateral vehicle deviation, assuming a driver of average driving skills) may be assessed using simulation results, and an exemplary controllability matrix, as shown in Table 1, thus generated. Exposure and severity may also be estimated from the simulation results in combination with statistical data. However, these assessments require a cross-functional team, composed of vehicle dynamics, powertrain, control systems and functional safety specialists, working together to use the system simulation results to support their deliberations, and to assist them in making expert judgments, to establish the appropriate ASIL.

5 Conclusions

The presented co-simulation toolchain permits hazard severity and vehicle controllability modelling in the presence of defined system failures, supporting improved hazard analysis and risk assessment by a cross-functional expert system engineering team. Thus the ASIL may be assessed with a higher confidence, from the early concept phase onwards. This co-simulation based methodology is available for trucks, buses, passenger cars and motorcycles, with conventional, hybrid electric and fully electric powertrains.

A refinement of the presented method relies on the definition of a broader spectrum of human driver types, with differing driving skills e.g. reaction times, steering angle rates, etc. These can be used to more precisely determine the controllability levels (C0 to C3) for each driving maneuver and powertrain failure.

Other advantages of system simulation support in the hazard analysis and risk assessment lie in the reproducibility it provides, and the higher coverage of different situations (which are composed from the countless variations of the parameters above) it permits.

The same co-simulation techniques may also be used to develop and validate fault mitigation strategies e.g. failure detection functions, improved hardware including sensors and communication networks.

Acknowledgements. The research leading to these results has received funding from the ARTEMIS Joint Undertaking under grant agreement number 295311 and from specific national programs and/or funding authorities.

References

[1] International Organization for Standardization, ISO 26262 Road vehicles - Functional safety, Geneva, Switzerland, 2011 and (2012)

[2] Jones, S., Böhm, H., Weingerl, P., Cheng, C.: Dynamic simulation of complex mechatronic systems: Torsional vibrations in powertrains, vehicle dynamics & safety. In: Systemanalyse in der Kfz-Antriebstechnik VII - Haus der Technik Fachbuch, vol. 129, p. 83. Expert Verlag, Renningen (2013)

[3] Jones, S., Ellinger, E.: Vehicle System Simulation for Electrified & Conventional Powertrains. In: SIMVEC - Berechnung, Simulation und Erprobung im Fahrzeugbau, p. 81. VDI-Bericht 2169, Baden-Baden (2012)
[4] Jones, S., Kural, E., Knoedler, K., Steinmann, J.: Optimal Energy Efficiency, Vehicle Stability and Safety on OpEneR EV with Electrified Front and Rear Axles. In: Fischer-Wolfarth, J., Meyer, G. (eds.) Advanced Microsystems for Automotive Applications 2013, Berlin, Germany (2013)

When Do We Get the Electronic Battery Switch?

Werner Rößler

Abstract. The battery pack of an electric or a hybrid vehicle is stacked to dangerous voltages up to 400 volts. It has to be disconnected during parking, for maintenance and very quickly in case of an accident. Special high voltage relays are State-of-the-art. As these components are easily damaged, expensive, heavy and bulky, a solution based on pure semiconductors is desired.

This paper describes the benefits of a purely solid state battery switch. They provide the potential for increased reliability and more than a thousand times faster switching time in the case of an external short circuit. The weight could be reduced to a third from 3 kg to 1 kg, and the volume could be reduced over 79% from 7 liters to 2 liters. Despite the enormous performance gain, the production costs are likely to be similar or slightly lower.

Keywords: Hybrid vehicles, EV, Battery, Main switch, Relay, Semiconductor, MOSFET.

1 Battery System Architecture of a Hybrid or Pure Electrical Vehicle

Electrically driven or boosted vehicles have a battery system as energy storage. High voltages in the range of 400 volts are used. The battery stack is composed of several modules (blocks) connected in series. Every block is a series arrangement of typically 10 to 12 single cells with cell monitoring circuitry. Fig.1 shows the block diagram of such a system. The battery terminal is connected to the vehicle chassis ground via a high resistance path to get a symmetrical output voltage. Each output terminal has a switching element to disconnect the battery

W. Rößler(✉)
Infineon Technologies AG, Automotive System Engineering,
Am Campeon 1-12, 85579 Neubiberg, Germany
e-mail: werner.roessler@infineon.com

J. Fischer-Wolfarth and G. Meyer (eds.), *Advanced Microsystems for Automotive Applications 2014*, Lecture Notes in Mobility,
DOI: 10.1007/978-3-319-08087-1_16,

system in case of parking, and for maintenance. Under these conditions, all the consuming elements are switched off. This allows disconnection with very low current load. In case of an accident or an external short circuit, the switch has to react very fast. These emergency disconnections have to be managed under high load currents.

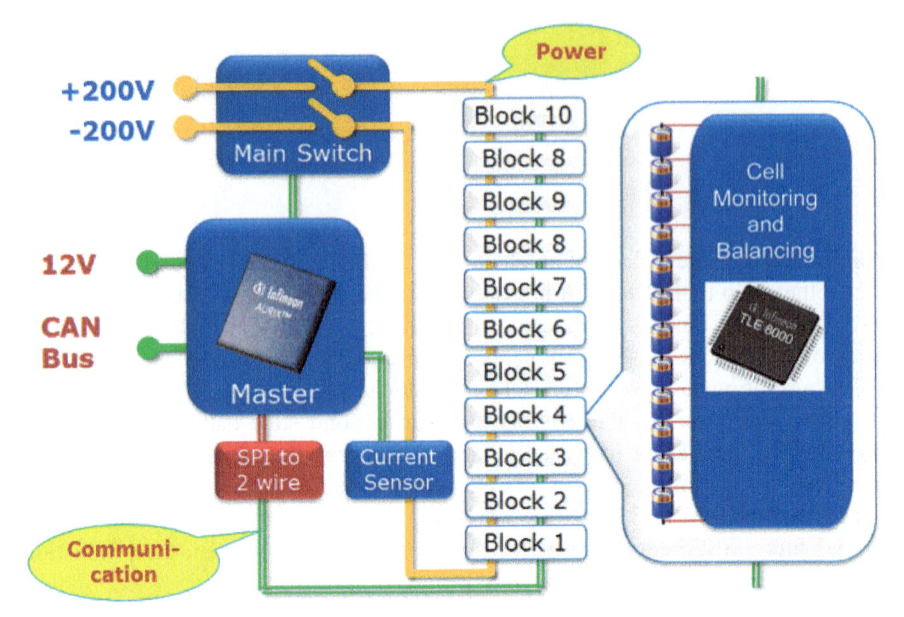

Fig. 1 High Voltage System of an electrically driven Vehicle

A powerful battery master unit has following tasks:

- *Monitoring of all battery cells*
  For lithium Ion cells, it is essential, that their voltage stays inside the allowed safe limits. If the cells are operated outside, damages like reduced capacity and increased self-discharge rates may occur. At higher voltages, overheat up to explosions can happen.
- *Balancing the cells in the case of different state-of-charge*
  This ensures that the state of charge of the single cells is equalized to reach the highest usable capacity. Active balancing provides the best performance.
- *Measure the system current*
  As all battery cells are linked in series, this will be the same for all cells. Therefore a single measurement spot with a current sensor is sufficient.
- *Control and supervise the main switch*
  The main switch is a very safety critical component. The possibility for a proper disconnection has to be guaranteed. Self-tests and a permanent monitoring of the functional parts is mandatory.

- *Calculation of the total state-of charge*
  The state-of–charge is important information for the driver. It indicates the remaining travelling distance. The value is calculated in a complex algorithm using the information about voltages, the current profile and the temperature.
- *Communicate with the vehicle control system*
  The battery main switch is embedded in the complete vehicle system. Besides the reception of switching commands, status messages are sent. In case of malfunction, the driver has to be informed about warnings and errors.

1.1 Requirements for the Battery Main Switch

The following collection of requirements is based on various discussions with different car makers (OEM) and Tier1 suppliers.

- *Safe disconnection*
  This is the most important task. If the vehicle is parked, the battery module has to be electrically disconnected from the rest of the vehicle. Both output poles of the battery string have to be decoupled. The leakage current has to be far below the perception threshold of one milliampere.
- *Operating Voltage*
  The output voltage range of the battery system depends on the type of the vehicle. The requirements are typically between 200 volts and 420 volts battery voltage.
- *Load Current*
  The max value of the load current depends on the type of vehicle. A mild hybrid car is less powerful than a pure electric sports car. A typical requirement is a continuous value of up to 150 amperes for high-speed cruising and a peak current up to 400 amperes. As these peak loads correspond to a power of about 160 kilowatts, they are limited to an acceleration phase below 10 seconds.
- *Charge Current*
  The charge current is an energy flow back to the battery. Fast high-power chargers can deliver up to 150 amperes. In a driving cycle, recuperation currents up to 350 amperes, limited for 10 seconds, are possible.

1.2 Solution in Existing Vehicles

Industrial solutions are often the reference for automotive implementations. For the disconnection of electrical circuits, relays are in use since many decades of years. Despite their later shown disadvantages, this technique was adapted for vehicle applications. The relays are available and their technology has been refined over the years. In many eyes, they are the only safe way to disconnect electrical circuit. In the past it was even mandatory to use a mechanical separation. The actual situation is that alternative solutions may be accepted, if their safety is proven.

2 Relay – An Ideal Switch?

At first glance, the relay fulfils the requirements perfectly. There are two metal contacts, which are pressed together or separated in a safe distance. But there is a difference between the ideal situation and the reality. As automotive OEMs are not happy with this situation, they are looking for alternatives.

- *Switching under load :* The most difficult task for a relay is a disconnection under load. Starting at voltages above 15 volts, an arc can be generated. Depending on the current value and the voltage, the arc can sustain the current flow over larger gaps. Therefore special high voltage DC relays are encapsulated and filled with special gases. Even if the arc is instantaneous, some metal contact material vaporizes and condenses at the walls of the relay chamber. Over the time, this leads to a better and better conducting layer, increasing the risk of leakage current in the Off state (see below).
- *On state :* If the relay contacts are closed and in a good condition, extremely low resistance values below 1 milliohm can be expected. Unfortunately, the switching process degrades the contact surface. Arcs caused by load switching and bouncing increase the contact resistance. Thus, the component ages.
- *Off state :* If the contacts are separated, a theoretically infinite resistance value is expected. But the reality shows differently. As explained above, metal particles at the walls of the chamber lead to an increasing leakage current, even if the contacts are disconnected.

3 Semiconductor Selection

3.1 Power Semiconductor Types

Semiconductor manufacturers offer a variety of different types for high power applications. Depending on the application requirements, a suitable technology has to be selected.

- *Thyristor :* Thyristors are widely used for power switches. Once turned on, they cannot be switched off via the control terminal. In AC applications, this is no problem, as the subsequent zero crossing interrupts the current flow. Because of this behavior, thyristors disqualify as DC main switch elements.
- *Bipolar transistor :* These already old-fashioned components are controlled by an inserted base current. For higher load currents, the amplification factor gets very low. For the required high loads, base driver stages with several amperes capability would be highly inefficient and expensive.
- *IGBT (Insulated-Gate Bipolar Transistor):* These power semiconductors are typically used in switching applications like motor inverters.

They are cheaper compared to MOSFETs with the same current capability. Compared to bipolar transistors, IGBTs need no driving current. Only the gate capacitor has to be charged. The big difference between a motor inverter and a main switch is the current direction. A Battery contactor is operated in both directions. As reverse current flows are not allowed, IGBTs have to be combined with an additional reverse diode. Even if these diodes are integrated in the package in most cases, the voltage drop cannot be avoided. As shown in Fig.2, the voltage drop of both, the IGBT and the diode rises above 1 volt even at low load currents. If IGBTs are connected in parallel, the curve above 1 volt gets flatter, but the 1 volt basic drop still remains.

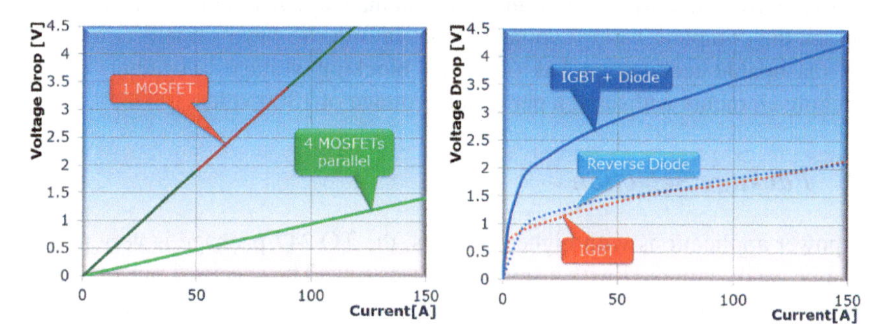

Fig. 2 Voltage drop: Left: of IGBT and Reverse Diode; Right: MOSFET

- *MOSFET (Metal Oxide semiconductor Field-Effect Transistor)* : Like IGBTs, MOSFETs need no control power, when they are switched on statically. Compared to the IGBTs, MOSFETs have no basic voltage drop. Their output behavior equals a resistor's characteristic. In case of parallel connected components, the total resistance value reduces proportional to the number of parts. MOSFETs may be operated with a reverse current. If they are switched off, the parasitic diode offers the path. The voltage drop is similar to a discrete diode. As this creates heavy losses at higher currents, it is not recommended to use MOSFETs in reverse mode over the diode. If the component is switch on with a positive gate voltage, the ohmic behavior is the same as in positive direction.

3.2 The Suitable Semiconductor Type

Regarding the arguments of chapter 3.1, thyristors and bipolar transistors have to be discarded. IGBTs can be used at least for lower current requirements and if the higher losses are accepted. Lower component costs are the benefit, but higher cooling effort defeats the advantage. Finally modern MOSFETs are the best choice. They can be easily connected in parallel. The minimum number of necessary components is given by the current capability of a single transistor. If more parts are used, the losses are reduced and the robustness is increased. So it is finally a trade-off between costs on the one side and efficiency and cooling effort on the other side.

4 Selection Criteria of the MOSFET Components

4.1 Blocking Voltage

The operation voltage can be up to 420 volts. For redundancy reasons, two switches are connected in series. Therefore their blocking voltages can be added, if both are switched off. If one transistor is damaged, each element has to be able to block the complete battery voltage alone. In order to have some headroom, the next reasonable technology has 650 volts blocking voltage. Together, the two switching elements withstand a permanent voltage of 1300 volts.

4.2 Package Selection

For power applications with higher voltages, the TO 247 package is very popular. It can be easily mounted with a screw to a heat sink. The special 4 pin version has two source terminals. Pin 2 carries the load current, while pin 3 connects the driver stage directly to the chip. This allows a safer switching behavior under extreme conditions.

Fig. 3 TO 247 Transistor Package

4.3 Current Capability

Modern MOSFETs with 650 volts blocking voltage in a TO 247 package can carry permanent currents up to 75 amperes at 25°C. A case temperature of 100°C reduces the allowed DC current to 62 amperes. A parallel combination of 7 MOSFETs would be theoretically sufficient. Pulsed currents of up to 496 amperes per transistor can be switched. In order to reduce the losses and to provide a safety margin, a number of 8 MOSFETs in parallel is recommended.

4.4 On Resistance (RDSon)

The on resistance is one of the most important key parameters of a transistor. It defines the voltage drop and herby the power losses in the on state. The up-to-date technology 'C7' has a maximum of 19 milliohms in a TO 247 package. The previous 'C6' technology could only offer 41 milliohms. To get the same performance, double the number of parts was necessary some time ago. The value of the on resistance is proportional to the temperature. Due to this, it is recommended to keep the temperature as low as possible. If the batteries for higher power are water cooled, this cooling system can be also connected to the main switch.

5 Feature Comparison Relay - Semiconductor

To make decision for a relay or a semiconductor based solution, it is useful to highlight different aspects.

5.1 Experience

Relays have existed longer than semiconductors. Relays have been optimized during this time. Power semiconductors are established in myriads of applications.

5.2 Lifetime

The lifetime is an important parameter for automotive applications. End customers expect an operation time of more than ten year without any trouble. Relays have a limited number of switching cycles. Semiconductors are able to switch many thousand times per second over several years

5.3 Reaction Time

As already shown later, relays are not able to handle an external short circuit satisfactorily. Instead of more than 2 milliseconds, semiconductors can be switched of in 100 nanoseconds.

5.4 Module Height and Volume

Volume of the components in vehicles is very important. The limited space should be optimized for passengers and luggage. The bulky relays require up to 7 liters volume for a switch box in existing solutions The height of an existing relay is more than 10 centimeters. As semiconductors are very small and flat, the modules can be designed with a height of 4 cm or even less. The height of a TO247 transistor is only 5 millimeters.

5.5 Module Weight

The weight is a key parameter for vehicles. As it directly influences the energy consumption, it should be as low as possible. The limited range of electrical vehicles makes the weight of each component critical. A reduction from approx. 3kg to less than 1 kg including a water cooled heat sink is possible.

5.6 Vibration Robustness

Vehicles are not operated stationary like many industrial applications. Depending on the quality of the road and the driving speed, vehicle components will be subjected to levels of vibration and shock. For semiconductors, vibration is no big issue.

5.7 Noise

Electric vehicles are very quiet. Therefore the clicking of a relay does not fit to the image and can be very obtrusive. To suppress this noise, some sound absorption measurements are necessary. This costs additional space, weight and money.

5.8 Component Costs

High voltage DC relays are often customer specific designs. There are many different types available. As they are not commodity products, it is impossible to find out the real costs as non-customer. Based on different indications, it seems that the pure component costs of semiconductors may be slightly higher. As car makers always try to reduce the costs, both variants are rated negatively.

5.9 System Cost

In case of the system costs, a semiconductor based solution could be cheaper. To calculate the real savings, a weighting of the benefits has to be done. However, survival of a short circuit would provide a saving of more than the cost of replacement in terms of vehicle repair and inconvenience.

Table 1 Comparison of relay and semiconductor

Relay	*Semiconductor*
• Cabling and joining technique for 5000 amperes max	• Cabling and joining technique for 600 amperes max
• Vibration damping required	• Robust against vibration
• Higher space demand	• Smaller
• Switchbox and eventually battery damaged after crash	• High protection of the battery, switch can be reset
• Noise suppression required	• No noise generation
• Limited life time	• Life time virtually unlimited
• High weight	• Lighter

6 Short Circuits Outside of the Battery System

A short circuit outside the battery can happen if the high voltage power cables are damaged during an accident and make an electrical contact. Another reason could be a defect in a load such as the motor inverter. Fig. 4 shows the current rise after an event at t = 2 milliseconds. Depending on the impedances of the battery system and the short circuit point, the current may rise to over 5000 amperes within 5 milliseconds.

Even if the accident is detected immediately, a relay needs several milliseconds to open. As these parts are not designed to open under loads of thousands of amperes, it is only possible to wait until the fuse blows after some tens or hundreds milliseconds. The high current value is applied to the complete system for this extended period. At least the switch box will have to be replaced. It is also almost inevitable, that the battery will be damaged and have to be replaced also.

Power semiconductors have possible switching times down to 100 nanoseconds, which is some ten thousand times faster. So there is enough time to evaluate the event carefully over some microseconds for a final decision. Only if a real short circuit occurs, the semiconductors will be switched off safely. The battery will not be damaged and the main switch can be switched on after the short circuit has been repaired.

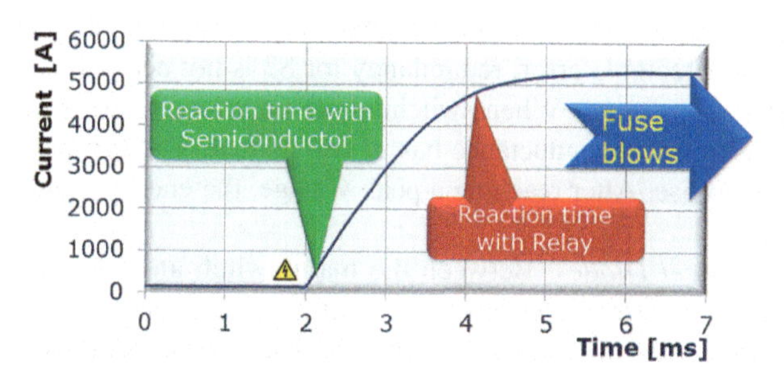

Fig. 4 Short circuit event

If it is guaranteed that the main switch can interrupt the current safely, the cabling system does not need to be designed for thousands of amperes in a crash situation. This can save money and weight. The costs for rescue of the battery and the resettable main switch should be included as part of the overall cost of ownership and added to the initial production costs. However, the philosophy of the OEM will determine if the total cost of ownership is taken into consideration.

7 Architecture of a Solid State Main Switch

Fig. 5 shows the architecture of a solid state main switch. To provide safety redundancy and bi-directional current flows, three independent MOSFETs are required.

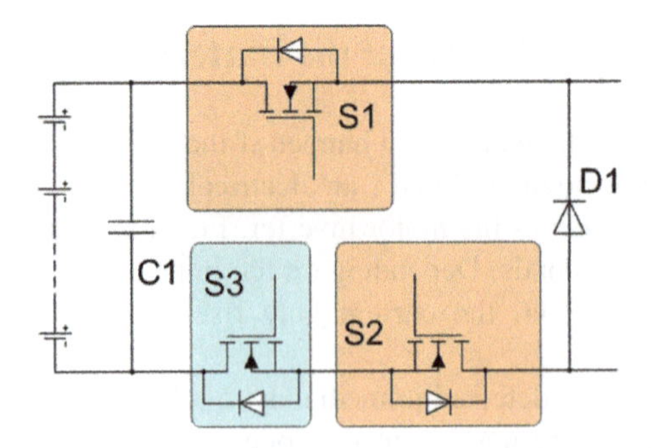

Fig. 5 Block Circuit Diagram

- *S1 (positive load path):* This switch disconnects the positive current path. Because of the parasitic reverse diode, the charge current cannot be interrupted.
- *S2 (negative load path) :* This switch disconnects the negative current path. The current flow can already be interrupted with S1 completely. Therefore S2 is fully redundant.
- *S3 (negative load path):* This switch disconnects the negative current path in reverse direction. Even if S1 and S2 are open, the battery still can be charged over the parasitic reverse diodes of the MOSFETs. To avoid unwanted overcharge, S3 is required. This is only necessary, if the charger or the recuperation system work erroneous and cannot be stopped. As this is already a detectable error, redundancy for S3 is not needed.
- *C1 Input Capacitor :* When switching off under load current, the energy in the parasitic source inductance has to be handled. The capacitor C1 reduces the voltage rise. After reaching a peak voltage, the energy swings back to the battery.
- *D1 Freewheel Diode :* Between the main switch and load (e.g. the motor inverter), the cables can have a length a few meters. In the event of an emergency interruption, the current needs a path to diminish. This path is offered by the freewheel diode D1. The output voltage of the main switch is clamped to the forward voltage of the diode until the energy stored in the cable inductance is dissipated.

8 Pre-charging of a DC-Link Capacitor

The secondary side of a battery main switch is connected to the energy consuming parts. The motor inverter is the biggest load. Fast switching semiconductors are used as choppers to convert the DC input voltage into a three phase AC motor voltage. A big capacitive input filter is required. These so-called DC-link capacitors have values of several hundred microfarads.

As a discharged capacitor is like a short circuit, it cannot be connected abruptly with the battery. Extremely high inrush currents have to be avoided. In existing solutions, a smaller additional relay together with a high load resistor formes the standard precharge circuit.

The fast switching capability of MOSFETs allows a pulsed precharge procedure. As the already existing high power switching elements can be used, the precharge function can be implemented without any additional material costs. The connection cables in a vehicle lead to unavoidable parasitic inductive parameters. This automatically reduces the current rise time. With defined pulses, a controlled precharge procedure can be achieved.

9 Example of a Prototype Implementation

In the Infineon automotive system laboratories, a first working prototype was developed. The performance measurements and tests are promising. The design is very modular, so that the components can be easily replaced by updated versions without touching the complete setup.

9.1 Power Modules

The switches S1, S2 and S3 (Fig. 5) are in the same current path. Therefore their architecture can be equal. The number of 8 MOSFETs in parallel is a good compromise between performance and cost. The transistors are mounted on an aluminum heat sink, which is also used as high current path. An 8 mm screw hole in the center is used as input respectively output terminal. The module has a size of 10 x 10 centimeters. By the arrangement of eight transistors in a circle around the central terminal, the current distribution is optimized.

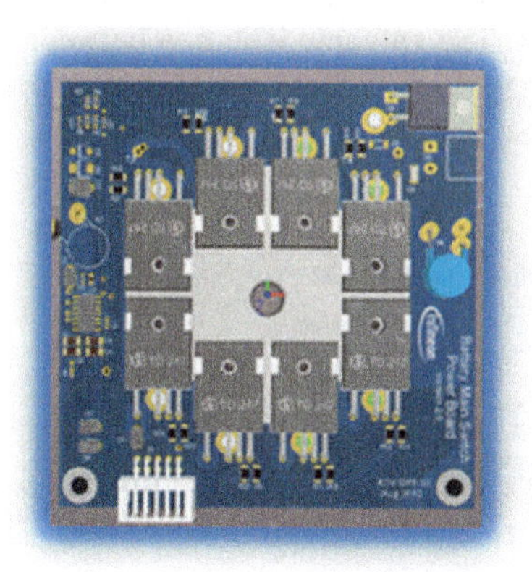

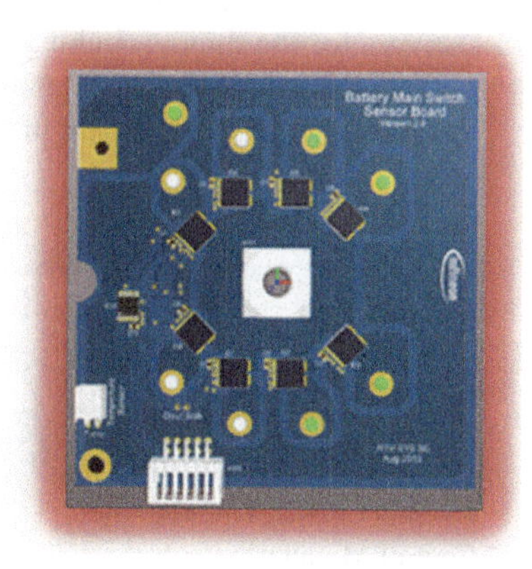

Fig. 6 Power Module (left) and Sensor Module (right)

9.2 Sensor Module

The sensor module carries eight hall sensors with a measurement range of 50 amperes each. Hall Effect sensors comprise of a galvanic isolated semiconductor chip. The communication path is already related to vehicle ground and need no further signal transmission over different voltage potentials. The actual current value can be read out over a serial SPI interface. Additionally a FOC (fast overcurrent) signal line can be used for a fast hardware based fuse function. As only the magnetic field around a conductor is used for the measurement, no voltage drop like in a shunt solution is produced. This minimizes the power losses in the system.

This design with eight parts is not economical for high production quantities, but is able to demonstrate the benefit of a Hall Effect based measurement. As soon as suitable parts for higher current values are available, the module will be updated. For redundancy reasons, a combination of two 200 amperes sensors could be a good solution.

9.3 Control Module

The control module acts as interface between the communication port and the MOSFETs. The key components are:

- *Controller:* Calculation and control part of the module is a simple 8 bit microcontroller (marked with ④ in Fig.7). The tasks are as:
  - Communication with the battery controller
  - Initialization and monitoring of the driver ICs
  - Read-out of the current sensor values
  - Measure voltages and temperatures
  - Control the switch states
  - Generate the precharge pulse timing
- *Gate drivers:* The three ICs marked with ①②③ (Fig. 7) are isolated gate drivers for the three switching modules. The galvanic isolation between input and output stage withstands voltages up to 2000 volts. Multiple supervision functions lead to a high safety level.

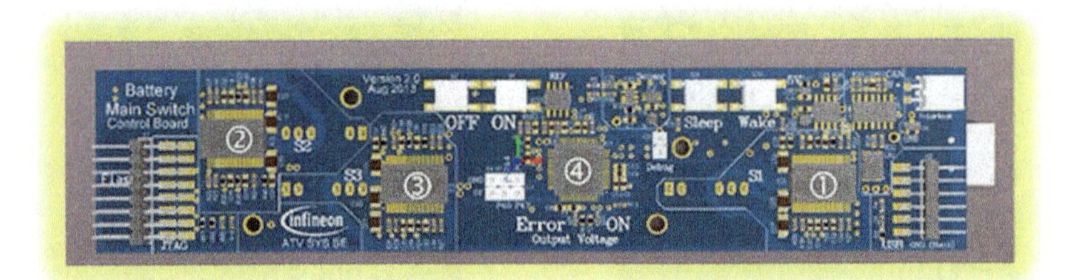

Fig. 7 Control Module

9.4 Complete Setup

The single modules are plugged together around a central heat sink. The cooling core is prepared be connected to a water circulation. As batteries for higher power applications need anyway a water cooling system, it can be used for the switch, too.

Fig. 8 Complete Main Switch Module

10 Conclusion and Outlook

Relays are the existing state-of-the-art solution for the battery main switches in electrically powered vehicles. Because of their substantial disadvantages, they will be replaced by semiconductors in the future. Permanent progress in the electrical parameters of new semiconductors will furthermore improve the performance.

The continuous improvements of the on resistance values lead to a reduction of the cost. A lower number of MOSFETs in parallel will give the same performance. It is also possible to reduce the losses and the cooling effort.

After the necessary paradigm change from the electromechanically solution to a fully semiconductor based main switch, in the future a solid state solution will be the standard.

On Board Energy Management Algorithm Based on Fuzzy Logic for an Urban Electric Bus with Hybrid Energy Storage System

Davide Tarsitano, Laura Mazzola, Ferdinando Luigi Mapelli, Stefano Arrigoni, Federico Cheli and Feyza Haskaraman

Abstract. Nowadays considerable resources have been invested on low emission passenger vehicle both for private and public transportation. A feasible solution for urban buses is a full electrical traction system fed by supercapacitor, that can be recharged at each bus stop while people are getting on and off. Moreover, in order to consider the worst operating condition for the bus (like traffic jam of higher distance to be covered), a conventional battery is also installed, obtaining an hybrid energy storage system. An energy management function, able to manage the two on board energy storage system based on fuzzy control logic, has been developed and validated by means of numerical simulations and compared to a previously presented one in order to evaluate its performances.

Keywords: full electrical vehicle, energy management logic, fuzzy control logic, low emission, supercapacitor.

1 Introduction

In order reduce urban transportation emission for public busses, a full electrical traction system is designed. Given that urban busses cover a short distance between consecutive stops, the designed traction system is primarily fed by supercapacitor: even if it can't store a big amount of energy, it appears to be the

D. Tarsitano(✉) · L. Mazzola · F.L. Mapelli · S. Arrigoni · F. Cheli
Department of Mechanics, Politecnico di Milano, via La Masa 1, Milan, Italy
e-mail: {davide.tarsitano,laura.mazzola,ferdinando.mapelli, stefano.arrigoni,federico.cheli}@polimi.it

F. Haskaraman
Department of Mechanical Engineering, Massachusetts Institute of Technology, Cambridge, MA, USA
e-mail: feyzahas@mit.edu

J. Fischer-Wolfarth and G. Meyer (eds.), *Advanced Microsystems for Automotive Applications 2014*, Lecture Notes in Mobility,
DOI: 10.1007/978-3-319-08087-1_17,

better storage device because it is fast enough be recharged at each bus stop and its life cycle of charge/discharge is much higher than the one of a battery. The present paper reports the analysis and validation of a bus hybrid energy storage system, constituted by a Li-ion battery and a supercapacitors bank [1, 2], designed for fast charging at bus stops. A numerical simulation model of the whole vehicle has been developed [3]. This model is suitable to define an energy storage management function for the on board energy system. The algorithm used is a fuzzy logic-based strategy able to split the power requested by the traction system between batteries and supercapacitor. This algorithm is fundamental and it is defined according to some peculiar purpose, e.g. to allow the battery for working in a steady state condition [4] or for maximizing the vehicle performance [5].

For the considered application the main purpose of the hybrid on board storage system is to extend the battery life. The proposed algorithm has been tested by means of a simulation model in order to verify the global performances. Results were eventually compared to the one obtained by means of the application of a heuristic approach [3].

2 Vehicle Description and Modeling

The powertrain of a urban electric bus, which is analyzed in this paper is schematically represented in Fig. 1: the charging station at bus stop is reported as well in the same figure. The energy storage system has been studied in order to allow the bus to cover the distance between two different bus stops using almost all of the energy in the supercapacitors; once the vehicle is stopped at the bus stop the on board supercapacitors bank (SC1) is charged using the energy stored in the charging station supercapacitor bank (SC2) through a 100kW DC/DC converter and a mobile contact.

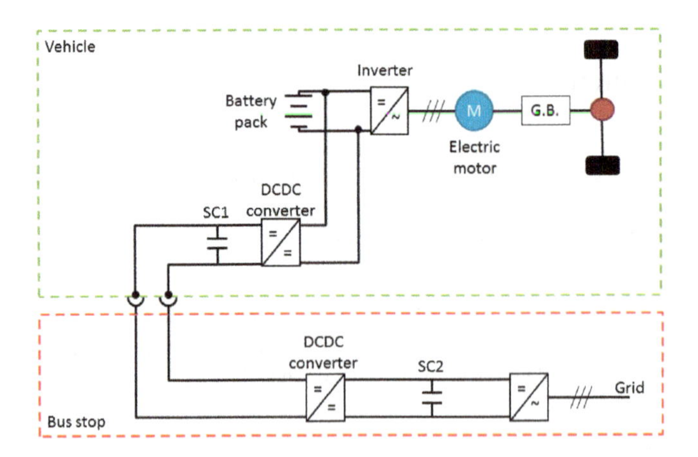

Fig. 1 Scheme of Electric bus power-train and storage system

Whereas the SC2 bank is charged from the grid through an AC/DC converter rated of lower power in order to have uniform loads for the power distribution grid. Some of the principal vehicle data are reported in Table 1.

Table 1 General data of the electric bus

Vehicle	
Mass	6500 kg
Wheel radius	0.356 m
C_D	0.573
Frontal area	5.4 m^2
Energy storage system	
Battery type	LiFePo4
Battery energy	24 kWh
Battery rated voltage	537.6 V
Supercap bank capacity	30.55 F
Electric power train	
Rated power	61 kW
Rated speed	4410 rpm
Rated torque	130 Nm

The simulation model has been developed using the object oriented approach [6-7]: every single device is modelled as an object with input signal and output state variables, and connected to the others. All the single devices represented in Fig. 1 (Lithium-ion battery, inverter and electric motor, gear box (G.B.), DC/DC converters, supercapacitors bank) together with pilot and longitudinal dynamic of the vehicle represents the overall system (see in Fig. 2, where T_{EM} and T_{Drive} are respectively the requested and the effective Electrical Motor torque and T_{Brake} is the mechanical brakes torque). Moreover in this case study a constant average power for auxiliary loads has been considered.

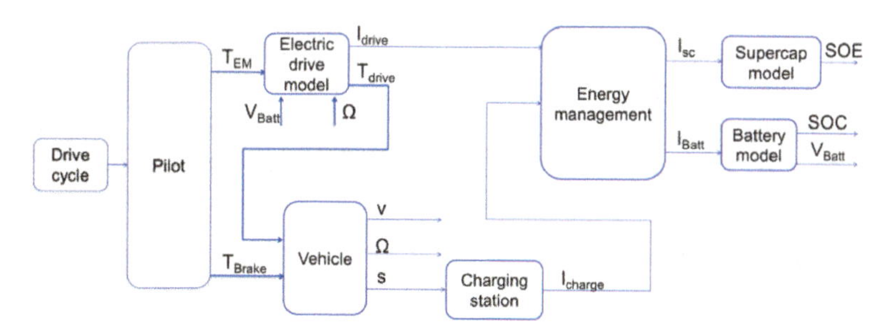

Fig. 2 General diagram of the simulation model

All the details of the numerical simulation model are available in [7-8], however, in the following each sub - model is briefly described.

First a drive cycle, replicating the mission of the vehicle, interacts with the pilot model transferring the information needed to assess the mission (e.g maximum speed). Then the pilot model, working in a closed loop compares the required speed to the instantaneous one and computes the necessary torque to reduce the error between the input function and the output one. Hence the necessary torque for the traction drive system and for the mechanical brakes is determined.

The input of the electrical traction system (composed by the induction motor and the inverter) is then the torque defined by the pilot model, the mechanical rotor speed Ω related to the longitudinal dynamic model and the battery voltage from the relative subsystem. It evaluates, using the steady state equations of the electrical machine [7-8], the overall power required by the inverter and, as consequence, the total current that the energy storage system has to provide.

Then it's defined an energy management model which splits the total power requested by the inverter between the power supplied from the battery p_{batt} and the supercapacitors through the DC/DC converter p_{scl},.

The battery model, which replicate the behaviour of the battery itself, computes, starting from, the battery current defined in the energy management model, the battery pack voltage and the battery Status Of Charge (SOC). Eventually an Equivalent Circuit Model (ECM), in order to consider also the dynamic behavior of the battery, i.e. first order Randles circuit, can be chosen as a good compromise between accuracy and computational effort.

The general model used to represent the supercapacitor receives as inputs the current required from the energy management and it provides the bank terminal voltage and the capacitor Status Of Energy (SOE). The supercapacitor bank is modeled as a single element with model characteristic (capacitance C, series resistance RS, dielectric leakage resistance RL) obtained from the available datasheet information. It is possible to determine the terminal voltage of the supercapacitor and, consequently, its SOE as a dynamic function of the current required by the energy management model. At last, since the supercapacitor voltage variation is considerable during the discharge, a DC/DC converter is needed (and modelled) to maintain constant the voltage for the traction inverter.

To reconstruct the energetic power flow between the components a simple longitudinal dynamic model of the bus is introduced. This model receives as input the torque given by the electrical traction system model and calculates the electrical motor speed Ω, the vehicle speed v(t) and the covered distance s(t).

3 Fuzzy Control

3.1 *Fuzzy Control Overview*

Fuzzy control, over recent years, has emerged as one of the most attractive research areas in control applications. Heterogeneous examples of fuzzy control are increasingly common in literature during recent years, so that it's impossible to present an exhausting overview of all applications that have been made. Among a wide variety of available application examples, it's possible to find fuzzy logic applications in energy management strategy. In [9] a fuzzy controller is implemented in an hybrid energy system, while in [10-11] are presented two applications of fuzzy control in automotive field: the controller aim is to split power request between the different kind of sources available onboard the hybrid vehicle.

Fuzzy logic (the logic on which fuzzy control is based) is very intuitive and similar to human thinking and natural language. Reasonably, it's able of capturing the approximate, inexact nature of the reality. In a very concise way, the Fuzzy Logic Controller (FLC) can be defined as a set of linguistic control rules related by the dual concepts of fuzzy implication and the compositional rule of inference. Then, the FLC provides an algorithm capable to convert the linguistic control strategy based on expert knowledge into an automatic control strategy. Experience shows that the FLC yields results comparable (in some cases superior) to those obtained by traditional control algorithms. In particular, FLC approach appears to be preferable when the processes are too complex for a conventional quantitative analysis or when the available sources of information are qualitative, inexact, or uncertain. Thus fuzzy logic control may be considered between a conventional precise mathematical control and human-like decision making, as indicated by Gupta [12].

Our investigation includes fuzzification and defuzzification strategies, the derivation of the database and fuzzy control rules, the definition of a fuzzy implication, and an analysis of fuzzy reasoning mechanisms.

3.2 *Fuzzy Control Implementation*

The fuzzy logic designed in this paper is four variables dependent:

- the State Of Energy of the supercapacitor (SOE);
- the State Of Charge of the battery (SOC);
- the overall power requested P_{req} at the on board storage system for traction purpose and for auxiliary loads;
- the speed of the vehicle.

The result of the logic implementation is the actual split factor of power demand fulfilled by the battery (k_{batt}).

In [13] a fuzzy logic energy management dependent from three variables (SOE, SOC, P_{req}) has been considered. In the present paper also speed, and hence kinetic energy, is considered as a power source, so fuzzy rules are extended in order to consider it.

In Fig.3 the overall logical functioning of the fuzzy controller is presented.

The main ideas behind the rules are below listed.

- Comparing the charge of battery and the supercapacitor: if supercapacitor is low compared to the battery and the power request is negative (battery is being charged) k_{batt} should favor the filling of the super capacitor.

- When SOC or SOE is in excess, recuperating the brake energy should be given to the other. Therefore, k_{batt} will be low and vice versa.
- When speed increases, kinetic energy of the system has increased and this acts as a power source. Therefore, we can decrease the load in the battery (both for charging and discharging).

In the following fuzzification, fuzzy rules for inference and defuzzification are briefly described. The membership functions shape used for the inputs and single output is a combination of trapezoidal functions at borders and triangular functions in the middle. As example, Fig. 4a shows membership functions of SOC.

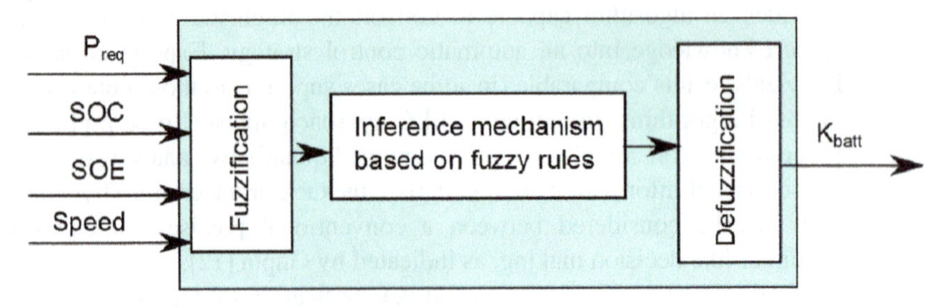

Fig. 3 Simplified logical functioning of fuzzy controller

The linguistic values "NB, NM, NS, ZE, PS, PM, PB, LE, ML, ME, MB, GE" represent: "negative big", "negative medium", "negative small", "zero", "positive small", "positive medium", "positive big", "little", "medium little", "medium", "medium big", "great".

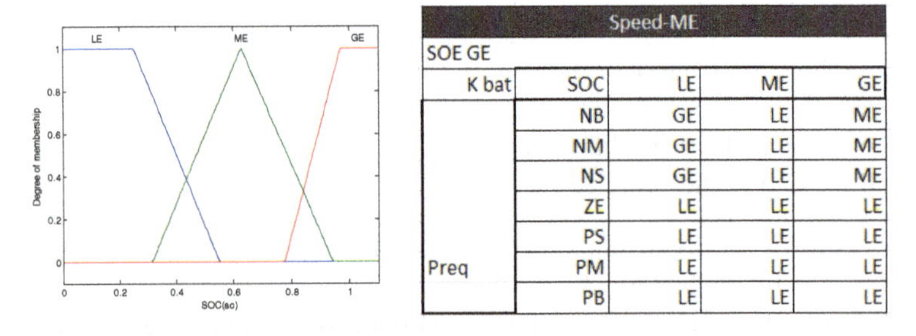

Speed-ME				
SOE GE				
K bat	SOC	LE	ME	GE
	NB	GE	LE	ME
	NM	GE	LE	ME
	NS	GE	LE	ME
	ZE	LE	LE	LE
	PS	LE	LE	LE
Preq	PM	LE	LE	LE
	PB	LE	LE	LE

Fig. 4 a) Membership Functions of SOC input and b) Inference rules table (1 of 9)

The inference rules are defined as a sequence of fuzzy sets of input and output variables. For example a typical rule has the form:

"If P_{req} is NB, SOC is LE, Speed is ME, and SOE is GE, Then k_{batt} is GE."

Because of a large number of inputs, it's impossible to list the rules in one table, so nine different tables are required to show the premises and consequents of all the initial rules. In Fig.4b one table is shown as an example: in this table Speed and SOE are defined as medium and great respectively and, in according to SOC and P_{req} values, a fuzzy value for output is defined.

The input for the defuzzification process is a fuzzy set of the results obtained applying the inference rules. The final desired output is a single number, so the fuzzy set must be defuzzified in order to resolve a single output value from it. The defuzzification method used is the centroid calculation, which returns the center of area under the curve.

3.3 Fuzzy Control Simulation

In order to validate the fuzzy logic controller and to evaluate its efficiency, the outcome of this research are compaered to the one of a previous one where a different control logic was applied [3]. To guarantee the reliability of the comparison a well-defined urban drive cycle is considered [3]

In particular the algorithms are tested on two partial drive cycles (long and short) as presented in Fig.5.

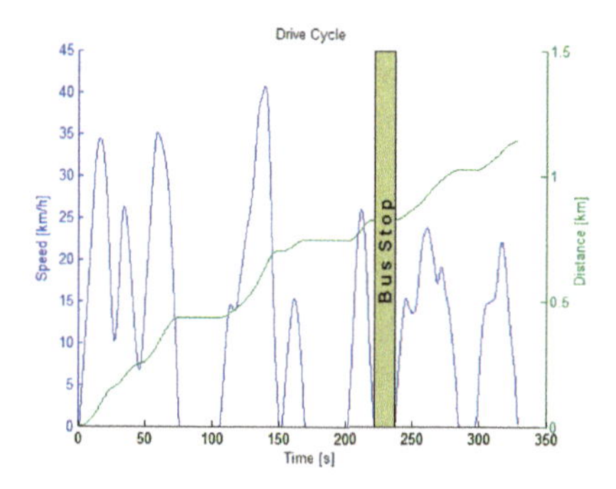

Fig. 5 Drive cycle

A comparison between an heuristic control strategy designed in [3] and the fuzzy controller is proposed in Fig.6 - 8. In particular Fig. 6 shows the comparison between the power repartition for the two differetn on board storage system, while in Fig.7 the energy levels are reported, and Fig. 8 the whole current delivered is compared.

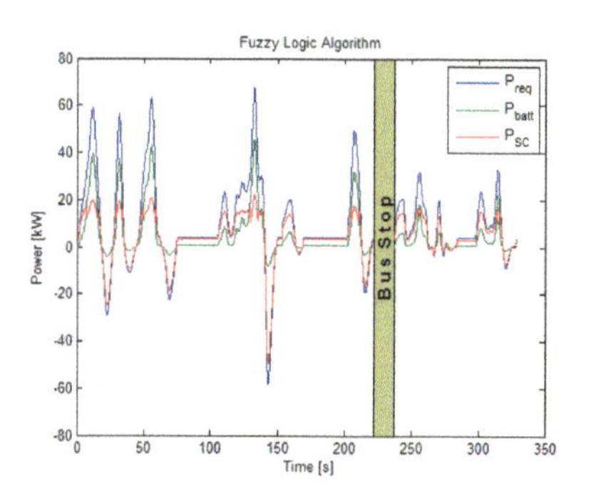

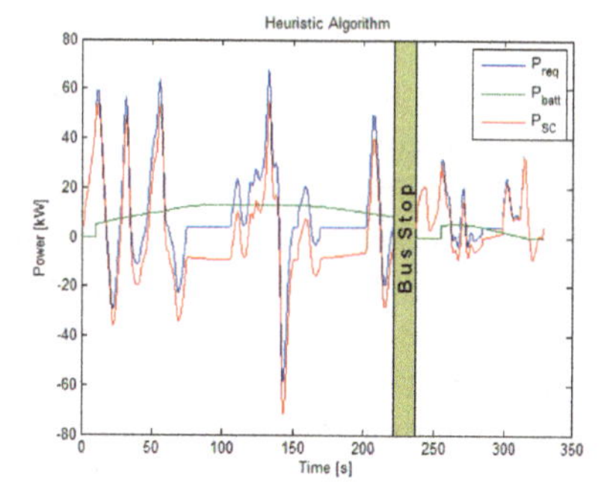

Fig. 6 Power distribution profile

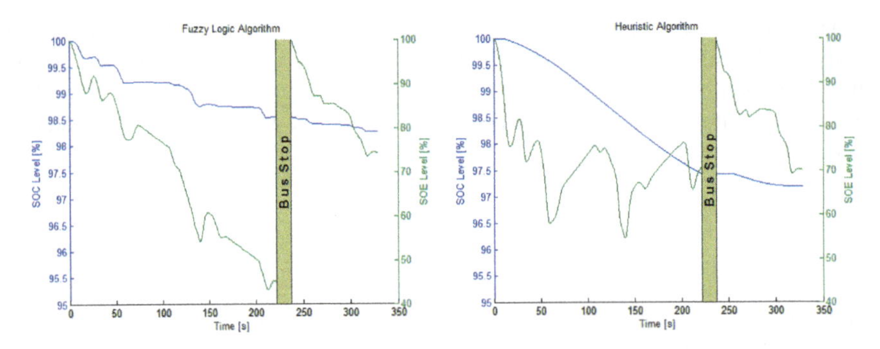

Fig. 7 SOE-SOC level profile

From the compared results it's evident that several benefits can be achieved with the implementation of a fuzzy logic in the management of the powertrain energy. In fact the overall amount of energy to complete the drive cycle is lower when the fuzzy logic is implemented. However some drawbacks are present: the power delivered by the battery appears to be more smooth when the heuristic approach is implemented.

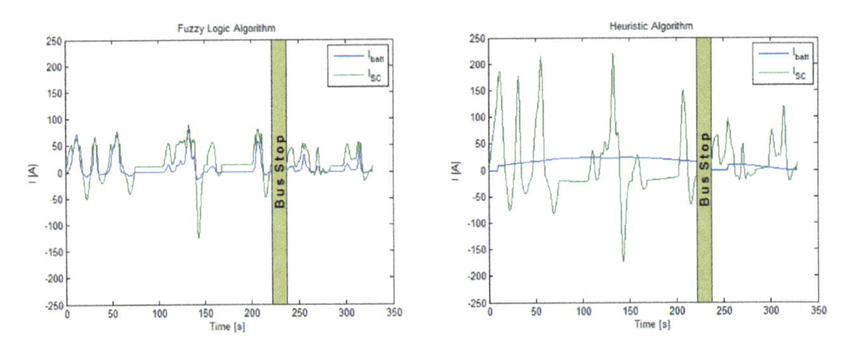

Fig. 8 Electric current profile

4 Conclusion

A novel fuzzy logic controller able to split power request of a full electric bus between superconductor and battery energy sources was presented and tested on a numerical simulation model. The fuzzy logic control algorithm generally allows to reach the bus stop with a lower amount of energy with respect to the previous heuristic algorithm, and, as consequence, with a lower depletion of the battery. On the contrary the current delivered by the battery results more smooth with the old heuristic algorithm, even if with the fuzzy algorithm the battery never work in overload condition, preserving its life.

Finally the new fuzzy algorithm has shown to be a feasible solution for the hybrid bus on board storage system: the benefits introduced on the extension of battery life have to be deeply investigated considering an appropriate aging model for this crucial component.

References

[1] Zhu, C., Lu, R., Tian, I., Wang, Q.: The Development of an Electric Bus with Super-Capacitors as Unique Energy Storage. In: IEEE Vehicle Power and Propulsion Conference, VPPC 2006, Windsor (2006)

[2] Khaligh, A., Li, Z.: Battery, ultracapacitor, fuel cell, and hybrid energy storage systems for electric, hybrid electric, fuel cell, and plug-in hybrid electric vehicles: State of the art. IEEE Transactions on Vehicular Technology 59(6), 2806–2814 (2010)

[3] Mapelli, F.L., Tarsitano, D., Annese, D., Sala, M., Bosia, G.: A study of urban electric bus with hybrid energy storage system based on battery and supercapacitors. In: ASME 2013 International Mechanical Engineering Congress & Exposition IMECE 2013, San Diego, USA (2013)

[4] Allegre, A.L., Bouscayrol, A., Trigui, R.: Influence of control strategies on battery/supercapacitor hybrid energy storage systems for traction applications. In: 5th IEEE Vehicle Power and Propulsion Conference, VPPC 2009, Dearborn (2009)

[5] Mapelli, F.L., Tarsitano, D.: Energy control for plug-in HEV with ultracapacitors lithiumion batteries storage system for FIA alternative energy cup race. In: 2010 IEEE Vehicle Power and Propulsion Conference, VPPC 2010, Lille (2010)

[6] Mapelli, F.L., Tarsitano, D., Agostoni, S.: Plug-In Hybrid Electrical Commercial Vehicle:modeling and prototype realization. In: IEEE International Electric Vehicle Conference (IEVC), Greenville, USA (2012)

[7] Cheli, F., Mapelli, F.L., Manigrasso, R., Tarsitano, D.: Full energetic model of a plug-in hybrid electrical vehicle. In: SPEEDAM International Symposium on Power Electronics, Electrical Drives, Automation and Motion, pp. 733–738 (2008)

[8] Mapelli, F.L., Tarsitano, D., Annese, D., Sala, M., Bosia, G.: A study of urban electric bus with a fast charging energy storage system based on lithium battery and super capacitors. In: ASME 2013 International Mechanical Engineering Congress & Exposition, ICEME 2013, Monaco, France (2013)

[9] Lagorse, J., Simões, M., Miraoui, A.: A Multiagent Fuzzy-Logic-Based Energy Management of Hybrid Systems. IEEE Transactions on Industry Applications 45(6) (November/December 2009)

[10] Schouten, N., Salman, A., Kheir, A.: Energy management strategies for parallel hybrid vehicles using fuzzy logic. Control Engineering Practice 11, 171–177 (2003)

[11] Martínez, J., John, R., Hissel, D., Péra, M.: A survey-based type-2 fuzzy logic system for energy management in hybrid electrical vehicles. Information Sciences 190, 192–207 (2012)

[12] Gupta, M.M., Tsukamoto, Y.: Fuzzy logic controllers - A perspective. In: Proc. Joint Automatic Control Conf., San Francisco, FAIO-C (August 1980)

[13] Chenghui, Z., Qingsheng, S., Naxin, C., Wuhua, L.: Particle Swarm Optimization for energy management fuzzy controller design in dual-source electric vehicle. In: Power Electronics Specialists Conference, pp. 1405–1410 (2007)

Part IV
Vehicle Electrification

COSIVU - Compact, Smart and Reliable Drive Unit for Commercial Electric Vehicles

Tomas Gustafsson, Stefan Nord, Dag Andersson, Klas Brinkfeldt and Florian Hilpert

Abstract. The EU-funded FP7 project COSIVU [1] aims at a new system architecture for drive-trains by development of a smart, compact and durable single-wheel drive unit with integrated electric motor, compact transmission, full silicon carbide (SiC) power electronics (switches and diodes), and an advanced ultra-compact cooling solution. The main goals of COSIVU is to increase performance, flexibility as well as safety and reliability of commercial hybrid and electric vehicles, which are even more demanding with respect to power, performance, durability [2], and availability than other types of vehicles. In addition, the new architecture will be adapted to other vehicle platforms such as passenger cars. The COSIVU solution is integration of the wheel motor and the inverter into one system package. During the first twelve months of the project the COSIVU system architecture concept has been developed and a highly modular packaging concept was chosen for the power stage, using "Inverter Building Blocks" (IBB). SiC bipolar transistors and diodes have been selected and production of packaged SiC devices has started. The design of a double sided cooling version of the modules has been initiated. A theoretical design of a new driver solution, which includes power saving and reliability enhancing features, has been done. A thermal investigation test bench has been designed for the thermal characterization of the SiC based power modules.

Keywords: Electric drive-train system, power electronics, silicon carbide, double-sided cooling, Inverter Building Block, heat exchanger, CFD-analysis.

T. Gustafsson · S. Nord
Volvo Group Trucks Technology, Advanced Technology and Research,
Götaverksgatan 10, 40508 Göteborg, Sweden
e-mail: {tomas.gustafsson.3,stefan.nord}@volvo.com

D. Andersson(✉) · K. Brinkfeldt
Swerea IVF AB, P.O. Box 104, 43122 Mölndal, Sweden
e-mail: {dag.andersson,klas.brinkfeldt}@swerea.se

F. Hilpert
Fraunhofer Institute for Integrated Systems and Device Technology IISB,
Drives and Mechatronics, Schottkystr. 10, 91058 Erlangen, Germany
e-mail: florian.hilpert@iisb.fraunhofer.de

J. Fischer-Wolfarth and G. Meyer (eds.), *Advanced Microsystems for Automotive Applications 2014*, Lecture Notes in Mobility,
DOI: 10.1007/978-3-319-08087-1_18,

1 Introduction

A state-of-the-art electrical drive-train unit usually consists of three more or less separated functional blocks: a control & communication module, a power module and a traction module. The control & communication module contains the microcontroller unit (MCU) performing all timing critical functions such as high-bandwidth current control loop, PWM signal generation above a carrier frequency of a few kHz. It also provides a communication interface (in vehicles typically a CAN interface) for both receiving torque commands as well as providing diagnostic information and error messages. The power module contains mainly the power switches, mostly insulated-gate bipolar transistors (IGBTs), with anti-parallel free-wheeling diodes, both based on silicon chips, that are fed with 600 V DC integrated with DC-link filters and also power switch drivers and protection electronics. The traction module consists of the electrical motor, either integrated together with a gear and/or integrated within the wheel. The novel approach of the COSIVU project is to bring the three functional blocks together as close as possible into one system package.

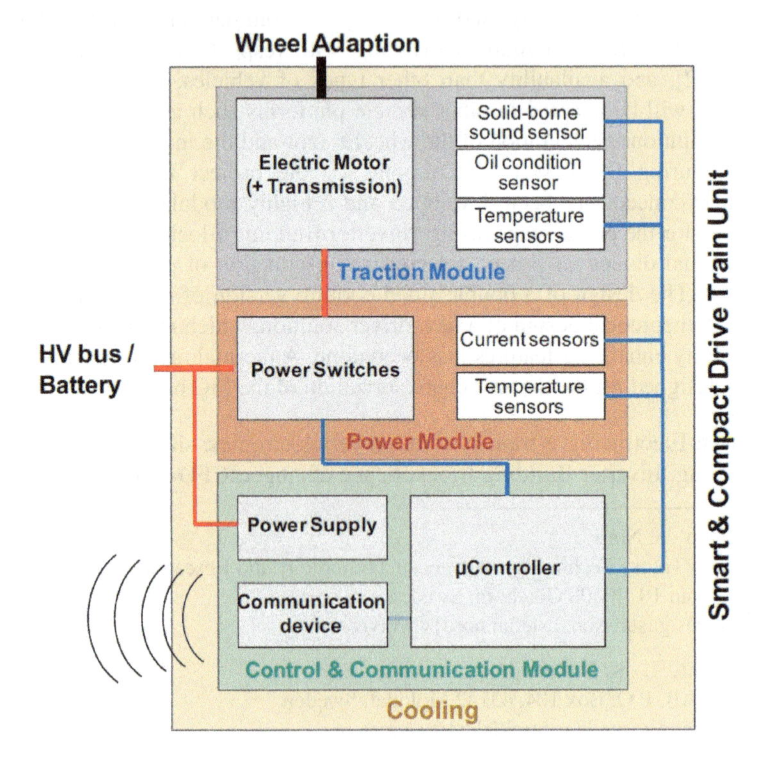

Fig. 1 Topology of the drive-train concept aimed by the COSIVU project

The main hurdle for an earlier realization was the fact that the power losses associated with state-of-the-art IGBT technology makes it hard to integrate them within the electric motor unit due to the high temperature environment to be found there, which makes the cooling for the IGBT modules much more challenging. COSIVU will overcome this bottleneck by using of advanced power switches (bipolar transistors) and Schottky diodes based on SiC technology from TranSiC, which are combining low conduction and switching power losses and thus being able to dramatically reduce power losses by -50...-70%. Consequently, the inverter efficiency is increased from approximately 98% to more than 99%. However, the real potential of this improvement is: As less than 50% of the losses occur, less than half of the heat must be dissipated from the semiconductor. Consequently, the complete cooling circle can be downsized in a similar way. Depending on the actual design, the mass of the heat sink can be reduced, less cooling fluid is needed, and a smaller pump suffices for the lower flux required. Thus, the exclusive usage of SiC for both, diodes and switches, allows realizing an even higher degree of integration. The power module is now small enough to be integrated right into the traction module.

2 COSIVU System Architecture

The picture below outlines the overall COSIVU system architecture consisting of the ICM (Inverter Controller Module), BDS (Base Driver Supply), IBBs (Inverter Building Blocks). For clarity the EM (Electrical machine), Gear and brake together with auxiliary systems (water cooling and oil cooling) is showed in order to give the complete system overview.

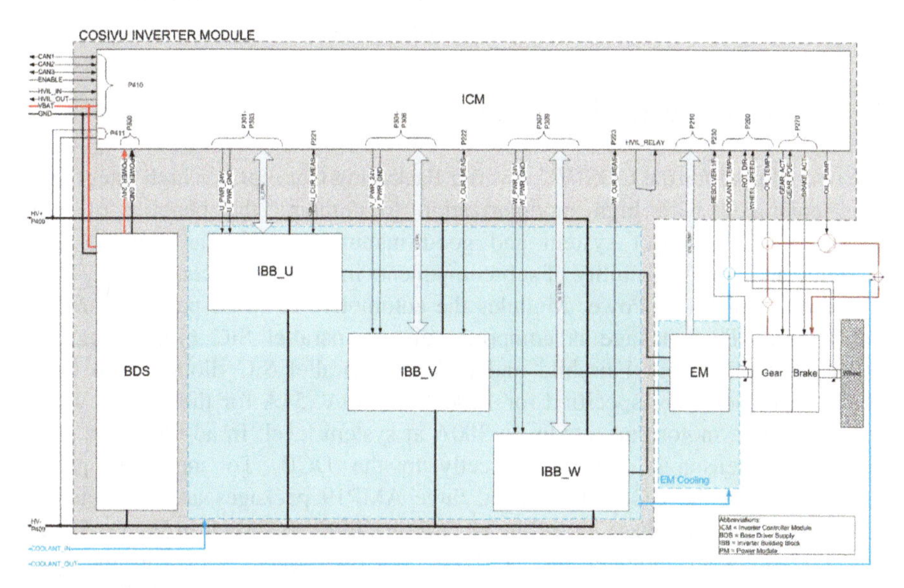

Fig. 2 COSIVU System Architecture

The strategy for the system architecture was to create a decentralized modular design that was both compact, easy to repair and cost effective. This resulted in a compact inverter unit integrating power electronics and control electronics. For ease of service and repair the whole unit is a fully pluggable solution including all electric and fluid connections. By decentralizing the power and control electronics there is less cables and thus copper that goes to the wheel hub and space is saved within the vehicle chassis. Due to the high integration level, the inverter and the electric machine will use one common water glycol cooling circuit. The inlet at the inverter is connected in series with the EM. Due to their higher operation temperatures the gears and the wet brake will use a separate oil circuit with a heat exchanger to the water glycol cooling circuit. This also saves space on the chassis as the oil cooling system is eliminated. On a 4WD vehicle there will be four COSIVU systems, one in each wheel, all systems are identical allowing less different parts and higher volumes and thus lower production and storage cost. Each COSIVU system can be controlled independently via a CAN bus by a master ECU that uses complex drive schemes such as torque vectoring, at local level each COSIVU system will control the electric machine with a local and fast control loop for highest performance and accuracy.

Internally the inverter unit consists of three different kinds of modules, ICM Inverter Control Module, BDS Base Driver Supply & IBB Inverter Building Block. Each of these modules has a unique function in the inverter and is equipped with state of the art proactive self-diagnosis and constructed in a modular way for easy service and replacement. Electric machine control is done by using space vector modulation. The internal ADC is almost exclusively used to measure the phase currents to achieve the best performance and predictability, other analog sensors that doesn't need fast response time, as temperature sensors, are read by a separate SPI ADC. The ICM also controls actuators like the oil pump and gear change mechanism.

3 Inverter Packaging Concept

The overall goal for the COSIVU Inverter Packaging Concept is a high integration level combined with high modularization to combine the benefits of both strategies – a compact system and good manufacturability as well as easy servicing and good reusability of sub-components in future projects.

For the half-bridge Power Modules the automotive certified package AMP19 was chosen. The package is composed of two parallel SiC bipolar junction transistors (BJT) from Fairchild and their anti-parallel SiC diodes from Cree. These components are specified for 1200V and 50A (54A for the diode), which allows to drive motor currents up to 300A at system level. In addition one NTC temperature sensor is soldered directly to the DCB. To meet the power requirements of the COSIVU Project, three AMP19 packages are connected in parallel to drive one phase leg of the electric machine, as shown in Figure 3. The Power Modules are mounted on one direct cooled baseplate, following the outer shape of the available round design space. The baseplate is sealed radially against the inverter housing. Using a direct cooled baseplate eliminates the need for an additional thermal interface material between baseplate and the cooling plate itself while obtaining the ability to pre-assemble and test the Power Modules before integrating them into the system.

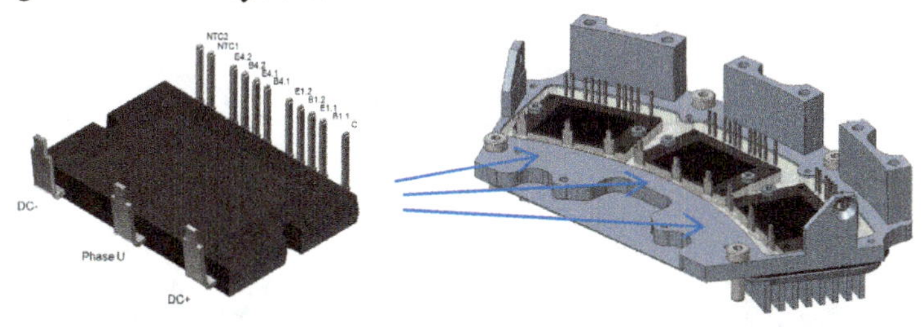

Fig. 3 AMP19 Package and Power Modules on direct cooled baseplate for one phase leg [FhG IISB]

The Power Modules are covered with an electrical isolating carrier and the Power Distribution PCB that contains electrical connections to the DC-Link Capacitor, the AC-Current-Sensor and connectors to the Base Drive PCBs. The shielded magneto-resistive Current Sensor is moulded directly to the copper bus bar linking the Power Distribution PCB and the AC Connector to the electric machine.

The DC-Link-Capacitor is a specially designed foil capacitor and obtains the functionalities of the capacitor, a low-inductive DC Power Link and a mechanical support structure for the Base Drive PCBs. The Inverter Controller Module is mounted on top of it. The relatively thick copper plates of the capacitor are used as the DC Power Link between the single capacitors and the DC Inlet of the inverter. This concept reduces significantly the overall System Inductance and will result in lower EMC disturbances compared to an external DC Power Link between the capacitors.

The Base Drive for one phase leg is split into three identical PCBs, each plugged into the connector on the Power Distribution PCB. The connector at the top side will be attached to the Inverter Controller Module when mounted on top of the capacitor. The Base resistors of each Base Drive are thermally connected to the cooling plate via aluminium blocks screwed to the direct cooled baseplate.

The described sub-system of the COSIVU Powerstage contains all necessary parts to drive one half-bridge of the Inverter and is therefore called Inverter

Fig. 4 Sub- and complete assembly of the Inverter Building Block (IBB) [FhG IISB]

Building Block (IBB). It is a mechanically independent sub-system, scalable in position and number to integrate three to six phase inverters axially to electric drivetrains, using identical system architecture, as demonstrated in Figure 5. With the center cutout even through-axle systems are possible.

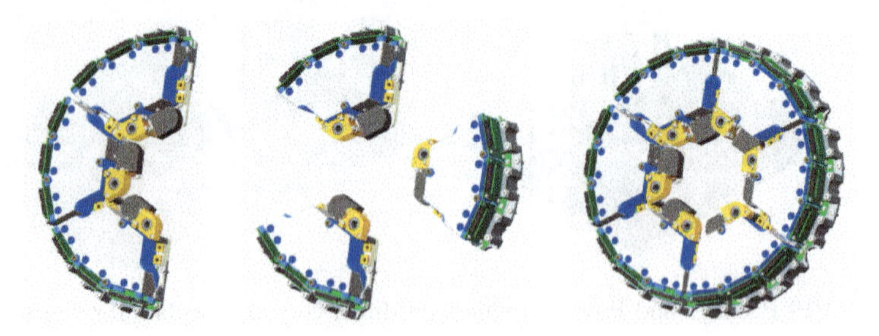

Fig. 5 Adaption of the system architecture to different designspaces or phase numbers [FhG IISB]

For the COSIVU Powerstage three IBBs are used to design a three phase inverter with the Inverter Controller Module mounted on top and being directly connected to the Base Drives. The symmetric DC connections at the IBBs give the ability to use the Powerstage in mirrored housings, like usually used for left and right wheel drivetrains. However, for the COSIVU Drivetrain one common system for left and right wheel is used to minimize spare parts and to maximize economy of scale effects.

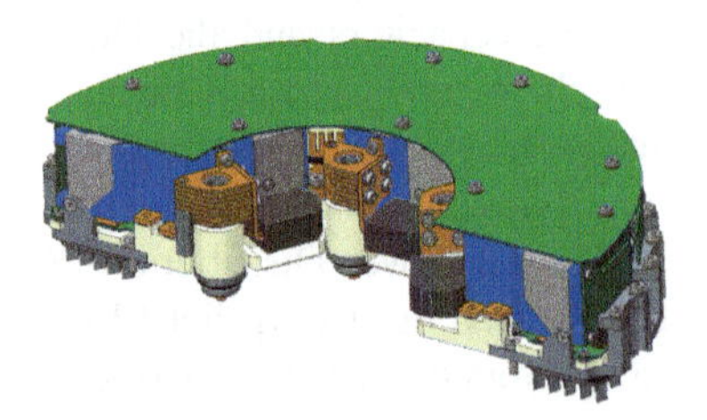

Fig. 6 The Powerstage of the COSIVU Inverter with the Inverter Controller Module (ICM) [FhG IISB]

The aluminum Inverter housing contains the Powerstage together with the ICM and the BDS. All connections to the electric machine are pluggable: for the AC connection three single pin connectors are used, internal temperature and rotor position sensors of the electric machine are connected via d-sub connectors. The water-glycol cooling circuit is linked with a leakage-free connector, allowing the COSIVU Inverter to be separated from the drivetrain without draining the cooling system. While plugging the Inverter to the EM, two alignment pins ensure the correct orientation of the system.

The connections to the commercial vehicle are also realized fully pluggable: one 2-pole High Voltage Connector, one combined Low Voltage Connector for 24V Supply and Signals and two leakage-free fluid connectors are used.

Fig. 7 COSIVU Drivetrain Unit for a commercial vehicle, left wheel (Inverter unplugged) and right wheel (Inverter plugged) System [FhG IISB]

When plugged, the whole Inverter fits inside the existing service opening in the trailing link of the commercial vehicles undercarriage. With the all pluggable COSIVU Inverter System easy servicing of the integrated drivetrain unit is possible. In addition to the commercial vehicle system, the COSIVU Inverter System will be integrated in a hybrid passenger car demonstration vehicle to show the flexibility of the COSIVU concept. The drivetrain architecture consists of two electric near-wheel direct drives at the rear axle and a combustion engine at the front axle.

To meet the differing design space constraints, the inverter housing is adapted, using identical system architecture inside. The whole system is pluggable as described for the commercial vehicle above, allowing easy servicing and testing of the prototype system.

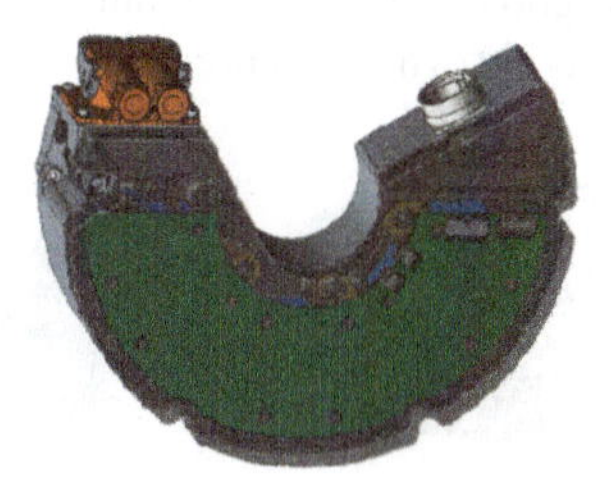

Fig. 8 COSIVU Inverter System for a passenger car application with through-axis design [FhG IISB]

4 Power Module Cooling Concept – ^{2}COOL

A new package cooling concept, which integrates the cooling system with the substrate will be verified in this project. The purpose is to improve the cooling effect per area compared to state of the art power systems. This means less volume is required for cooling in the drive unit, which gives more freedom in the system integration.

The cooling concept is based on a direct manufacturing process, which gives a high degree of freedom in geometry design and thereby also allows the heat exchanger outer surfaces to be fully integrated with the drive unit. The material used is Cu and laboratory tests show that the thermal conductivity of the directly manufactured Cu material is currently around 70 % of bulk Cu thermal conductivity. Further improvement of compactness and efficiency of the printing process is expected during the project.

Initially, two different cooling structures have been investigated using thermal CFD analysis. One was a simple inline pin-fin structure (shown in Figure 9a). The other was a sponge-like structure (shown in Figure 9b).

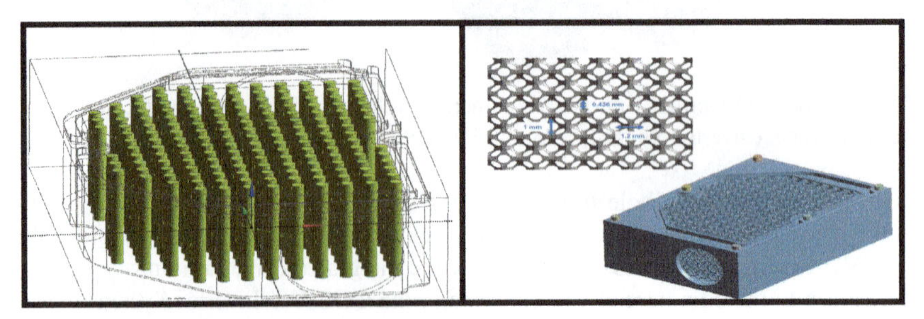

Fig. 9 The two different cooling structures investigated

A coolant flow between 5 - 15 l/min was used and only single side cooling was applied. Five hotspots resembling the active switching devices were assumed and each of these were loaded by 30 W. The results show that the maximum temperature increase for the pin-fin version was around 12 C° near the inlet and outlet, and 8-9 C° at the center of the module. For the sponge-like structure the temperature increase was slightly less at the coolant inlet and outlet, while slightly higher at the center hotspot as shown in Figure 10.

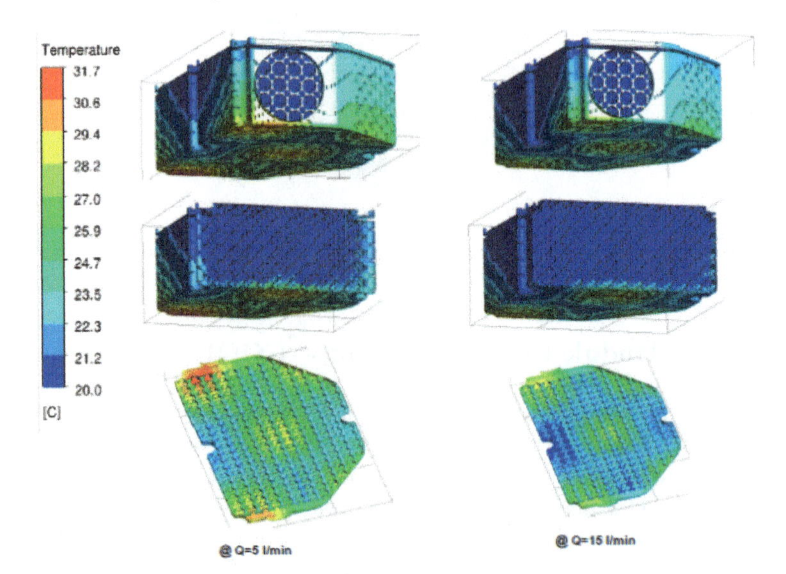

Fig. 10 Results for sponge cooling structure

The flow for both the inline pin-fin and sponge structures was predominantly in straight lines from inlet to outlet with little mixing in the vertical direction. It was also found that the sponge-like structures produce a significantly higher pressure drop (Figure 11).

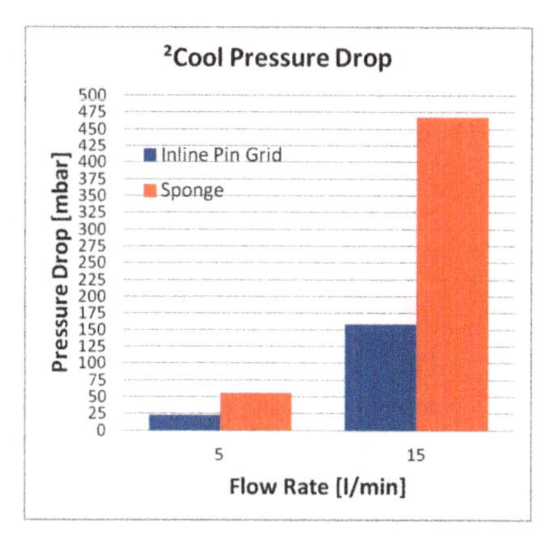

Fig. 11 Pressure drop comparison between the different structures

Following the initial analysis it is clear that at least half of the structure height could be removed and that more vertical mixing of the coolant is needed. Therefore, a tilted sponge structure, which forces the coolant to flow more in the vertical directions will be evaluated. The new design is shown in Figure 12. Some improvements to the pin-fin will also be made (staggering of the pins and height reduction).

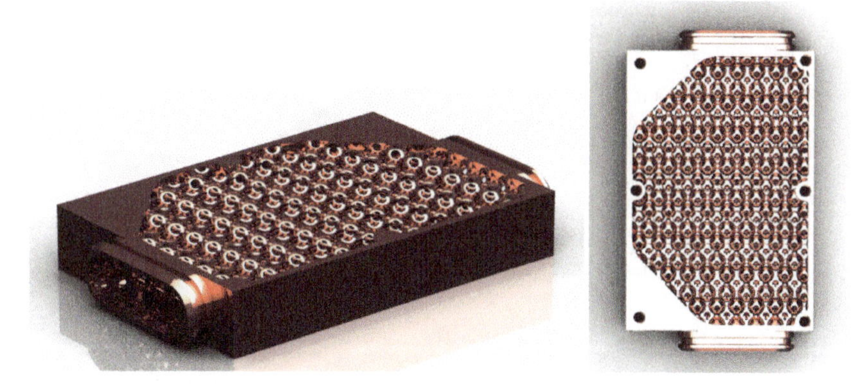

Fig. 12 New design of the sponge-like cooling structure

5 Conclusions

This paper presented a Compact, Smart and Reliable Drive Unit for Commercial Electric Vehicles within the frame of the currently running project COSIVU.

Furthermore the strategy for the system architecture was outlined and the Inverter Packaging Concept with its high integration level combined with high modularization was described in detail. Finally, a new package cooling concept, which integrates the cooling system with the substrate, was discussed and simulations on different cooling structures were presented.

Acknowledgement. The Authors would like to acknowledge the European Commission for supporting these activities within the project COSIVU under grant agreement number 313980.

References

[1] Rzepka, S., Otto, A.: COSIVU – Compact, Smart and Reliable Drive Unit for Fully Electric Vehicles. Micromaterials and Nanomaterials (15), 116–121 (2013)

[2] Nord, S., Cortes, A.: Electro-mobility: Key Technologies for Sustainable Transport Solutions. Micromaterials and Nanomaterials (15), 16 (2013)

Development of a Solid-Borne Sound Sensor to Detect Bearing Faults Based on a MEMS Sensor and a PVDF Foil Sensor

Jurij Kern, Carsten Thun and Jernej Herman

Abstract. Vibration analysis is an effective method to determine the health of a rotating mechanical machine. A test is set-up with two different sensors, MEMS and PVDF-Foil based. Experimental data is presented and a test to detect undamaged and damaged bearings has been performed. The MEMS sensor shows a good performance with clear indication of failure frequencies. The PVDF Foil in this configuration also shows the ability to detect the difference, but the natural properties leading to mechanical amplification of bearing vibrations are limiting the performances for weak mechanical failures. Monitoring algorithms are employed under standard conditions: RMS value of the signal, Kurtosis, Power spectrum density and Envelope analysis.

Keywords: Condition monitoring, Bearing faults, PVDF foil, MEMS, Kurtosis, Envelope detection, Hub bearing, Sensor, Vibration detection, Power spectrum density.

1 Introduction

Vibration analysis is an effective method for monitoring the condition and fault diagnosis in bearings of rolling elements. Bearings are used in nearly all rotating

J. Kern
Elaphe Propulsion Technologies Ltd., Luznarjeva 3,
4000, Kranj, Slovenia
e-mail: Jurij.kern@elaphe-ev.com

C. Thun(✉)
Hella Fahrzeugkomponenten GmbH,
Dortmunderstr. 5, 28199 Bremen, Deutschland
e-mail: carsten.thun@hella.com

J. Herman
Elaphe Propulsion Technologies Ltd.,
Babno 20a, 3000, Celje, Slovenia
e-mail: jernej@elaphe-ev.com

J. Fischer-Wolfarth and G. Meyer (eds.), *Advanced Microsystems for Automotive Applications 2014*, Lecture Notes in Mobility,
DOI: 10.1007/978-3-319-08087-1_19,

machinery. The reliability of the bearing depends on the type of bearing selected, precision of the components i.e. shaft, housing spacer and the operating conditions. The manufacturing technology of today is not the limiting factor for the service life time of a bearing. Subsurface stress that occurs due to fatigue accounts for 3% of all bearing failures in their service life time. [1].

In the COSIVU project [2] a new compact and highly integrated electrical drive is developed. One of the goals in this project is to increase the reliability of the drive's life-time. The increased mechanical density and closer integration of all systems components leads to the need of having a life-time and functional monitoring of crucial components.

The components are Silicon Carbide-Transistor [3] based power electronics, mechanical parts and technical fluids. The mechanical parts are the bearing of the electrical engine and the gearbox. Within the COSIVU project an algorithm for monitoring functional and health state was proposed from Hella and Elaphe. Polyvinylidenfluorid (PVDF) foil is used for the evaluation of vibrations and for the development of a health monitoring system in the COSIVU project. This technology is already used in another Hella project [4], where capturing the sound of scratches and dents introduced by minor accidents are explored. In the project the foil offered a good chance of having a good and cost effective sensor instead of other technologies like a MEMS accelerometer. The PVDF offers a wide variety of applications for instance the ability to glue the sensor to nearly any surface. The application of using the PVDF foil for detecting vibrations was also used in [5]. For the above mentioned reasons the PVDF foil is also used for the evaluation and development of a health monitoring system in the COSIVU project.

In addition to being the first choice within COSIVU this technology is widely explored at Hella in various internal projects. This paper presents a comparison of PVDF and MEMS technology as well as describing a mechanical improvement proposal on the PVDF signal detection. We report here the first results when performing measurements with a simple PVDF foil on a real bearing.

2 Physics of Solid-Borne Sound and the Source of Bearing Vibration

The movement of solid materials produce solid-borne sound within the material. This sound can be transmitted inside the material in various wave types [6] either as longitudinal or transversal waves. The source of movement can be an excitation from outside, like a mechanical force by a punch, or from inside (intrinsic excitation) in the material's atomic structure forced by temperature differences. The frequency range of solid-borne sound has no upper limit, but starts from seismic vibration 0.1Hz- 10Hz, over the acoustic sound frequencies 16Hz to 20 kHz until it reaches the ultra-high vibrations (200 kHz) in the material itself.

The type of excitation and the outer physical dimensions of a solid material determine the main wave type. A comparison of the physical dimension with the calculated longitudinal wave length defines the dominant wave type. Young´s modulus, the density and the Poisson`s ratio are three material constants. These are the main parameters for the structure borne sound transmissions behaviour. Because bearings are mainly made out of steel, the speed of a longitudinal wave is quite fast (around 6000m/s). In spite of the bulkiness and thickness of the bearing material, fast rotational surface waves with a small material displacement are created for which the main wave form is transversal. Thus bending waveforms are not very strong in a healthy roller bearing and audible sound is not present.

A mechanical rolling-element bearing is one of the most common ones used in various mechanical applications. It presents a vibrational system by rolling elements, outer raceway and cage. All elements together with the rotational period interact to form a complex vibration generator. The sources of vibrations in a bearing are surface roughness, waviness, geometrical imperfections and variable compliance [7].

Depending on mechanical size and type of the bearing each bearing produces its natural frequency from each element.

3 Bearing Basic Defect Frequencies

Every bearing exhibits certain characteristic frequencies. As the inner or outer ring rotates the balls moving over the running surface excite the bearing structure and produce vibrations which are detectable by a sensor. If damage occurs on either the outer ring, inner ring or the balls the amplitude of the vibrations are amplified. When observing the frequency spectrum of a bearing's vibrational signal, specific frequencies are amplified when a fault is present. These are calculated by the formulas [8] below where f_r is the rotating frequency, D is the pitch diameter, d is the ball diameter, n is the number of balls and ϕ is the contact angle from the radial plane:

$$BPFO\left(\text{Ball pass frequency pass outer ring}\right)=\frac{nf_r}{2}\left(1-\frac{d}{D}\cos\phi\right) \tag{1}$$

$$BPFI\left(\text{Ball pass frequency pass inner ring}\right)=\frac{nf_r}{2}\left(1+\frac{d}{D}\cos\phi\right) \tag{2}$$

$$FTF\left(\text{Fundamental train frequency}\right)=\frac{f_r}{2}\left(1-\frac{d}{D}\cos\phi\right) \tag{3}$$

$$BPRF\left(\text{Ball pass roller frequency}\right)=\frac{Df_r}{2d}\left(1-\left(\frac{d}{D}\cos\phi\right)^2\right) \tag{4}$$

4 PVDF-Foil

Polyvinylidenfluorid is a partly crystalline polymer with piezoelectric properties. In general the piezoelectric material generates an electric charge when mechanically stressed. Most of the structures borne sound sensors are made from piezoelectric ceramics (PZT). These PZT elements have to have a mass spring system in order to work properly. Function and behaviour is comparable to the PZT Sensors.

The advantages of foils are high mechanical flexibility and low thickness as well as low cost. They are also able to produce a much higher output signal compared to PZT Sensors. The frequency response of a PVDF foil extends up to the megahertz range. A disadvantage of the foil in the present application is the fact that that the sensor is responding best on bending material waves while the bearing transmits in transversal sound modes. Also the EMC behaviour is inferior due to the relatively wide area a sensors covers [9].

5 MEMS Sensor

MEMS or micro electromechanical systems are small devices that enable measurement of acceleration forces. MEMS sensors were first widely used in the automotive-airbag applications. They are extensively used in consumer electronics, such as smartphones and game controllers. They also offer huge industrial potential due to their low cost and a wide range of applications.

The main difference between the piezo-effect and the capacitive effect sensors is in the frequency response. Piezoelectric accelerometers have a higher frequency response which is beneficial for bearing fault diagnostics, since early warnings of bearing faults are usually present in the high frequency spectrum and masked in the low frequency spectrum by noise [10]. It is difficult to get good data with envelope analysis of a low cost MEMS signal, because its frequency response is limited to 1,6 kHz (ADXL325 used in this paper). The bearing faults usually excite the frequencies that are in a few kHz range[11], but still the whole spectrum should be carefully examined for possible fault signatures. Piezoelectric sensors, however, have a higher price and require a charge amplifier to give an output signal that can be processed further.

MEMS sensors are often made from the same materials and processed with the same processes as Si chips and electronic circuits, but they have holes, channels and membranes that imitate their large mechanical parts, so they do not share the same characteristics. Not all MEMS are made from silicon, since metal, polymer, ceramics and glass can be used as well[12].

6 Analysis of Signal

Signal analysis was done in four steps. The signal from the MEMS data is passed through a low pass filter, which is a 4th order Butterworth low pass filter with a cutoff frequency of 1 kHz. For the PVDF data the signal was low pass filtered at 20 kHz. After filtering RMS (Eq. 5) and Kurtosis(Eq. 6) are calculated for each measured sample. The sample with the highest Kurtosis was chosen for envelope and power spectrum density analysis.

$$X_{RMS} = \sqrt{\frac{1}{N}\sum_{n=1}^{N}|X_n|^2} \tag{5}$$

$$KURTOSIS = \frac{1}{N}\sum_{i=1}^{N}\left(\frac{x_i - \mu}{\sigma}\right)^4 \tag{6}$$

Both Kurtosis and RMS are good indicators of faults and impulses in the acceleration signal. RMS has been used to detect faulty gearboxes [13] and Kurtosis for detection of faulty bearings [14]. Kurtosis and RMS calculation are simple statistical indicators and do not require very high computing power. Also since both are single parameter value it allows for easy continuous monitoring for an extended time without the need for large data space storage. By evaluating the time history trends of the signal the prognosis can be performed.

The power spectrum density (Eq. 7) is calculated from the sample with the highest Kurtosis.

$$P_{xx}(f) = \mathrm{E}\left\{\frac{2f_s}{TC_a^{\,2}NB_e}|\mathrm{X}(f)|^2\right\} \tag{7}$$

Envelope analysis is done by viewing the frequency spectrum of an analytical signal and then selecting the optimal pass band to filter the unwanted noise frequencies. A Hilbert transform is then done on the band-passed time domain signal from which the envelope is calculated. Then the envelope(Eq.8) is Fast-Fourier transformed and shifted to the frequency spectrum.

The envelope of an analytic signal is defined as

$$\hat{x}(t) = x(t) * \frac{1}{\pi t} = \frac{1}{\pi}\int_{-\infty}^{\infty}\frac{x(\tau)}{t-\tau}d\tau \tag{8}$$

7 Measurements and Testing

The measurement of the two sensor types took place at Hella Fahrzeugkomponenten (PVDF) and Elaphe (MEMS). The measurement conditions were kept the same, as both sensors were applied to the same bearing type. The implemented mechanical damages to the bearings were the same in both cases since identical bearings were used. The measurements were done at 600 RPM and the sample rate of the raw data was 80 kHz. The sample interval was 10 samples at Hella and 100 samples at the Elaphe test bench. The placement of the sensors is shown in Fig. 1.

The bearings used in the experiments are standard hub wheel bearings for a city car. The bearing BAR 0051 is a double row angle contact bearing with its characteristics described in Table 1.

Table 1 Calculated fault frequencies

Bearing characteristics	*Bearing fault frequencies*
Number of balls [n] = 16 (2 rows)	BPFO = $6.877^* f_r$ [Hz]
Diameter of ball [d]= 9.525 mm	BPFI = $9.1224^* f_r$ [Hz]
Running diameter[D] = 53mm	FTF = $0.43^* f_r$ [Hz]
Contact angle[ϕ] = 38°	BPRF = $2.75 * f_r$ [Hz]

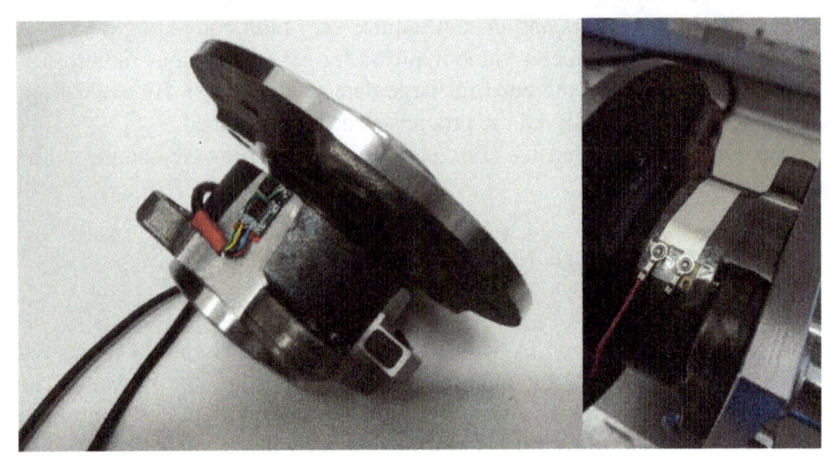

Fig. 1 Bearing with MEMS sensor (left) and PVDF foil (right)

The raw signals of both sensor measurements are shown Fig. 2 in and Fig. 3

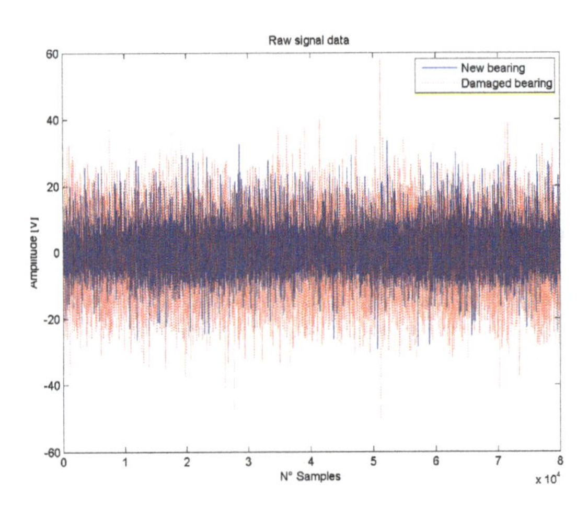

Fig. 2 Raw signal good and bad bearing with PVDF foil

The performance of the PVDF foil regarding the signal to noise ratio is not as good as the MEMS sensor set-up. This is mostly due to the EMC environment. A no-load measurement (0 rpm) showed that environmental noise level influences the measurement. The raw signals of PVDF foil were consequently treated with the inverse noise addition via software to improve the raw signals.

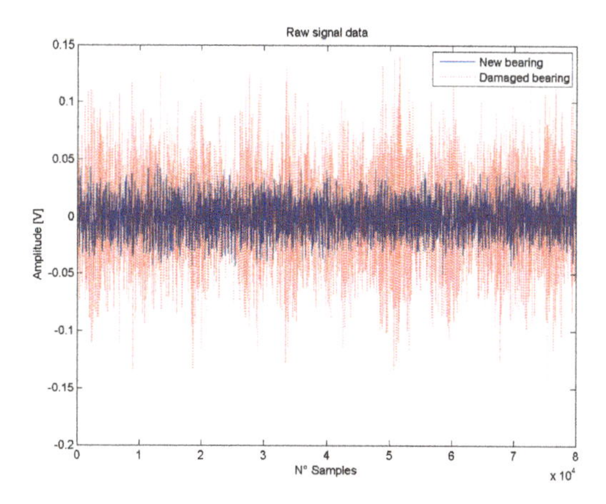

Fig. 3 Raw signal of good and bad bearing of the MEMS sensor

8 Results and Discussion

8.1 Results from MEMS Data

From Fig. 4 and Fig. 5 there is a clear indication of the difference between the signals from a new or healthy bearing versus a damaged one. RMS of the vibration of the damaged bearing is on average 3 times higher than in a healthy one. The same can be seen in the value of Kurtosis. Kurtosis is a measure of impulses in the signal. The value of Kurtosis for a healthy bearing, where there are none of very few faults is equal to 3. This value is the same for a Gaussian white noise where all frequency components are equally distributed [15], meaning no peaks for a certain set of frequencies. The value of Kurtosis for a damaged bearing is higher than 3 and reach as high a value as 7 in some samples.

Power Spectrum density analysis reveals both BPFO (67 Hz) and BPFI (89 Hz). The difference in the values comes from the bearing slip [7] and a non-precise speed measurement, which is estimated from the motor inverter data. There is also an increase in amplitude at BPRF 27 Hz. The rotating frequency is also visible at around 10 Hz. The envelope spectrum also shows higher BPFI and BPFO, but not as pronounced as in the PSD image. The limitation of the low cost sensor is obvious; as its frequency response is too low to conclusively state that the fault is on either the inner or outer ring. There is a clear indication of the BPFO being the prevailing fault from the PSD analysis.

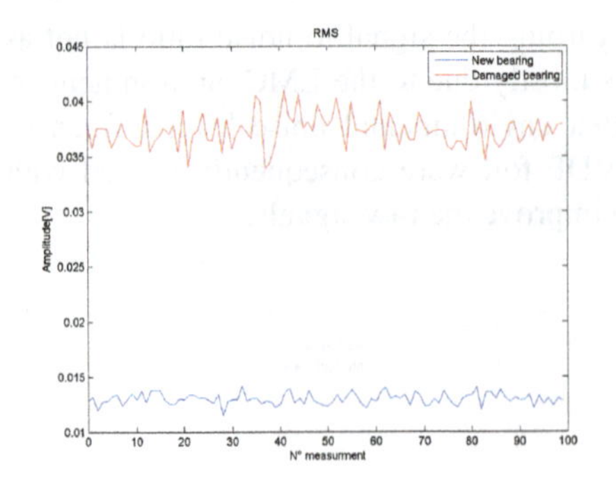

Fig. 4 RMS values form the MEMS sensor

Fig. 5 Kurtosis of MEMS sensor

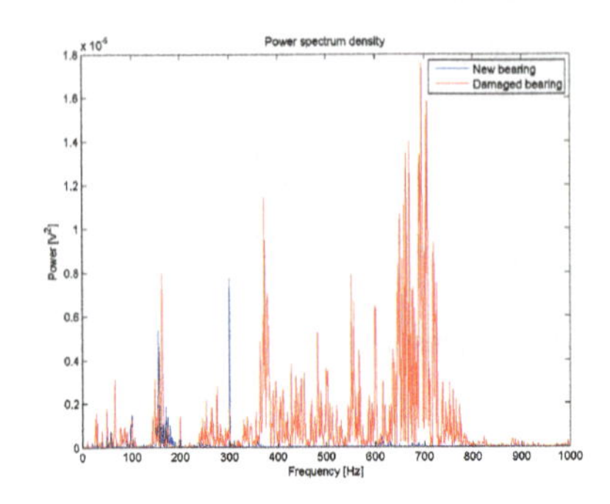

Fig. 6 Power spectrum density of the MEMS measurements

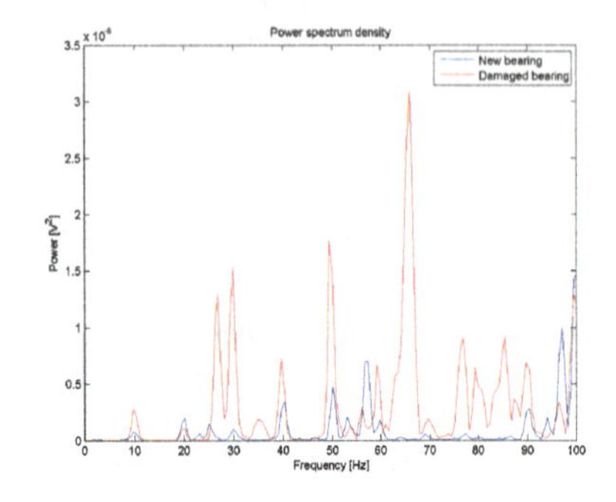

Fig. 7 Power spectrum of MEMS, Detail: 1Hz-100Hz

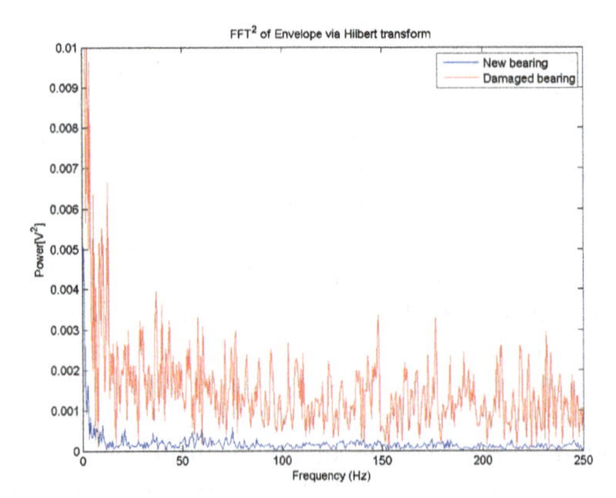

Fig. 8 FFT analysis of the MEMS sensor

8.2 Results from PVDF Foil Data

The analysis of the Kurtosis and RMS of the filtered signal at 20 kHz from Fig. 2 gives following values:

- Kurtosis
  - New bearing: 2.9
  - Damaged bearing: 3.3
- RMS
  - New bearing: 3.74
  - Damaged bearing: 7.27

Both Kurtosis values are close to the common threshold of 3. The RMS values differ with more than 50%. Both statistical indicators show there is difference between a new and a damaged bearing.

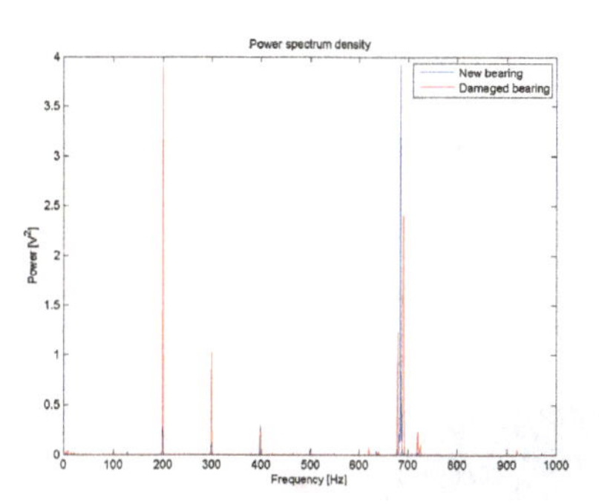

Fig. 9 Power spectrum density of the PVDF Foil

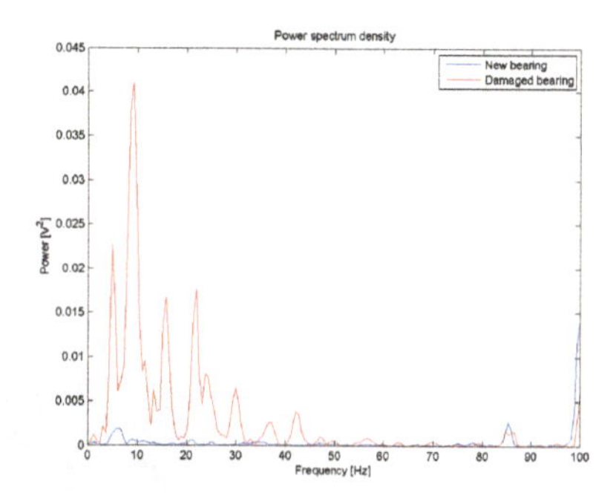

Fig. 10 Power spectrum density (1-100Hz) of the PVDF Foil

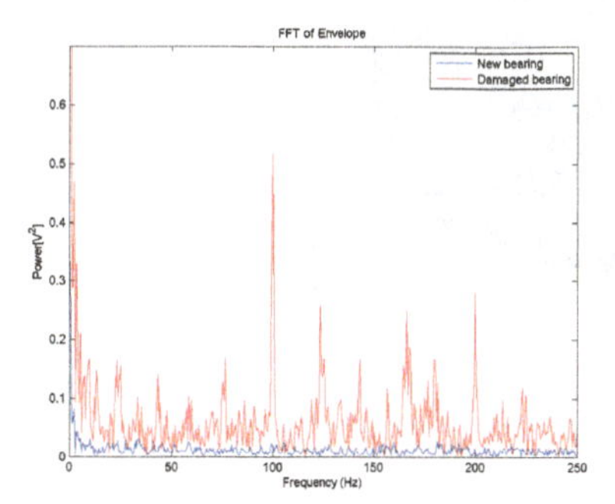

Fig. 11 Envelope analysis of the PVDF Foil

In the Power spectrum density Fig. 10 the rotating frequency of 9.15Hz could be identified. The damaged bearing shows the indication at 24Hz of the BPRF error frequency at the given rotating frequency.

The envelope structure is shown in Fig. 11: Envelope analysis of the PVDF Foil shows peaks at 2 x BPFI (166Hz) and 2 x BPFO (126Hz). The major peak at 99Hz (4 x BSF) indicated a massive error on the bearing balls.

To sum up, the difference between a good and a bad bearing can be detected through the high amplitude differences in the raw signal, but the indication in the spectra are not so clear, due to mechanical coupling and electromagnetic noise influence.

9 Conclusion and Further Steps

The analysis of the low cost MEMS sensors for use in an integrated condition monitoring system has shown that it is possible to detect faults. Basic signal processing techniques, like RMS, Kurtosis, PSD and Envelope analysis were performed on a filtered vibration signal to diagnose characteristic faults in a bearing. The performance of the low cost MEMS sensor is good enough for a rough estimation of the bearing state of health without the need to use high computing power. The limitation of the sensor is that only the low frequency spectrum can be observed and the chance of detecting the faults in early stages is inherently low. The low frequency spectrum is also usually masked by noise.

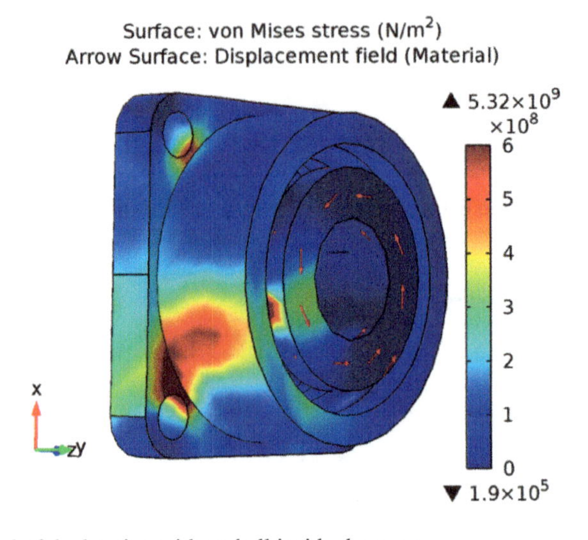

Fig. 12 FEM model of the bearing with on ball inside the cage

The PVDF offers a good opportunity for a low-cost vibrational sensor, but the physics of the sensor is not favourable for the suggested application space. Therefore important necessary design steps have been identified. The first results of the comparison with a MEMS sensor module show that the basic electrical properties of the pure sensor element are only sufficient to detect strong bearing damages. A good signal conditioning with proper hardware filter is mandatory in order to gain the best signal to noise ratio. The basic failure frequency could be detected and a distinction between a good and bad bearing is possible. The mechanical properties have to be adapted in order to change and amplify the transversal wave to a longitudinal bending wave form. To do this, a model of the bearing with a FEM simulation has been built up. A one ball bearing is modelled and the rotation is put on the inner ring, then the outer displacement of the outer ring is measured. Different structures will be applied on the ring to simulate the bearing and then transferred displacement e.g. vibration will be measured. First simulation results show a potential in the mechanical amplification.

References

[1] Lacey, S.J.: An overview of Bearing Vibration Analysis, Schaeffler UK (2007)

[2] FP7 COSIVU web page, GC.ICT.2012.6-8: -PPP GC 2012, http://www.cosivu.eu

[3] Zhang, H., Tolbert, L.M., Member, S., Ozpineci, B.: Impact of SiC Devices on Hybrid Electric and Plug-In Hybrid Electric Vehicles 47(2), 912–921 (2011)

[4] KESS project web page, BMBF-No.: 16N11 780 (2012), http://www.kess.uni-bremen.de

[5] Luber, W., Becker, J.: Application of PVDF foils for the measurements of unsteady pressures on wind tunnel models for the prediction of aircraft vibrations. In: Structural Dynamics - Proceedings of the 28th IMAC 2011. Conference Proceedings of the Society for Experimental Mechanics Series, pp. 1157–1176 (2011)

[6] Cremer, L., Heckl, M.: Structure-Borne Sound: Structural Vibrations and Sound Radiation at Audio Frequencies (2005)

[7] Randall, R.B.: Vibration Based Condition Monitoring: Industrial, Aerospace and Automotive Applications

[8] Randall, R.B., Antoni, J.: Rolling element bearing diagnostics—A tutorial. Mech. Syst. Signal Process. 25(2), 485–520 (2011)

[9] Natu, M.: Bearing fault analysis using frequency and wavelet techniques. IJCEM Int. J. Comput. Eng. Manag. 15(6) (2012)

[10] Sawalhi, N., Randall, R.B.: Localized fault detection and diagnosis in rolling element bearings: A collection of the state of art processing algorithms. no. Hums (2013)

[11] Konstantin-Hansen, H.: Envelope analysis for Diagnostics of Local Faults in Rolling Element Bearings, Denmark

[12] Andrejašič, M., Poberaj, I.: MEMS ACCELEROMETERS, Ljubljana (2008)

[13] Abouel-seoud, S., Ahmed, I., Khalil, M.: An Experimental Study on the Diagnostic Capability of Vibration Analysis for Wind Turbine Planetary Gearbox. Int. J. Mod. Eng. Res. 2(3) (2012)

[14] Dube, A.V., Dhamande, L.S., Kulkarni, P.G.: Vibration Based Condition Assessment Of Rollingelement Bearings With Localized Defects. Int. J. Sci. Technol. Res. 2(4), 149–155 (2013)

[15] de Lorenzo, F., Calabro, M.: Kurtosis: A Statistical Approach to Identify Defect in Roller Bearings. In: 2nd International Conference on Marine Research and Transportation, pp. 17–24 (2007)

Compact, Safe and Efficient Wireless and Inductive Charging for Plug-In Hybrids and Electric Vehicles

André Körner and Faical Turki

Abstract. Conventional charging systems for electric and plug-in hybrid vehicles currently use cables to connect to the grid. This methodology creates several disadvantages, including tampering risk, depreciation and non-value added user efforts. Loose or faulty cables may also create a safety issue. Wireless charging for electric vehicles delivers both a simple, reliable and safe charging process. The system enhances consumer adoption and promotes the integration of electric vehicles into the automotive market. Increased access to the grid enables a higher level of flexibility for storage management, increasing battery longevity.

The power class of 3.7kW or less is an optimal choice for global standardization and implementation, due to the readily available power installations for potential customers throughout the world. One of the key features for wireless battery chargers are the inexpensive system costs, reduced content and light weight, easing vehicle integration.

This paper demonstrates a wireless charging design with minimal component content. It includes a car pickup coil with 300 mm side length and low volume and mass 1.5 dm³ power interface electronics. After an overview of its hardware requirements, power transfer and efficiency benefits are presented, providing the anticipated horizontal and lateral deviations.

An intense magnetic field is required to transfer the target power at low volumes between the transfer units. This field heats up any metal object over the transfer coil, similar to an induction oven. Consequently, the system should be powered down whenever a metal object is detected in this area. A Foreign Object

A. Körner
Hella KGaA Hueck & Co.
Beckumer Str. 130, 59555 Lippstadt
e-mail: andre.koerner@hella.de

F. Turki(✉)
Paul Vahle GmbH & Co. KG
Westicker Strasse 52, 59174 Kamen
e-mail: faical.turki@vahle.de

J. Fischer-Wolfarth and G. Meyer (eds.), *Advanced Microsystems for Automotive Applications 2014*, Lecture Notes in Mobility,
DOI: 10.1007/978-3-319-08087-1_20,

Detection (FOD) design has been developed to continuously monitor the critical high field area. Device testing results are also provided.

Field characteristics are verified alongside the vehicle, ensuring system safety for living beings; compliant with all applicable standards reference limits which is more challenging than the basic limits [13].

Keywords: electric vehicle, wireless charging.

1 Introduction

Contactless energy transfer has been available since the late 90's as a standard technology [2] where a mobile load must be supplied with energy. It is widely accepted in assembling automation. Since the advent of automobile electrification, this technology is becoming increasingly more popular. However the first systems that came onto the market had differing concepts, since the manufacturers were focusing on the actual function and not their compatibility with other grid connected power equipment and charging applications [5]. The first systems, being strictly unidirectional, will be focused on vehicle charging. Already with such systems, power grid operators might get some control about the distribution of the network load by charging using the concept of time control.

Fig. 1 Wireless induction charging System

The energy transfer chain consists of two sub-systems: (1) The stationary infrastructure producing the required primary magnetic field (2) An on-road vehicle that transforms the primary field into electric power.

With the assumption of the systems complexity as constant, the optimization toward better market acceptance and lower pricing leads to a solution involving shifting the complexity to the infrastructure side and keeping the vehicle components as simple as possible. This involves the regulation of the power flow from the stationary primary side of the system.

All the systems available on the market may be represented by the structure in Fig. 2:

A standard wireless power transfer system usually consists of a primary stationary inverter fed by a supply grid via a one phase or three phase grid front-end. The primary inverter is responsible for rectification, power factor correction and converting the resulting DC-link voltage into AC with a normal switching frequency in the kHz-range. This process drives a resonant induction coil. In most systems, the frequency is within the range of the resonant circuit connected to the inverter output. The primary and secondary coils are responsible for magnetic coupling. The reactive power required to transfer the power over a large air gap is provided by a combination of compensation capacitors. The capacitor values are tuned to a nominal resonant frequency of the induction coils. The induced voltage in the secondary coil is rectified and optionally converted to the required voltage/current levels. This power can then feed the HV (high voltage) battery directly since there is already galvanic separation from the grid, or may feed the DC-link of an existing on-board charger. The latter option is a less complex scenario for resonant circuit topologies, but produces a lower efficiency.

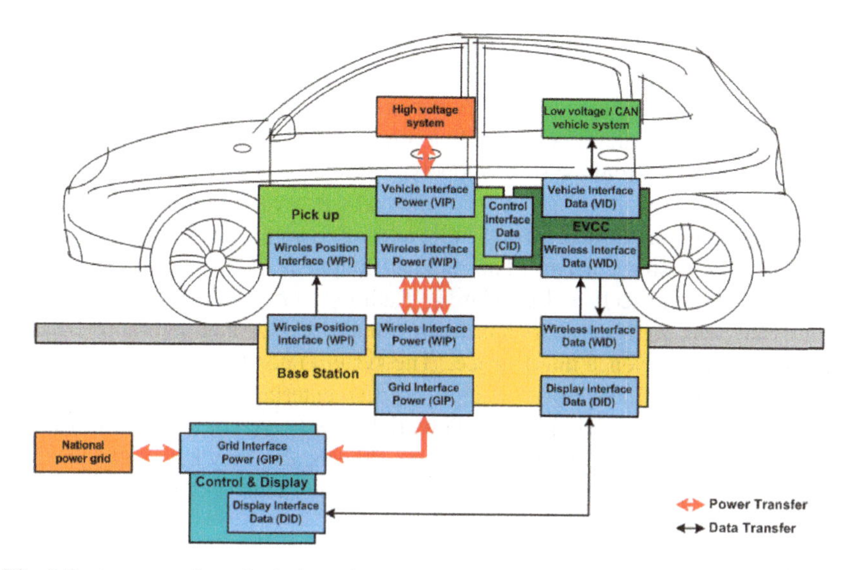

Fig. 2 System overview of wireless chargers

2 Vehicle Integration

The primary objective to introduce contactless charging systems to vehicle manufacturers is to increase the ease of charging for the end customer/user. For example: picking up a charging cable in the morning, where it has sunken deep into the snow and frozen solid to the ground by Joule heating after completing the prior charging and cool down cycle. Rain and dirt are also a concern for both the charging cable and vehicle connector interface. Plug-in hybrid vehicles possess smaller batteries that potentially require charging more frequently throughout the day.

Required cable and plug handling prerequisites may lead to a “forget it” style attitude when time is minimal.

The charging process must be automated, as much as possible, when compared to conventional charging, providing the ideal scenario where the driver only parks the vehicle on the charger. Charging automatically takes over and controls the process. The system provided, supports the driver, positioning the vehicle over the device. The system would possess a tolerance for misalignment between the vehicle coil and stationary floor coil (charging pad). It uses a wireless interface to authenticate the car, direct the driver´s steering and braking until the vehicle coil is well aligned with the floor coil and then initiate and continuously control the energy flow.

Another important objective is the electrical safety. Cables generate the risk of trip hazards and vandalism. Electrical power contacts on the EVSE (Electric Vehicle Supply Equipment) and vehicle are known to be one of the major causes of charging related failures. All this is avoided or significantly reduced inherently by implementing an inductive charging scheme.

A successful system integration into a vehicle requires a lightweight compact design, including a minimal, low profile for the underbody. A safe and simple power control process is required. Power transfer efficiency must be 90% or greater from grid to battery, comparable to current developments, charging with a power cable and electronic system support. The electrical system’s performance must meet the battery charging requirements. The process begins with a charging phase at a constant maximum current, followed by a shorter phase with continuous voltage and decreasing current. The radiated magnetic fields must comply with legal regulations and technical standards. Radio systems, including vehicle keyless entry/keyless go systems, tire pressure monitoring or non-automotive based methods cannot experience interference. The system must be innocuous including inductive heating influences on foreign objects entering air gaps between the charging pad and the vehicle pickup. Potential objects are not allowed to heat-up due to the risk of injury or fire hazards. Live objects also may have to be protected, even if no influence from low frequency magnetic fields is known today.

3 Magnetic Circuit

The magnetic circuit supporting a wireless power transfer system is equivalent to a transformer with a large air gap in its main area resulting in a relatively large leakage inductance on the primary and secondary sides.

A standard set-up for the magnetic circuit, with an air gapped transformer configuration, consists of a soft-magnetic material, usually ferrite (like most inductive components for high kilohertz frequencies) which contains the fields, and a copper winding made from litz wire to reduce the dissipation power caused by skin and proximity effects (See Fig. 3).

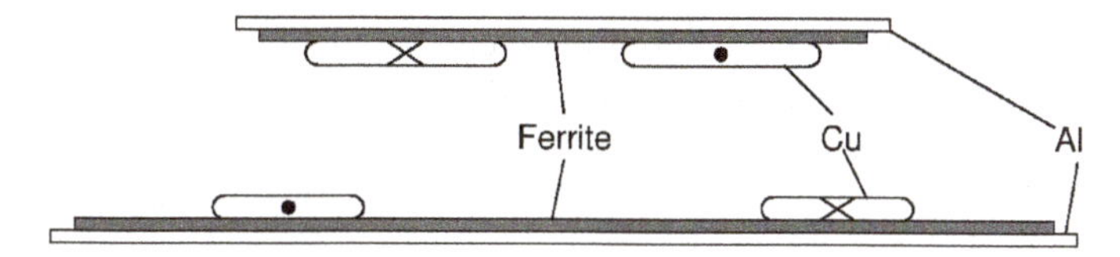

Fig. 3 Magnetic design of circular coils

This type of, circular configuration has an advantage, containing the magnetic field thus preventing it from reaching the vehicle underbody. For additional protection, an aluminum plate can be added to improve the shielding at high kilohertz frequencies.

The bulk of the magnetic flux is located in the middle of the air gap, with only leakage fields produced at the sides. The value of magnetic coupling in the nominal working area of the transformer varies from about 0.1 to 0.3 on a typical vehicle with air gap clearance up to 250 mm. The value 0.5 can be reached only for very large coils.

In contrast to the symmetric arrangement of the induction coils, vehicle manufacturers prefer to keep the vehicle side (secondary) coil as small as possible. Making both coils smaller would disturb the positioning tolerance in such a way as to reduce the power transfer efficiency below acceptable levels [3]. Such asymmetric magnetic circuit is shown in Fig. 4, where the stationary primary coil is on the left and vehicle secondary coil is on the right.

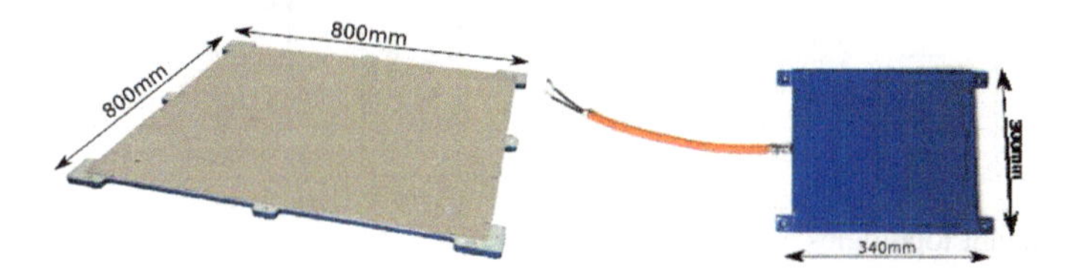

Fig. 4 Tested circular coils

The chosen primary and secondary coils (Fig. 4) differ in dimensions: 1m x 1m for the base plate on which is mounted an 800mm x 800mm primary coil, then a 300mm x 340mm vehicle side secondary coil. It will be shown that this system is able to work with asymmetric magnetic coupling. The dimensioning of the larger primary coil (800mm x 800mm) is based on the German VDE application guide and prevents excess heating of metallic foreign objects [4].

4 Resonant Circuit Topology

Depending on the decision of which side the air gap is magnetized during the light load condition and on the power electronics topology, the resulting resonant circuit may be any combination of parallel and series configurations on primary or secondary side. This leads to four combinations (Fig. 5).

From a physical perspective, all of these combinations are feasible, without the need to choose one arrangement as a standard, due to the same nominal power rating; the mechanical designs for every topology are equivalent. A rule must exist for the system behavior to handle various scenarios by adapting the control loop.

The compensation strategy is a major issue, due to the major impact it has on the behavior of the transfer system. In this process, only the two-sided resonance combinations are relevant, due to inductive power transfer systems and only one resonant circuit not feasible for low coupling applications like electric vehicle charging via the underbody.

As shown in Fig. 5, the magnetization current may be delivered with a parallel circuit on both the primary and secondary sides (Fig. 5c), but in a series resonant circuit on the secondary side (Fig. 5d), the magnetization can't be utilized by the load side, if it has a simple rectifier.

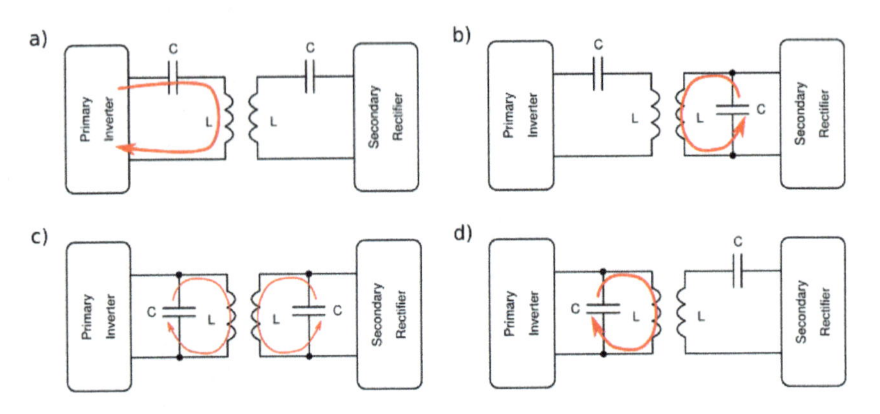

Fig. 5 Resonant topologies

The most desirable and accurate topology for the primary resonant circuit has to be selected. To realize a universal primary arrangement, a DC-link voltage has series topology and is easier to implement with a popular full bridge converter, delivering the appropriate level of reactive power to magnetize the air gap by controlling its output current. See Fig. 5a.

The function and behavior of the resonant circuit can be explained by first reviewing the combination of a series to series layout, (Fig. 5a) which is a symmetric system with series oscillating circuits on both sides. The following diagram is the equivalent circuit of the T-System with gyrator-function.

The gyrator-function is shown in Fig. 6: The capacitor C in Fig. 6a compensates the total inductance of the respective windings completely. If they are split into main (mutual) inductance L_h and leakage (stray) inductance L_s, C can be theoretically separated into two capacitors in series. One of them compensates the leakage inductance L_s and the rest has the same reactance X of the main inductance L_h with opposite sign so they can compensate each other. This way, a

T-system is generated with different signs as shown in Fig. 6b. It follows the load resistance, transformed into X^2/R on the feeding side.

The disadvantage of this topology is the feeding inverter needs to maintain a nearly constant magnetizing current over the entire operating power range, leading to reduced efficiency during light load-operation.

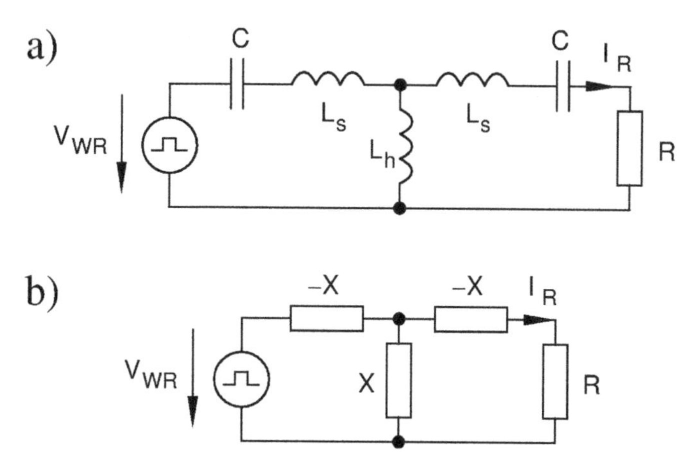

Fig. 6 Gyrator circuit

This is a preferred method, but has a smaller number of components and the charging load profile on the grid connection operates mainly in the high power range. As a result the counter argument for reduced light load efficiency becomes less significant. This solution is not vulnerable to a variable air gap, changing the inductances of the system and the resonant frequency. This provides the ability for the inverter to obtain ZCS (zero current switching) / ZVS (zero voltage switching) conditions, leading to low switching losses.

A scenario to consider is a parallel pickup coupled with the universal primary system. In this case, the equivalent circuit can be assumed to be a series resonant circuit possessing a variable load resistance that does not vary from the series pickup with the exception of the gyrator effect not present, due to a current source transformed into a current source. A detailed description of the load behavior is provided in [5].

The equations governing the circuit impedance and current of the gyrator circuit are shown in Equation 1.

Equation 1: Gyrator impedance and current as a function of passive component values

$$X = 2\pi f L_h$$

$$I_R = \frac{V_{WReff}}{jX}$$

5 Power Flow Control

As shown in Fig. 7, the charging process of an electric vehicle battery, especially lithium based, is separated into two control modes that should be supported: constant current and constant voltage. This requirement is generated by the on-board vehicle BMS (Battery Management System).

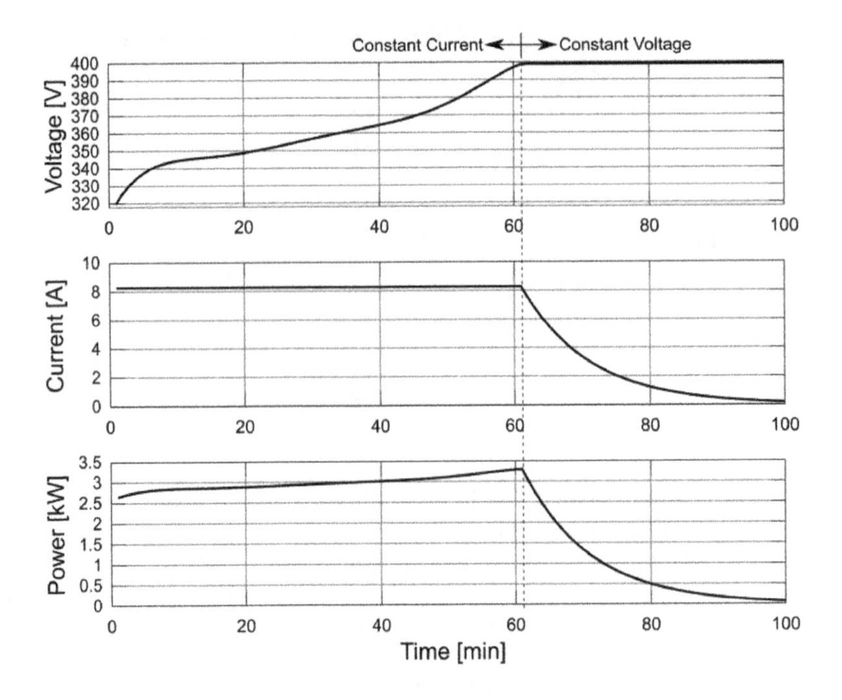

Fig. 7 Typical 1C-Charging process of a Lithium-based battery

Due to the relatively low availability of mounting space of E-vehicles, with associated high costs of on-board power electronics and components; a simple rectifier solution is preferred versus that of an active SMPC (Switch-Mode Power Controller). This requires the information from the BMS to be transmitted via a wireless data transfer system to the primary base station in order to control the charging process and power fed to the primary coil.

Fig. 8 shows the complete control system using a wireless data transfer system. The wireless component is assumed to be readily available since authentication and infrastructure metering data is required to ensure a safe and efficient charging process.

The consequence of this control structure is that the primary inverter has to control the power flow by a power semi-conductor switching strategy.

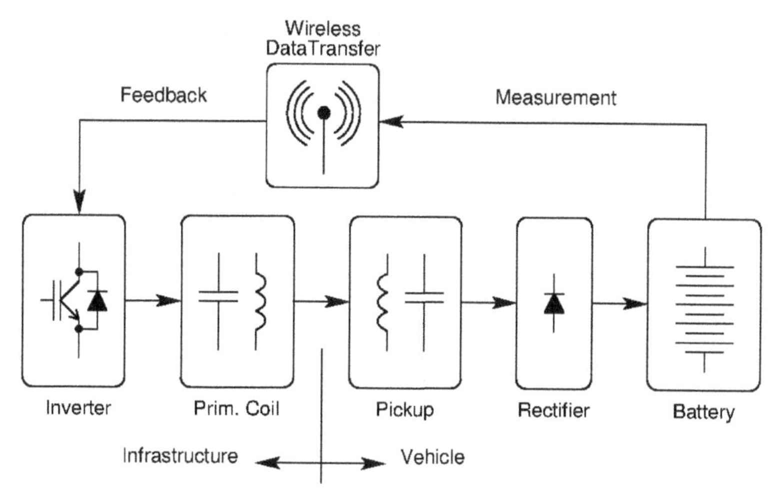

Fig. 8 Closed-loop control of power flow through wireless data transfer

6 Power Electronics

Fig. 9 shows a grid-front end consisting of an EMC filter followed by full-wave rectifier feeding a boost style PFC. The PFC supplies a DC-link inverter which drives the primary coil.

The primary inverter generates an AC current in the high kilohertz range. This is required to generate the primary field, especially for the magnetization of the transformer, for the series pickup.

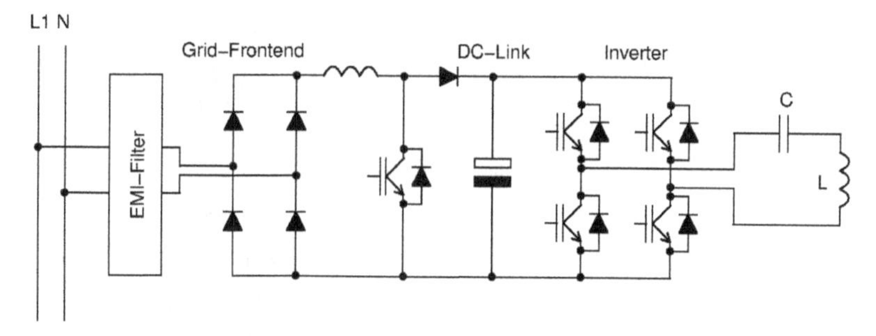

Fig. 9 Stationary power electronics

If the inverter is required to control its output current or voltage, a possible method is continuously varying the DC-link voltage, but this imposes an additional buck converter or a less conventional grid front end [6] [7]. There are other possible methods for controlling the output, by influencing the following:

- Semi-Conductor, Pulse Width: This leads to high switching losses and more EMI due to capacitive commutations of voltage inverters and inductive commutation of current inverters.
- Switching Frequency: Even if light loads are fed with frequencies higher than the resonant frequency; to behave as an inductive commutation at series resonance, the semiconductors have to switch at non-ZCS. This hard switching leads to higher switching losses and more EMI than resonant converters.
- Tuning of the Primary Resonant Circuit: This is done by switching additional capacitors in series or in parallel with the compensation capacitors by means of bidirectional switches. This method is equivalent to the previous scheme of adjusting switching frequency which leads to higher switching and conduction losses.
- The mean value of the output voltage by synchronized full or half wave switching.

The final strategy, varying the mean value of the output voltage, synchronizing full or half wave switching, is selected as the preferred method. The output current is controlled by a pulsed voltage produced by switching the respective diagonal power semiconductor legs of the full-bridge inverter. On the other hand, if a zero-voltage is required, either the high side pair or low side pair of semiconductors is switched on. Only whole periods or half periods may be switched (Fig. 10) resulting in either ZCS or ZVS operation. Power is then regulated by pulse skipping as opposed to duty-cycle modulation resulting in higher DC-AC inverter efficiency.

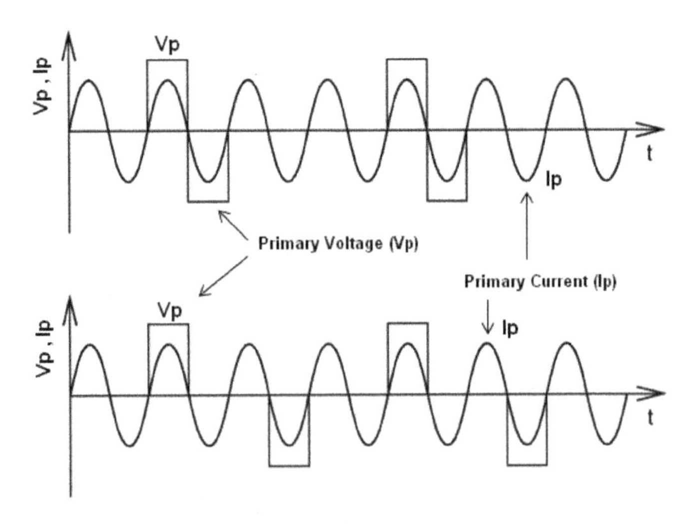

Fig. 10 Switching modes: full period and half period

The synchronization of the voltage V_p is accomplished by measuring the output current I_p. This kind of control algorithm allows the use of a constant DC-link voltage therefore minimizing the switching losses by turning off the semiconductors only at very low current and turning them on at zero voltage. This synchronization leads to a positive feature of the inverter, which is its ability to follow the resonant frequency of the system; even if it drifts from the nominal value. This is a key advantage which eliminates the system sensitivity to position tolerances caused by the inductances of the coils changing by the proximity of the ferrite used in the opposing coil. A similar effect is caused by temperature drift or variation in the tuning capacitor values.

The command for the output current is given by the data transfer to the micro-controller inside the inverter [8]. This allows the secondary side of the power electronics, located in the vehicle, to be as simple as possible:

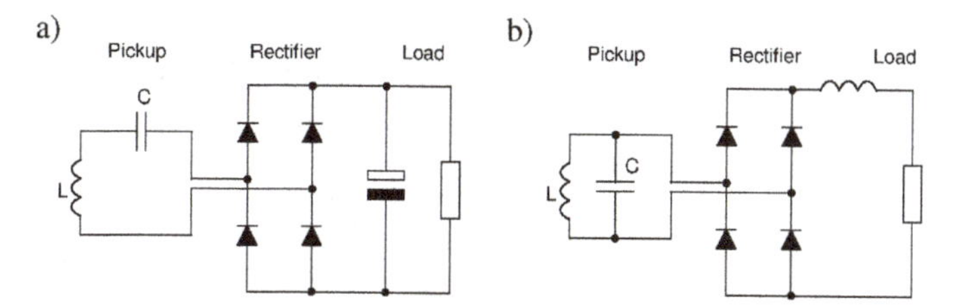

Fig. 11 Mobile vehicle power electronics

As shown in Fig. 11, the secondary (vehicle side) consists of a full-bridge rectifier and filter. Filtering may be performed by a parallel capacitance in the case of a series pickup (Fig. 11a) or by a series inductance in the case of a parallel pickup (Fig. 11b).

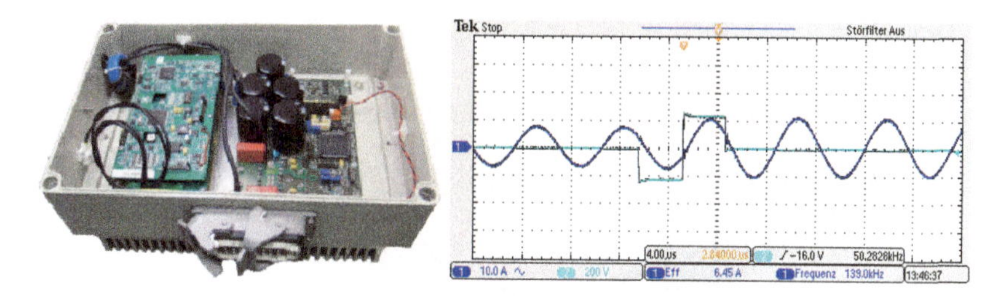

Fig. 12 Inverter package including the grid front-end (left) and its output current and output voltage (right)

7 Series / Parallel Performance

To test the primary hardware, the two pickup topologies for series and parallel must be investigated. Due to the almost constant power during the constant current phase of the battery charging process in Fig. 7, the systems are, rather, tested at the constant voltage condition which results in more power variation.

In nominal position (air gap Z- about 135mm with X- and Y-deviation equal zero) the comparison of the system is not realistic because in most cases the position of the car will not be optimal. Thus, a second scenario has to be taken into account, having a bad coupling resulting in a lower induced voltage. Fig. 14 gives an impression of the change in voltage, from a drop in the magnetic flux density, as a result of moving in one of the deviation directions.

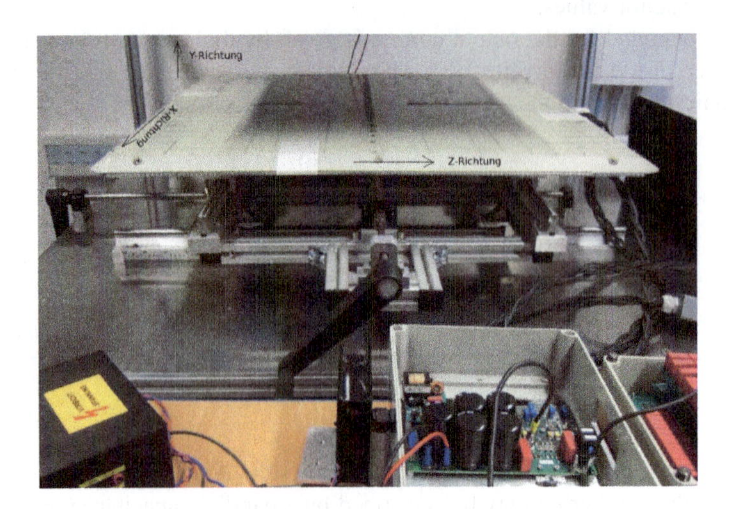

Fig. 13 Hardware test-bench

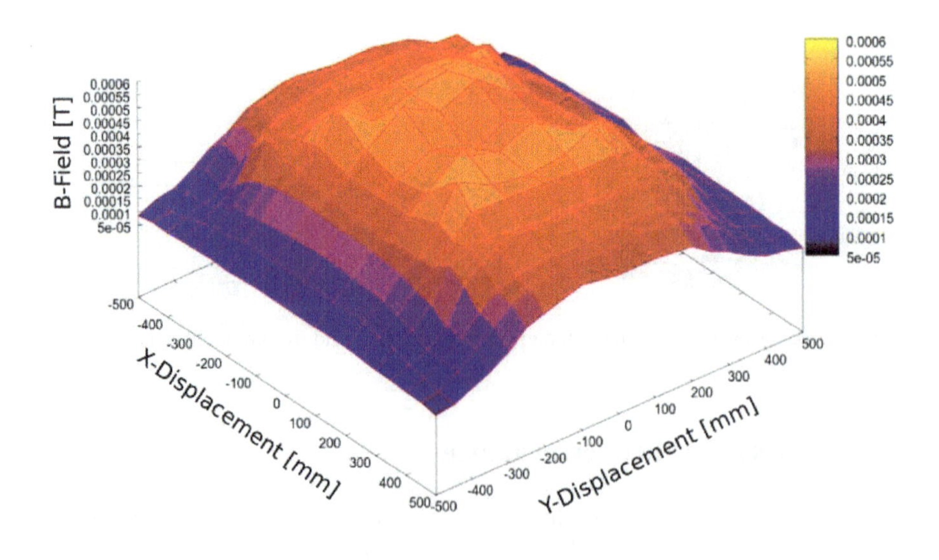

Fig. 14 Measured flux density in the air gap with respect to X and Y displacement

The efficiency characteristics shown in Fig. 15 were measured using a nominal air gap and zero displacements (X=Y=0). These are the optimal parameters for which the system was designed. Both coil pickup versions delivered good performance. However, the parallel pickup exhibited a more unstable current due to additional resonances.

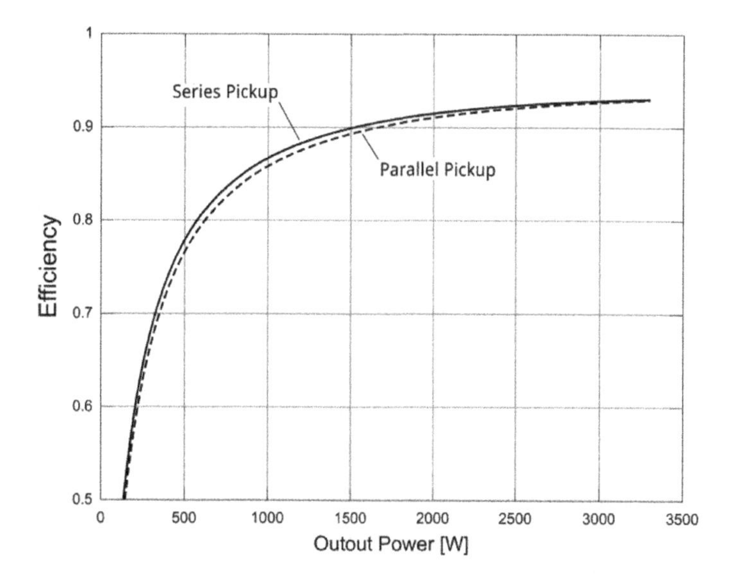

Fig. 15 System efficiency with series and parallel pickup at nominal position

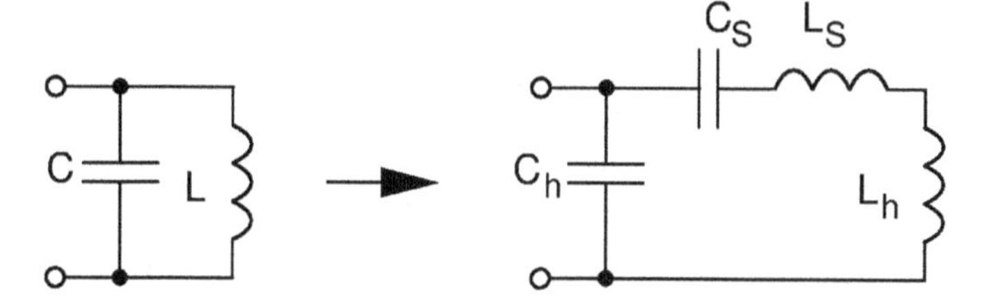

Fig. 16 Parallel circuit of the pickup with capacitor split

Fig. 16 shows the actual circuit tested in hardware. The resonant capacitor is split into two parts: C_h compensates the main inductance L_h and C_s compensates the stray inductance (leakage inductance) of the pickup, leading to a current source if the system is fed with a current source; but leads to a voltage source if fed with a voltage source. Obviously, this combination works well only for one value of coupling factor and therefore is sensitive to misalignment.

Due to the self-adjusting feature of the working frequency of the inverter to the resonant frequency, there must be one definite resonant circuit that determines this frequency. In Fig. 16, two possible paths exist for the load current (via L_h and L_s) which leads to two resonances and thus an unstable feeding frequency.

8 Bidirectional Power Transfer Capability

An additional possible feature of the contactless system is the ability to feed energy from the vehicle battery into the power grid via a bidirectional configuration.

Regardless of the powers feeding direction, which must be controlled, the power command results from the charging mode of the vehicle and is either generated by the battery management system (BMS) or from a voltage measurement if a DC-link voltage of a present charging device is fed.

On the other hand, if the system works in grid support mode, the power command is generated by the system operator. However, the BMS has the higher priority since it controls the energy content of the vehicle battery. Therefore the existing wireless data interface can be used.

According to the circuit topology in Fig. 17, two possible methods to control the power flow may be used:

- Both primary and secondary inverters switch with exactly the same frequency and the same duty cycle near 50% but with different phase angles. The phase difference between the inverters (usually between -90° and 90°) sets the direction/sign and amount of power. Therefore both inverters have to be synchronized by an additional signal. The advantage of this solution is that the control device is on the side which requires power (consumer).
- Only one inverter works for one feeding direction. The second inverter works as a passive inverter by deactivating the semiconductor switch gate signals. To close the loop, the system power requirements of the consumer are fed back to the feeding side via the wireless data interface. The latter has to control the power flow by employing a semi-conductor switching strategy. This happens via pulse width modulation, pulse packet control by switching whole numbers of intervals or the variation of dc link voltage by an additional DC/DC converter.

The second (latter) variant was chosen because of the low switching losses of the passive inverter, and the reduced complexity since synchronization of both systems is not required. The control of the inverter is done by switching whole periods of the resonant frequency.

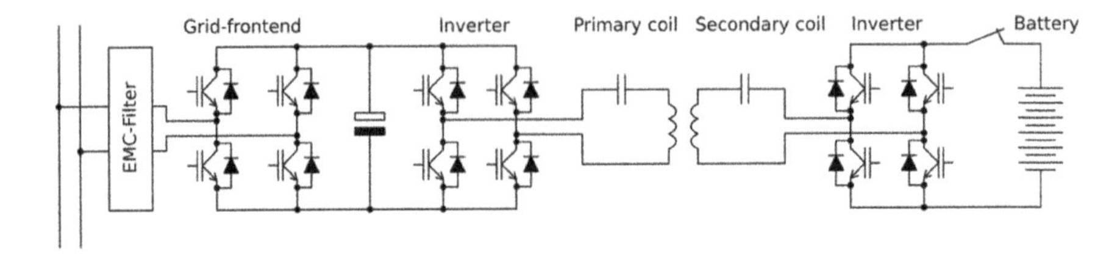

Fig. 17 Bi-directional energy transfer topology

9 Coil Displacement Experiments

The configuration was tested at a constant output load power level of 3kW and controlled output voltage of 400VDC. The current required to transfer 3kW was controlled by an adjustable electronic load on the secondary side. Measurements of primary current and efficiency were measured with respect to X-, Y- displacement and air gap distance Z. Measuring equipment used to gather test data included: oscilloscope outfitted with current probes for inverter supply current, primary current and load current. High voltage probes were used for measuring DC-link voltage and load voltage. Data was logged by a PC with analysis software in order to calculate inverter efficiency over multiple periods of the coil resonant frequency.

Table 1 Tested configuration: single layer bipolar planar primary + bipolar solenoid secondary

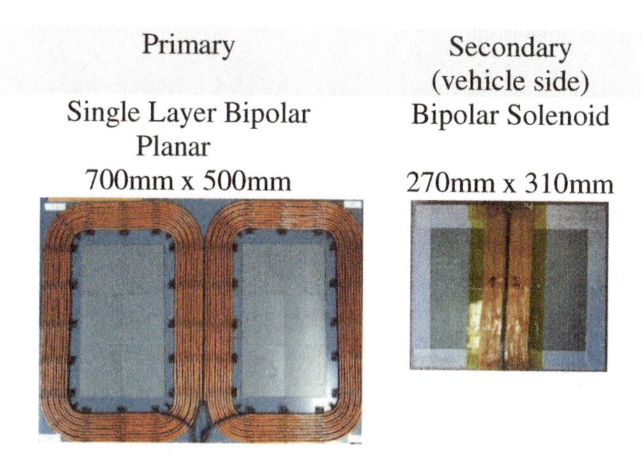

Primary	Secondary (vehicle side)
Single Layer Bipolar Planar	Bipolar Solenoid
700mm x 500mm	270mm x 310mm

Table 1 shows a bipolar arrangement due to two symmetric magnetic poles with opposite flux directions. It is built from a dual-coil on the primary side, where there is enough room to use a large area for the coil and thus realize the required positioning tolerances in X- and Y-directions (see Figure 18).

The secondary side uses a bipolar coil with a geometry equivalent to a solenoid. This is also a bipolar coil, with symmetrical field distribution for both flux directions making it compatible with the double-coil variant.

The current flowing through the windings on the top of the secondary coil makes the shielding layer mandatory.

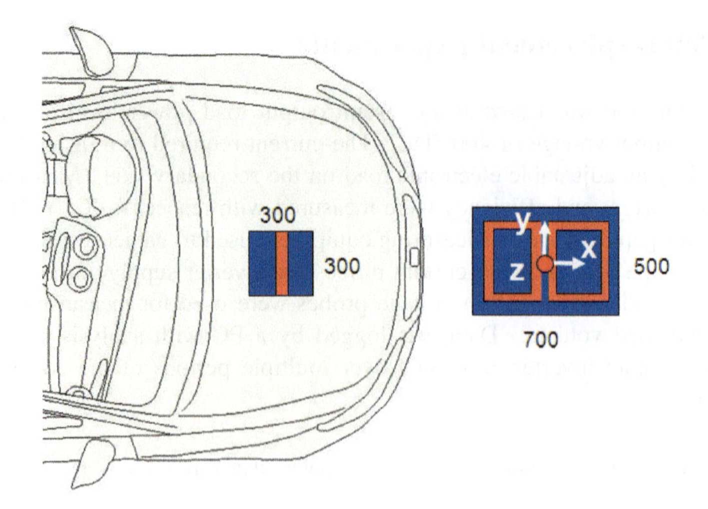

Fig. 18 Application alignment

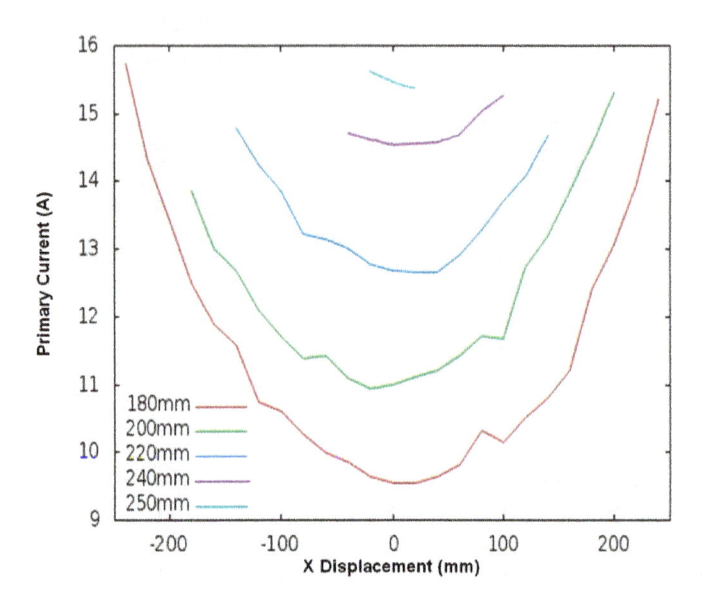

Fig. 19 Primary current versus X-displacement at 180mm, 200mm, 220mm, 240mm and 250mm air gaps

Fig. 19 and Fig. 20 show the variation in required primary current in order to achieve the 3kW power level with respect to X- and Y- displacements, while Fig. 21 and Fig. 22 show the inverter efficiency with respect to X- and Y- displacements.

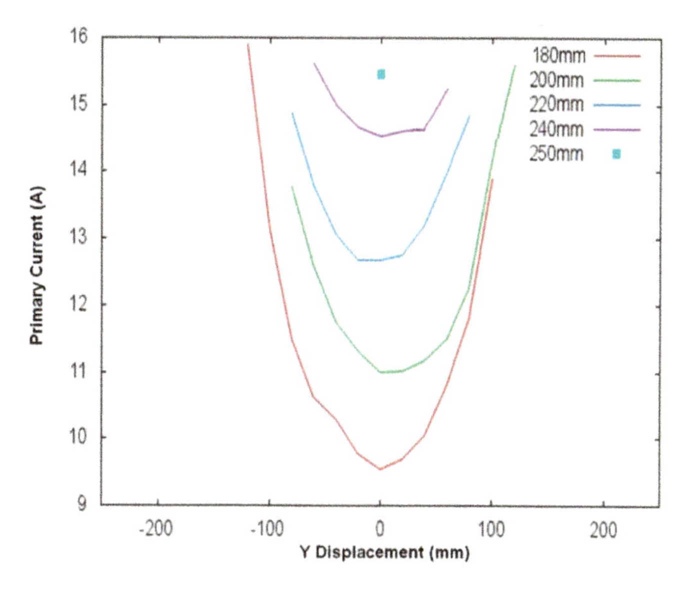

Fig. 20 Primary current versus Y-displacement at 180mm, 200mm, 220mm, 240mm and 250mm air gaps

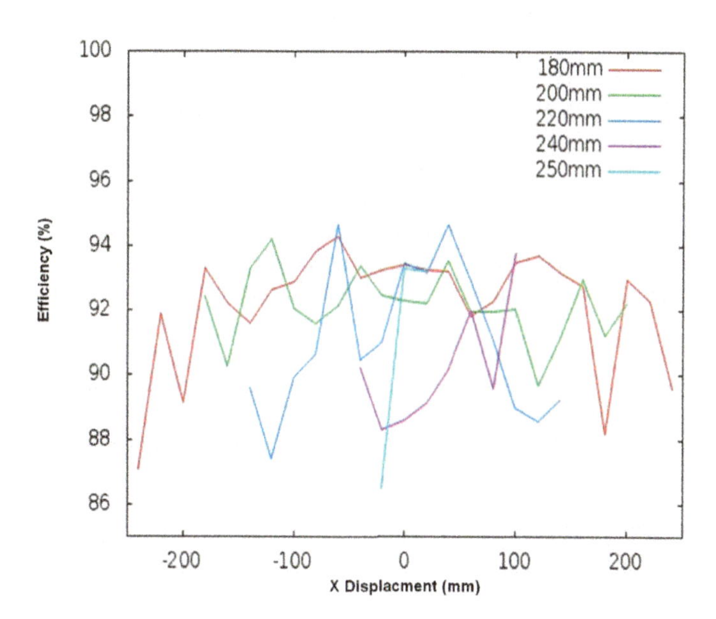

Fig. 21 Efficiency versus X-displacement at 180mm, 200mm, 220mm, 240mm and 250mm air gaps

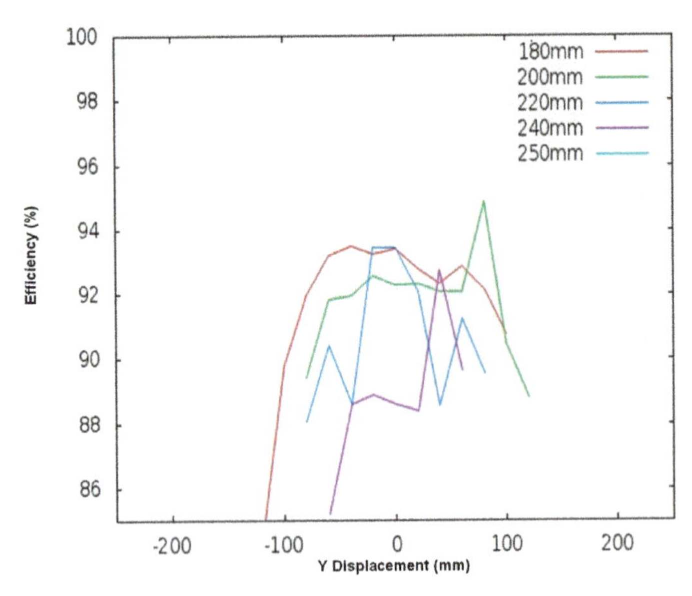

Fig. 22 Efficiency versus Y-displacement at 180mm, 200mm, 220mm, 240mm and 250mm air gaps

10 Foreign and Living Object Detection

The primary target for vehicle manufacturers is using a small lightweight secondary coil, offering a lower cost and easier system integration. However, this causes a high intensity magnetic field, potentially causing any metal objects exposed to the field to warm as a result of induced eddy currents and subsequent joule heating. This warming effect possesses similarities with induction cooking, and must be prevented to thwart safety risks with flammable liquids and materials. A Foreign Object Detection (FOD) system is installed on the inductive charging system to detect metallic objects around and on the surface or in proximity to the stationary coil. The FOD can be activated in advance, of high frequency power, detecting hazardous conditions and preventing injury or damages. In the event of real danger, power is not transmitted or activated; shutting off within 10 milliseconds after a risk is identified.

Objects that have been successfully detected by the system range from small to large and massive including iron and non-ferrous objects:

- Cigarette box
- House door key
- Wrench

More objects will be tested in future, especially based on standardization of test procedures and customer requirements and could include objects like:

- Coins
- Beverage cans
- Common garage tools

Live objects cannot be exposed to magnetic fields, based on liabilities, government mandates, customer requirements and industry based standards, including SAE J2954, or the ICNIRP recommendations. The magnetic field underneath the vehicle's body can potentially exceed legal limits. Additional systems must be installed to detect living beings, in order to switch off the power transfer, or prevent access to these areas. Presently, object specifications are under review. These vary between different requirements, for instance regarding the minimum size of considered animals. In spite of the increased safety measures, it still must be maintained to accomplish the charging process as expected by the car owner, in order to prevent flat batteries for the next car start. Therefore, the requirements need to be defined carefully and precisely, and the systems must be designed for precise object detection and quick repowering after a disturbing object is removed from the observation area. If a safety shut-off lasting for more than a few minutes occurs, the vehicles owner must be notified to ensure operation.

11 Keyless Entry / Go Systems Interoperability

Inductive charging systems create magnetic AC fields in low frequency range. The power control of the power transfer may create some kind of modulation, as shown in Fig. 12. Radio systems in cars or any other commercial, public or private application may interfere with that. The coil units are not designed to emit radio waves, but nevertheless the magnetic AC field is considerably detectable in the near field and even at distant locations, according to the general magnetic field strength characteristic as shown in Fig. 23.

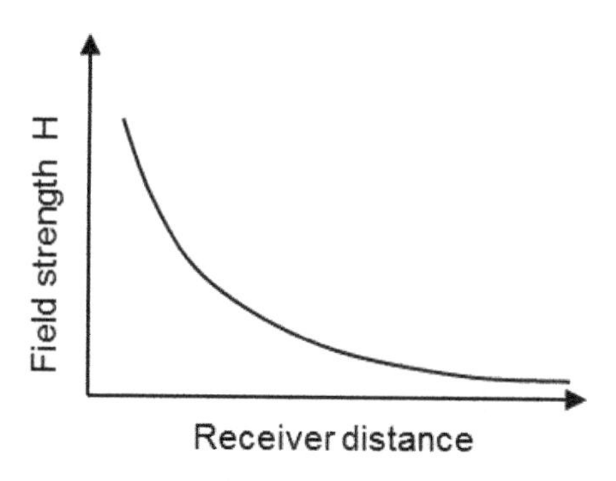

Fig. 23 Receiver field strength characteristic

While most known car radio systems meanwhile have shown that they are not disturbed by LF inductive charging systems, Keyless Entry & Go Systems show significant functionality limitations regarding their operation distances, up to complete disfunction. During the initialization of such systems, e.g. when a car driver touches the door handle, the car sends wake-up and identification data to the ID transmitter, what the car owner has in its hand, in his pocket or anywhere else. This data transmission uses LF signals, commonly at frequencies like 20kHz, 125kHz, 132kHz or 135kHz. The sender antennas are located e.g. in the door handle or at the sill. The LF receiver of the ID transmitter needs a field strength of about 70dBμA/m. Its typical frequency selectivity is wide enough for the required bandwidth of about 4kHz for the data transmission. Inductive charging systems both at 85kHz or 140kHz create easily 100dBμA/m field strength around the car at their resonance frequency. 85kHz systems tend to create higher field strength, as for a given charging power the field strength must be higher. For evaluation of the ID transmitter disturbance, the 100dBμA/m field iso surface, what results from all points with 100dBμA/m field, is a good indication for the disturbed space, when also the receiver selectivity and necessary signal /noise ratio is taken into account. This is valid at the same levels for 85kHz and 140kHz, due to the typical data receiver selectivity characteristic. A visualization of the simulated 70dBμA/m isosurface from a 125kHz keyless entry door handle antenna at a compact car is shown as top and side views in Fig. 24.

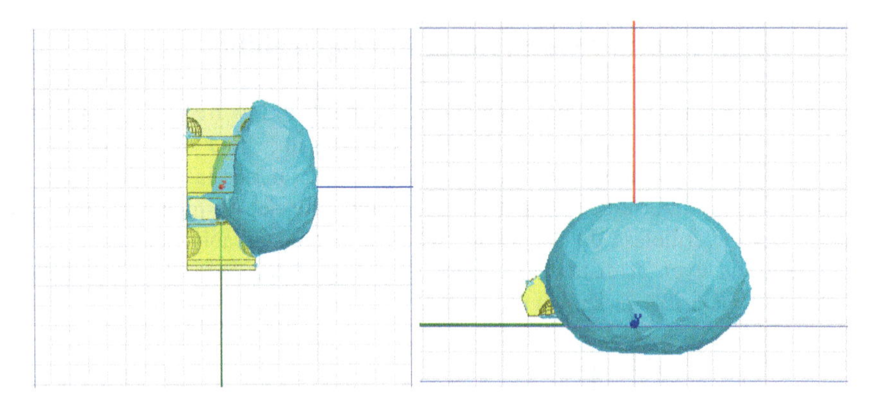

Fig. 24 Keyless entry door antenna 70dBμA/m iso surface

A simulation of the 100dBμA/m iso surface from a 140kHz circular inductive charging system at 3,3 kW power with 150mm air gap and no lateral misalignment is shown for comparison in Fig. 25. Obviously close to the floor there will be no data transmission possible, so e.g. opening the door with the ID transmitter in the briefcase standing on the ground will not work. A LF antenna in the door sill would be much more critical.

Of course a different charging coil geometry, power, frequency, location, or shielding will affect the picture and are known and used levers for optimization of the Interoperability of Keyless Entry/Go Systems with Inductive charging systems. But with 3,3 kW charging power and some xW data transmission power at similar frequencies it will remain a serious engineering task to maintain undisturbed LF data transmission for the charged car and for the vehicles around it.

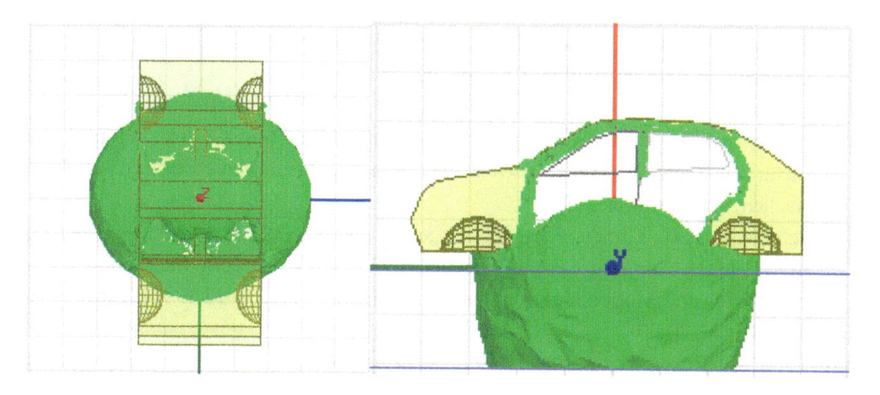

Fig. 25 Inductive Charging 100dBμA/m iso surface

12 Conclusions

The operation of a series resonant coil fed by an inverter adapting its working frequency to the resonance of the load was proven with two possible pickup topologies: parallel and series resonant. The investigation concludes both topologies, when paired with a secure wireless communication interface, are feasible and feature identical performance at nominal position without lateral displacement. Reduced coupling, caused by mechanical displacement, may create several variances in the efficiency of the system. This is due to the power losses at the provided reactive power lowering on the secondary side. However, parallel pickup provides the inverter control loop potential instabilities from increased resonances from the entire passive circuit.

The coil configuration was able to operate with the highest performance to contain the magnetic flux which increased the energy transfer efficiency and reduced the system's sensitivity to (X-,Y-) displacements. This configuration is also the most desirable providing the secondary coil which can be produced smaller, reducing weight and delivering the end application greater flexibility for mounting and packaging.

Wireless charging, combined with foreign and living object detection, is safe, convenient and an efficient system for charging plug-in hybrids and electric vehicles. Methods have been developed to detect common foreign objects as mentioned in this study and will be revealed in a future paper.

References

[1] VDE-AR-E 2122-4-2, Elektrische Ausruestung von Elektro-Strassenfahrzeugen Induktive Ladung von Elektrofahrzeugen Teil 4-2: Niedriger Leistungsbereich (March 2011)

[2] Meins, J., Buehler, G., Czainski, R., Turki, F.: Contactless Inductive Power Supply. In: Maglev Conference 2006, Dresden, pp. 527–535 (2006)
[3] Turki, F., Vosshagen, T., Schmuelling, B.: eCPS Ein induktives Ladesystem für Elektrokraftfahrzeuge. In: Tagung Elektrik/Elektronik in Hybrid und Elektrofahrzeugen und elektrisches Energiemanagement. Haus der Technik RWTH Aachen, Miesbach (2012)
[4] Schmuelling, B., Cimen, S.G., Vosshagen, T., Turki, F.: Layout and Operation of a Non-Contact Charging System for Electric Vehicles. In: 15th International Power Electronics and Motion Control Conference, EPE- PEMC 2012 ECCE Europe, Novi Sad, Serbia (2012)
[5] Turki, F., Reker, U.: Further Design Approaches of the Standardization: Inductive Charging of Electric Vehicles. In: 2nd International Electric Drives Production Conference and Exhibition, Nürnberg (2012)
[6] Chow, M.H.L., Lee, Y., Tse, C.K.: Single-Stage Single-Switch Isolated PFC Regulator with Unity Power Factor, Fast Transient Response, and Low-Voltage Stress. IEEE Transactions on Power Electronics 15(1) (January 2000)
[7] Boys, J.T., Huang, C.-Y., Covic, G.A.: Single-Phase Unity Power-Factor Inductive Power Transfer System. In: Power Electronics Specialists Conference, PESC 2008. IEEE (2008)
[8] Turki, F.: Bidirektionale induktive kontaktlose Energieübertragung zur Ankopplung von Elektrofahrzeugen ans Versorgungsnetz. VDE-Congress 2012, Stuttgart (2012)
[9] Turki, F.: A Wireless Battery Charger Concept with Lightweight and Low Cost Vehicle Equipment. In: Conference on Electric Roads & Vehicles, Utah, USA (2013)
[10] Turki, F., Guetif, A.: Supporting the low-voltage distribution network with static and mobile energy storage systems. In: IEEE International Multi-Conference on Systems, Signals and Devices Hammamet, Tunisia (2013)
[11] Turki, F., Vosshagen, T., Kürschner, D., Kratser, A.: Unified Supply Concept for Standardized Contactless Inductive Battery Charging of Electric Vehicles. In: Tagung Elektrik/Elektronik in Hybrid- und Elektrofahrzeugen und elektrisches Energiemanagement. Haus der Technik RWTH Aachen, Bamberg (2013)
[12] Schmuelling, B., Turki, F.: A SEPIC fed inductive charging device for electric vehicles. IEEE International Conference on Power Engineering, Energy and Electrical Drives, Istanbul, Turkey (2013)
[13] ICNIRP 2010, http://www.icnirp.de/documents/LFgdl.pdf

Reliability of New SiC BJT Power Modules for Fully Electric Vehicles

Alexander Otto, Eberhard Kaulfersch, Klas Brinkfeldt, Klaus Neumaier, Olaf Zschieschang, Dag Andersson and Sven Rzepka

Abstract. Wide-bandgap semiconductors such as silicon carbide (SiC) or gallium nitride (GaN) have the potential to considerably enhance the energy efficiency and to reduce the weight of power electronic systems in electric vehicles due to their improved electrical and thermal properties in comparison to silicon based solutions.

In this paper, a novel SiC based power module will be introduced, which is going to be integrated into a currently developed drive-train system for electric commercial vehicles. Increased requirements with respect to robustness and lifetime are typical for this application field. Therefore, reliability aspects such as lifetime-limiting factors, reliability assessment strategies as well as possible derived optimization measures will be the main focus of the described work.

Keywords: Electric drive-train system, power electronics, silicon carbide, double-sided cooling, reliability.

A. Otto(✉) · S. Rzepka
Fraunhofer Institute for Electronic Nano Systems ENAS,
Department Micro Materials Center, Technologie-Campus 3, 09126 Chemnitz, Germany
e-mail: {alexander.otto,sven.rzepka}@enas.fraunhofer.de

E. Kaulfersch
Berliner Nanotest und Design GmbH, Volmerstrasse 9B, 12489 Berlin, Germany
e-mail: eberhard.kaulfersch@nanotest.org

K. Brinkfeldt · D. Andersson
Swerea IVF AB, Argongatan 30, 431 53 Mölndal, Sweden
e-mail: {klas.brinkfeldt,dag.andersson}@swerea.se

K. Neumaier · O. Zschieschang
Fairchild Semiconductor GmbH, Technology Development Center,
Einsteinring 28, 85609 Aschheim, Germany
e-mail: {klaus.neumaier,olaf.zschieschang}@fairchildsemi.com

J. Fischer-Wolfarth and G. Meyer (eds.), *Advanced Microsystems for Automotive Applications 2014*, Lecture Notes in Mobility,
DOI: 10.1007/978-3-319-08087-1_21,

1 Introduction

Power electronics is gaining more and more importance in the automotive sector due to the slow but steady progress of introducing partially or even fully electric powered vehicles. The demands for power electronic devices and systems are manifold, and concerns besides aspects such as energy efficiency, cooling and costs especially robustness and lifetime issues. This is in particular true for commercial vehicles such as lorries or construction equipment, where in comparison to passenger cars higher performance requirements and harsher environmental conditions facing increased needs for total driving range (lifetime) and up-time.

The European joint research project COSIVU (project duration: October 2012 - September 2015) addresses these issues by developing a novel electric drive-train system architecture. The goal in this project is to realize a smart, compact, and durable single-wheel drive unit including an integrated electric motor, a 2-stage gear system, an inverter with SiC based power electronics, a novel control and health-monitoring system with wireless communication, and an advanced ultra-compact cooling solution (Fig. 1) [1].

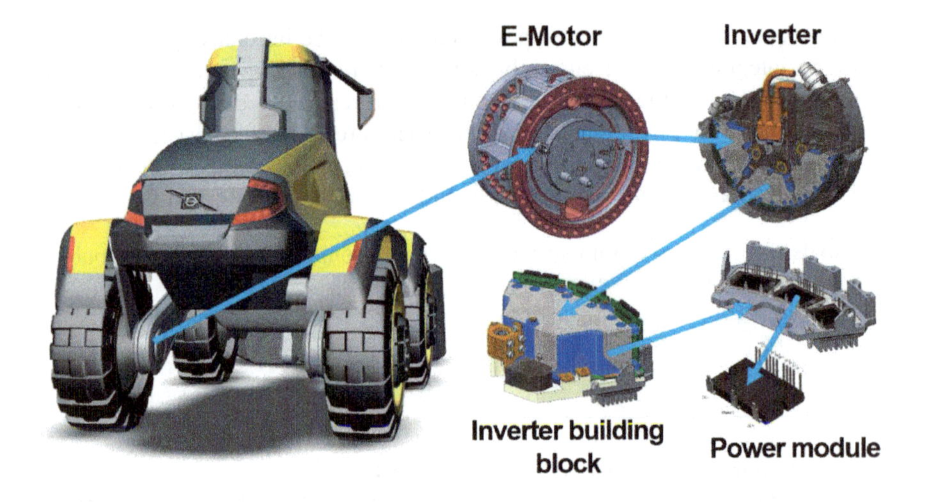

Fig. 1 Overview of the currently developed COSIVU drive-train system including inter alia single-wheel e-motor, inverter with SiC power modules, cooling system and control electronics (VOLVO CE, Fraunhofer IISB)

Furthermore, reliability assessment and optimization of critical components such as the power modules for the inverter will be addressed in this project. The underlying concepts as well as first results will be discussed more in detail in the following chapters.

2 New SiC BJT Power Module

The main components of the inverter unit are three equal inverter building blocks (IBB), a power supply for the base driver, an inverter controller module and an inverter housing. Each IBB in turn consists of a cooling plate, DC-link capacitors, current sensor, base drivers and three paralleled half-bridge modules based on an automotive qualified power module package from Fairchild Semiconductor (FSC), as shown in Fig. 2. The half-bridge modules are composed of SiC bipolar junction transistors (BJT) from Fairchild (part number reference FSICBH017A120) and their anti-parallel SiC diodes from Cree (part number CPW2-1200S050). These components are specified for 1200V and 50A (54A for the diode), which allows to drive motor currents up to 300A at system level.

The power module construction is based on a direct bonded copper (DCB) substrate with aluminum nitride (AlN) used as the ceramic isolator due to its superior thermal conductivity. For the die attach a lead-free soldering process has been used, whereas the electrical connection of the chip topside pads is done with aluminum wire-bonds (300μm for emitter, 150μm for base).

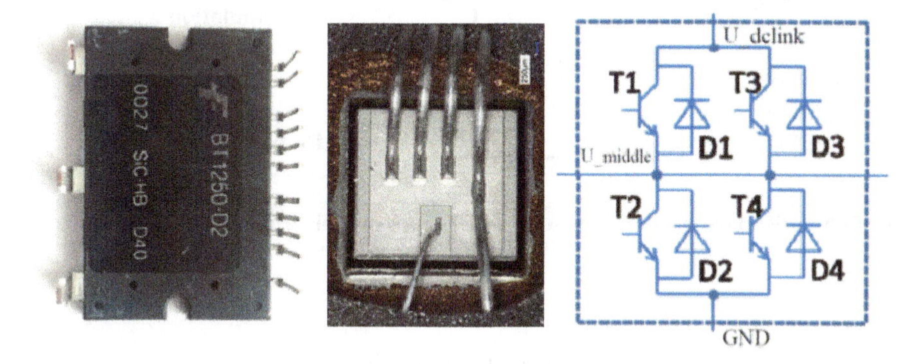

Fig. 2 Photography of the FCS SiC BJT power module (left) and of a SiC BJT (middle) and schematic plan (right)

For the encapsulation an epoxy molding compound (EMC) was applied. It has openings for threaded fasteners to assure an appropriate clamping force for an optimal heat transfer to the cooler. The overall size of the package is 44 mm by 29 mm by 5 mm. Furthermore, a thermistor for temperature indication is integrated into the module. However, in the COSIVU project the virtual junction temperature will be determined additionally to allow thermal impedance spectroscopy measurements for health monitoring purposes [2].

3 Reliability Aspects

3.1 Principle Approach for Lifetime Assessment

The harsh environmental conditions (passive and active temperature cycles, vibrations, shocks etc.), to which commercial vehicles are usually exposed to, clearly pose high risks for the power modules within the traction module. In addition, these systems have high safety requirements, particularly in the present case of multi-motor solutions. For this reason, the potentially critical reliability issues need to be assessed already during the concept and design stages, i.e., when real samples and demonstrators are not available yet. To face this challenge, a lifetime estimation process will be applied for the SiC power modules, as shown in Fig. 3.

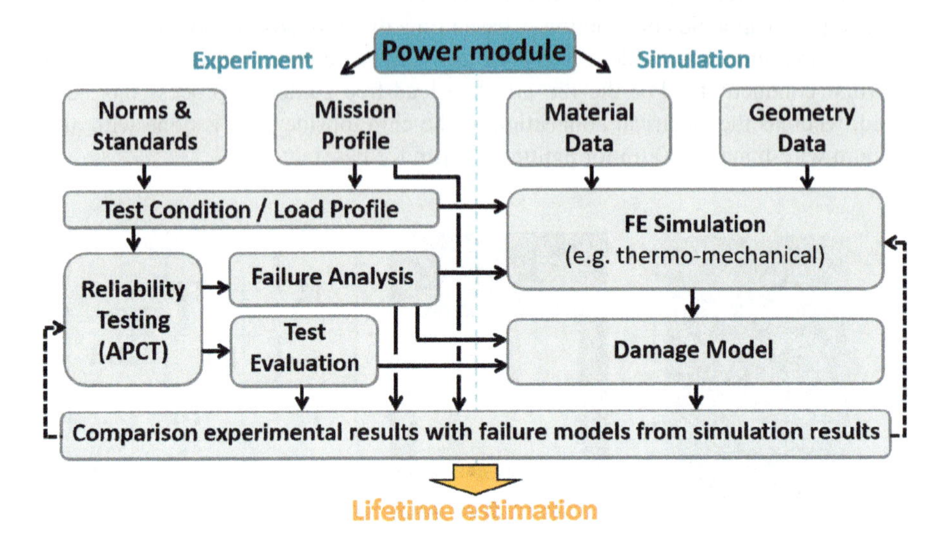

Fig. 3 Overview on the lifetime assessment methodology

Mission profiles, derived from the respective application scenarios for the power module together with relevant norms and standards, provide information about the frame conditions to perform reliability tests as well as the accompanying finite element (FE) simulations. The FE simulations are based on models covering the geometric, material, and load conditions. Practical reliability tests, which in the case of power electronics are conducted in terms of active power cycling tests, in turn provide information about failure mechanism types and their location as well as the statistical failure probability. Lifetime or damage models are linking these results with the simulation results to provide a lifetime estimation of the investigated product. The lifetime models, which need to be chosen or developed for the specific case, are usually based on the underlying physics of failure. Failures in the investigated power modules are mainly expected to be thermo-mechanically induced due to mismatches in the coefficients of thermal expansions (CTE) of the involved materials, such as degradation in the die-attach (solder joint) or in the chip metallization, wire-bond lift-off as well as delamination and cracks within the DCB substrate. However, further failure types such as chip cracking or gate oxide breakthroughs are also possible. Finally, with the knowledge about the weak point's, design rules can be set up to improve the power module reliability and to finally allow the COSIVU system to fulfill its mission under all condition.

3.2 Active Power Cycling Test Preparation

Active power cycling (APC) test are state-of-the-art for performing accelerated lifetime tests on power electronics. They are performed by means of an internal heating of the device under test (DUT) due to a targeted power loss insertion at a high cycle rate. For the SiC power modules, the available APC test bench [3] needs to be adapted in terms of mechanical sample fixation and electrical connection as well as in terms of the LabVIEW based control software.

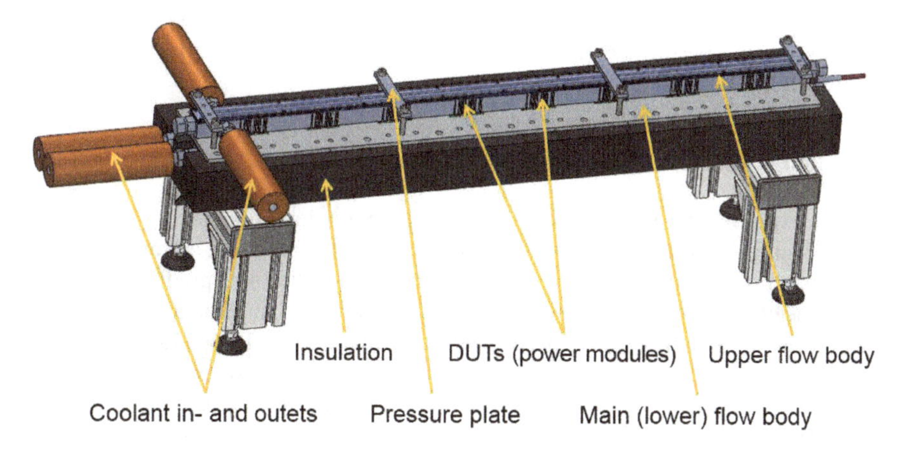

Fig. 4 3D CAD design of temperature controllable sample holder for performing ACP tests on COSIVU SiC modules (for single-sided as well as for double-sided cooling versions)

In Fig. 4 the 3D CAD design of the temperature controllable sample holder system is shown, which is extended in a way to emulate also a double-sided cooling environment by using two independent cooling bodies with the DUT being placed in-between (cf. chapter 4). The thermal simulation results for the aluminum cooling body in single-sided cooling mode are shown in Fig. 5. The coolant temperature was set to 60°C and the power modules are for simplicity reasons represented as simple heat sources (130W at 50A), representing the worst-case scenario which will also be used in the subsequent APC tests.

It can be noticed that with the developed design an even temperature distribution over the cooling body can be achieved with only minor deviations for the outer DUT places, which is important for comparison of the test results of each DUT among each other. Next step within the project will include thermal simulation for the double-sided cooling constellation and adaption of the existing

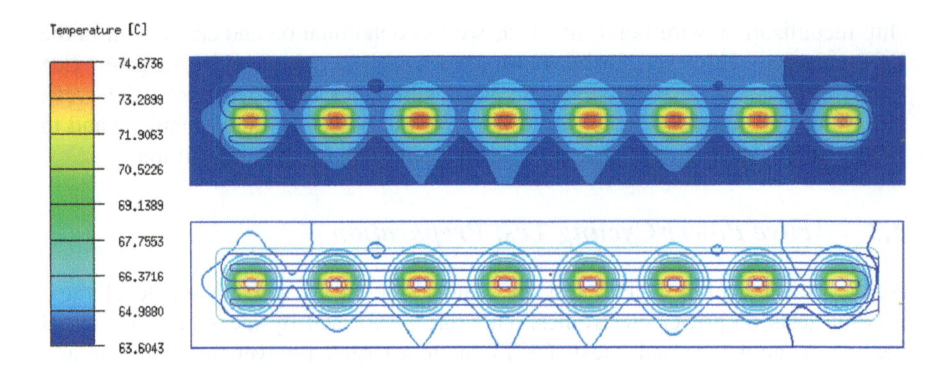

Fig. 5 Thermal simulation of APC cooling body

junction-temperature determination methods to the SiC power modules. Subsequently, power cycling tests on single-sided as well as on double-sided power modules will be performed for benchmarking purposes and for comparison with the numerical simulation results.

3.3 First Thermo-mechanical Simulation Results

Besides experimental investigations numerical simulation is necessary to systematically analyze and evaluate the response of the device under given boundary conditions in order to generate design guidelines for lifetime prediction. These models need to reflect the physics behind the failure mechanisms and have to be reproduced consistently by experiments and simulations. FE simulations have been performed to investigate the stresses and strains induced by processing and operational internal and external thermal loads relevant for the envisaged field of automotive application. Prior to modeling, relevant materials were analyzed regarding temperature and process dependence. An FE model of the module has been generated and material data have been implemented.

To evaluate influences of thermo-mechanical stresses and strains, extensive simulations of the package behavior and non-destructive evaluation have been performed. Simulation runs allow location and monitoring of mechanical stress concentration and accumulating plastic and creep strains over the process steps and through several thermal cycles. A typical process flow with soldering on DCB substrate and transfer molding as well as thermal cycling (TC) between -40 and +150°C were subject of investigation by thermo-mechanical simulation based on FE models for the molded module, generated as depicted in Fig. 6 with colors denoting property regions. The model allows a process simulation with individual materials added analogue to the real technological process to evaluate the individual stress state due to thermal mismatch at various steps of production.

Results of the warpage measurement and simulation for the surface profile across the DCB surface of a SiC power module are depicted in Fig. 7. Warpage is the primary information coming out of the calculations to evaluate production and test cycle results. It is obvious from the simulations that cycling is resulting in

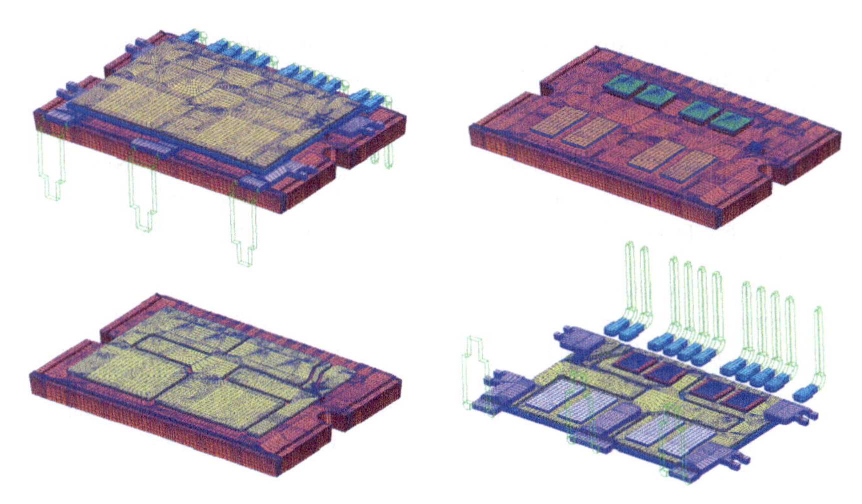

Fig. 6 Geometry and FE model of the SiC power module

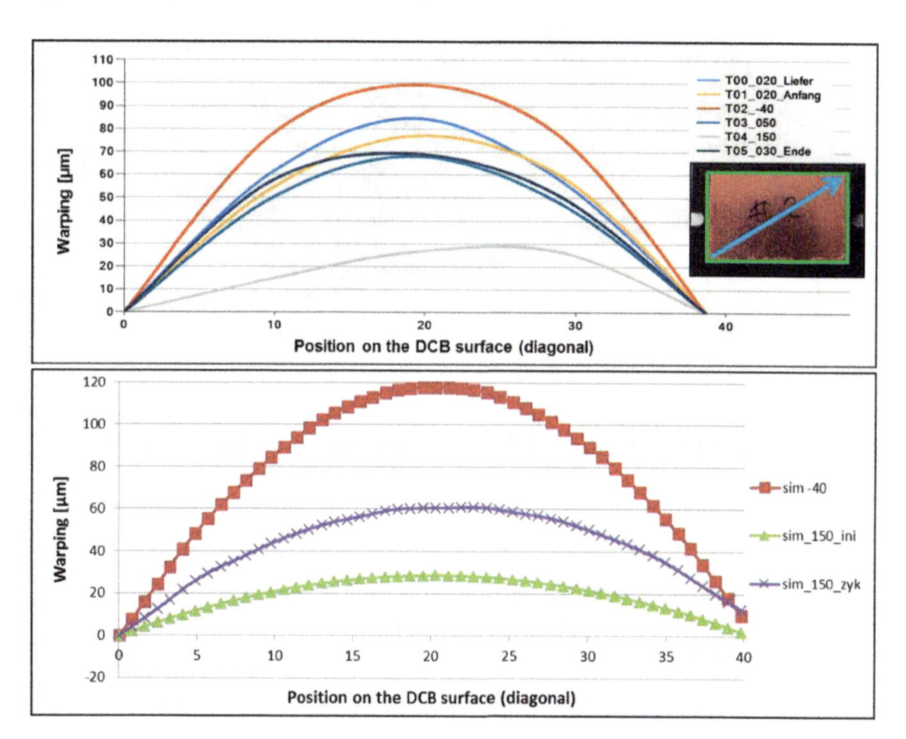

Fig. 7 Warpage measurement (top) and simulation results (bottom) on SiC power module for calibration of FE simulation model

remaining module deformation (sim_150_zyk) originating from mold compound creep. Additionally, TC changes the warped profile for the module (not fixed to any cooling plate). The simulation slightly overestimates the warpage at low temperatures but still reproduces the deformation behavior well.

Furthermore, thermo-mechanical reliability investigations have to address mechanical stress concentrations as well as accumulating equivalent creep strains, the latter serving as a failure criterion for low cycle solder fatigue. Significant strains and stresses evolve in the die attach (Fig. 8). The results indicate that the creep strains are primarily influenced by the high CTE mismatch between DCB and SiC dies.

Accumulating die attach creep will lead to solder fatigue and therefore increased thermal resistance to the substrate. To visualize weak points of the design, parametric FE models have and will be successfully used in the regarding reliability investigations. By identifying process steps and loading conditions mainly contributing to stresses and strains in the module and altering respective parameters, deformations and intrinsic stresses will be minimized in order to avoid excessive stressing of the device.

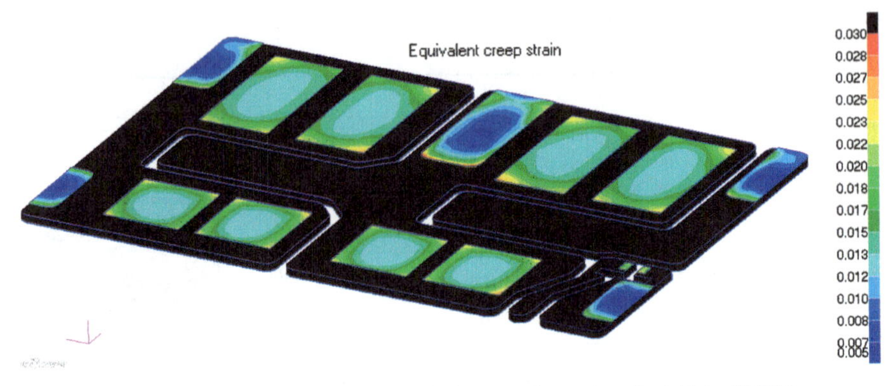

Fig. 8 Equivalent creep strains accumulated after one thermal cycle (-40 to 150°C)

4 New Cooling Concept Based on Double-Sided Cooling

To gain further improvement in the power module lifetime a double-sided cooling concept (^{2}COOL) is investigated within the COSIVU project. In this process, two different types of cooling structures have been evaluated: a simple inline pin-fin structure as well as a sponge-like structure (Fig. 9, left).

Thermal computational fluid dynamics (CFD) analyses have been performed for different coolant flows (5 – 15 l/min at 20°C) for single-sided heating scenario with five heat sources (30W each, four on the corners, and one in the middle). The simulation results for a flow rate of 5 l/min are shown in Fig. 10. For the pin-fin structure the temperature has increased by around 12K at the in- and outlet and by 8…9K in the center, whereas for the sponge-like structure the increase in temperature was slightly lower at the in- and outlet and slightly higher in the center. Further simulation results have also shown a higher pressure drop for the sponge-like structure (approx. 50 mbar vs. 25 mbar for 5 l/min and 470 mbar vs. 155 mbar for 15 l/min) and a predominant flow direction from in- to outlet in both cases.

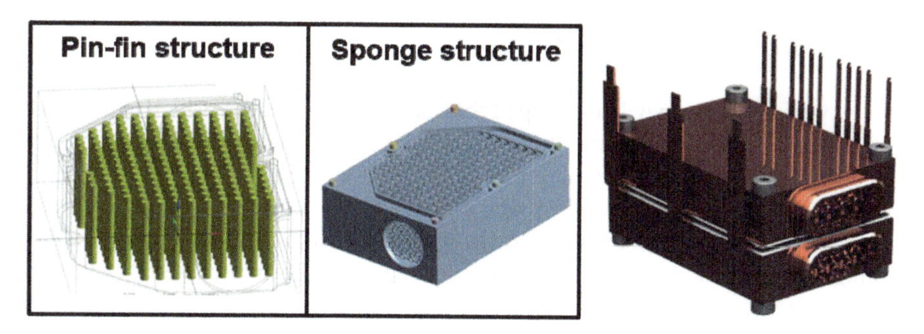

Fig. 9 The two investigated cooling structures (left) and the final module assembly (right)

The simulation results of this initial analysis clearly pointed out that further improvements are required, such as a reduction of the structure height as well as an increase of the vertical mixing of the coolant. To reach the latter one, a tilted sponge structure to force the coolant flow more in the vertical direction will be realized. The double-sided cooling assembly with the new sponge structure design is depicted in Fig. 9, right.

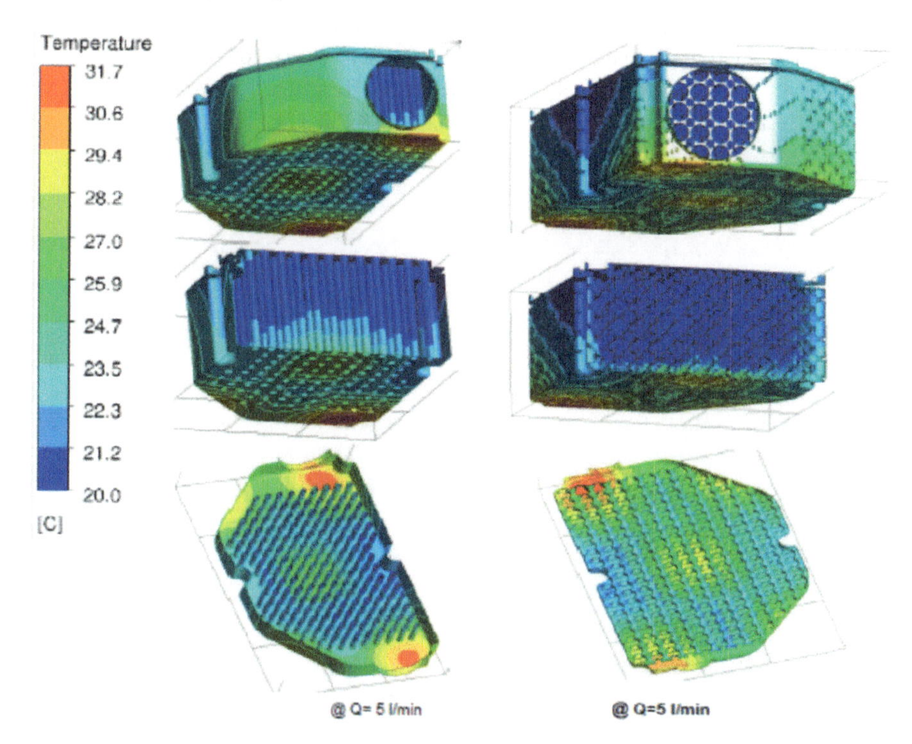

Fig. 10 CFD results for inline pin-fin (left) and sponge-like structure (right)

5 Conclusion

This paper presented a newly developed SiC based power module, which is going to be integrated into a novel drive-train system for commercial vehicles within the frame of the currently running project COSIVU. Furthermore, reliability measures, including the general lifetime assessment process, latest status on power cycling test preparation and first simulation results, as well as investigation results on different cooling structures for double-sided cooling concepts have been discussed.

Acknowledgement. The Authors would like to acknowledge the European Commission for supporting these activities within the project COSIVU under grant agreement number 313980.

References

[1] Rzepka, S., Otto, A.: COSIVU – Compact, Smart and Reliable Drive Unit for Fully Electric Vehicles. Micromaterials and Nanomaterials (15), 116–121 (2013)

[2] Hensler, A., Wingert, D., Herold, C., Lutz, J., Thoben, M.: Thermal impedance spectroscopy of power modules. Microelectronics Reliability (51), 1679–1683 (2011)

[3] Otto, A., Vohra, A., Rzepka, S.: Newly Developed Test Bench for Active Power Cycling Tests. Micromaterials and Nanomaterials (15), 132–134 (2013)

A Framework for Design, Test, and Validation of Electric Car Modules

Mehmed Yüksel, Mohammed Ahmed, Benjamin Girault, Timo Birnschein and Frank Kirchner

Abstract. This paper presents a practical framework and workflow for development and implementation of vehicle hardware and software components. It covers all activities from unit testing of single components to field experiments with the final constructed car. The described framework is based on the rapid control prototyping approach that is used for development and enables modularity in the design of subsystems. The framework was successfully used for the development and implementation of our new electric car concept (EO smart connecting car 2). Performance is analyzed through detailed simulations and experiments. The Framework reduces time and costs significantly for implementing component prototypes of the target system.

Keywords: Electric Vehicle, Hardware-in-the-loop, Software-in-the-loop, real-time control, rapid control prototyping, framework, path following, simulation, x by wire.

1 Introduction

In recent years, tremendous progress has been made in the field of intelligent vehicles for regular traffic. Research on vehicle automation for different automatic driving related tasks and assisted driving has increased during the last years.

Some early developments in electric vehicles include the "Lunar Roving Vehicle" (for Apollo programs 15, 16, and 17) for astronaut mobility on the moon [11].

M. Yüksel(✉) · M. Ahmed · B. Girault · T. Birnschein
DFKI GmbH - Robotics Innovation Center, Robert-Hooke-Straße 5,
28359, Bremen, Germany
e-mail: {mehmed.yueksel,mohammed.ahmed,benjamin.girault,timo.birnschein}@dfki.de

F. Kirchner
DFKI GmbH - Robotics Innovation Center, Department of Mathematics and Computer Science, University of Bremen, Robert-Hooke-Straße 1, 28359, Bremen, Germany
e-mail: frank.kirchner@dfki.de

J. Fischer-Wolfarth and G. Meyer (eds.), *Advanced Microsystems for Automotive Applications 2014*, Lecture Notes in Mobility,
DOI: 10.1007/978-3-319-08087-1_22,

In that vehicle, maneuverability is provided with all wheel steerable (AWS) and all wheel electric drive (AWeD) features. Hiriko [13] is another electric car, which also addresses urban mobility with its foldable construction (2.5–1.5 m), AWeD, and AWS features. Its design is a realization of the *CityCar* concept [12]. ROboMObil from the German Aerospace Center is one of the few autonomous electric car platforms with robotics features [9]. Another direction for autonomous car implementation is using a conventional car as technology carrier platform (e.g. Google Car [8] and BRAiVE car [7]).

In this work, we present a practical framework and workflow for development and implementation of vehicle hardware and software components. This covers all activities from unit testing of components to field experiments with the car. The presented platform was successfully used for the development and implementation of low, middle, and high level control layers.

Fig. 1 EOscc2 (exterior design by David Grünwald) and its *SujeeCar* test platform [3]

2 EO Smart Connecting Car 2

EO[1] smart connecting car 2 (EOscc2) (Fig.1) is a high maneuverable all-electric micro-car designed for crowded cities with very confined parking spaces designed and constructed with robotic features. Its design is based on the fully functional technology demonstrator EO smart connecting car 1 (EOscc1) [1][2]. EOscc2 is developed to demonstrate the possibilities of a sideways driving, turn on the spot (Fig.3), shrink and dock to charging stations. EOscc2 is designed to be a *real* car with respect to practicality, safety, meeting the burden of regulatory compliance, and cost. Its chassis is a welded lightweight very high precision steel tube frame. The car has a modular active AWS. Four actuators are used for steering (between 32° to −92°) and changing the vehicle's ride height (within 16 cm). In addition, each wheel is equipped with a brushless DC (BLDC) wheel hub motor with integrated brakes for maximum efficiency. This results in a drive by wire vehicle that has none of the massive and bulky components. The shrink/fold feature results in a car with footprint of 3.4 m² (~36.6 ft²) through 1.4 m² (~15 ft²) for a minimum parking space. It is design for driving on test areas and public roads safely with a

[1] "EO" means in Latin "I go". The name belongs also to the fully electric vehicle family with robotic features, developed by DFKI since 2010, that includes EO smart connecting cars 1 and 2.

maximum speed of 70 km/h (~44 mph). To enter and exit the car in any folding position, the car is equipped with two *scissor-style* doors.

Autonomous functions are supported with several sensors and cameras. EOscc2 can dock to power outlets, other cars, or extension modules. EOscc2 is planned to park itself, dock to charging stations, undock, leave the parking space safely and realize a pick-up service for the driver within a parking lot. It is highly adaptive with a coupling mechanism for Car2Car, Car2Extender, or Car2ChargingStation [1][2][3].

To have a feasible system implementation for such a complex system, it should be kept as clearly arranged as possible. Therefore, the control layers "perception and planning" (high–level) and "actuation" (middle/low–level) of EOscc2 are separated. The actuation layer is developed in MATLAB/Simulink, tested on a rapid control prototyping (RCP) unit, and runs later on an embedded vehicle control unit (VCU). Precise multi-body dynamics simulation is used for rapid and cost efficient development [3][14]. The high-level control will run on the Robot Construction Kit (ROCK) framework[2].

3 Framework and Workflow Description

The presented framework used for the development and testing of electric cars consists of several tools. These tools are described as follows:

***Computation, modeling, and simulation software*:** it is used for the numerical computations (e.g. kinematics), control of the motors, modeling of the car's state machine. It is also used to analyze and visualize log data. For EOscc2 we used MATLAB/Simulink.

***Multi-body dynamics simulation software*:** to simulate the dynamics of the car and force/torque estimations. We used Adams/View that can also be interfaced with MATLAB/Simulink to perform co-simulations [4] using a detailed 3D model of the car.

***Rapid control prototyping solutions*:** For real-time applications, some software and hardware components are used to control different systems directly. In most of the commercially available RCP components, a Simulink generated model can be directly uploaded. The developer can monitor and record the parameters and modify them online. In our development, we used the dSPACE RTI-Libs and ControlDeskNG software components on the MicroAutoBox II RCP unit.

***Software development tools*:** used for the micro-controller programming. CooCox CoIDE was selected. It is specific for ARM Cortex MCU based microcontrollers and provides debugging tools.

***Documentation system*:** combining the use of *doxygen* for code documentation and the wiki and tracking platform *trac*.

---

[2] www.rock-robotics.org

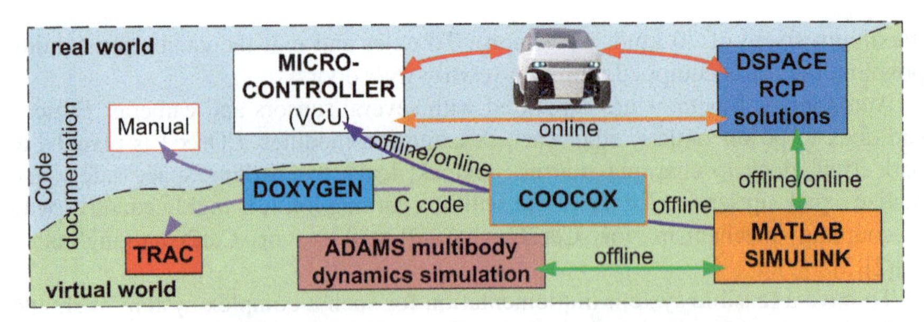

Fig. 2 Tools used within the framework divided into virtual (simulation) and real (hardware)

The main advantage of the chosen tools is their interactivity (Fig.2). They can re-use the code generated by other tools or allow online interaction between them or with the user/developer. Moreover, software-in-the-loop (SIL) tests can be performed. The code that will be later used for the car can be connected to simulation for qualitative validation and also for optimization. Hardware-in-the-loop (HIL) tests are made possible as well.

Thanks to its various I/O interfaces, the RCP unit can be easily connected to motors or input devices. However, the car- in its final configuration- is so complex that the provided interfaces (especially CAN interfaces) are not enough to control the whole system. Despite its many advantages, it was decided not to use the RCP unit in the final version of the car but to use an in-house designed vehicle control unit (VCU) running on a microcontroller.

The chosen RCP (and other most commercially available) components generate code directly for microcontrollers. In theory, this would allow skipping the software development and program the microcontroller as it is done with RCP unit. However, the functionalities are limited and do not cover the hardware specific interfaces, which are the most demanding parts of the programming.

3.1 *Workflow Description*

Car development can be split into three major levels: the low–level software and driver modules that are close to hardware (e.g., motors and electronic devices), the middle–level software that controls the whole car and finally the high–level algorithms for the autonomous and intelligent features. Those levels do not necessarily happen sequentially. Thanks to the SIL and HIL possibilities, high fidelity tests can be performed, independent from each other. This makes the whole development time shorter and in addition reduces the costs.

3.1.1 Low–Level Control of Hardware Components

The low–level software modules manage the communication with the BLDC wheel hub motor controllers and the electric actuators used for the steering, lifting, morphology, or the docking interface. The interfaces, using different CAN protocols (device specific, CANopen), were modelled and developed in MATLAB/Simulink with integrated RCP tools for each device separately and tested on a specific test bench. The required CAN database container (dbc) files were written and imported to the model. The properties of the hub motors and actuators were measured and their control parameters were tuned. Because of its modularity and scalability, the same software module could be re-used within the main model controlling the whole car. Thus, the RCP unit could be used for device control and parameter tuning. The last step is adapting the protocols for the micro-controller, reusing some parts of the RCP unit's code that were already written in C.

3.1.2 Middle–Level Control Components

The middle–level software components cover the computation of the wheel kinematics and the vehicle drive modes. Based on the steering wheel position, accelerator, and brake values steer angles and wheel speeds are computed according to desired drive mode (double Ackermann, diagonal steering, sideways or turn on the spot) (Fig.3).

The wheel suspension kinematics are computed for actuator values to reach the desired steer angle. The algorithms for kinematics and the drive modes were implemented in MATLAB/Simulink. The correctness of the algorithms was verified through co-simulation and later with unit testing. The Simulink blocks were connected for the simulation model of the whole car (in Adams/View) and driving trajectories were tested for each drive mode. These simulations validated the algorithms but were also the base for high-level simulations (e.g., path following) and evaluations of the dynamics of the vehicle. The developed Simulink blocks were then integrated in a larger model and tested directly on SujeeCar (Fig.3) test platform (using the RCP unit). Finally, the RCP unit code was ported to the micro-controller and verified using unit testing. The micro-controller will replace the RCP unit but there will be still the possibility to combine both of them, and having for example the RCP unit in *silent* mode for data logging.

Fig. 3 Ackermann, sideways and turn-on-the-spot drive modes

3.1.3 High Level Algorithms

To provide the driver–assistance systems and go towards the car autonomy, several high level algorithms need to be developed and tested. This includes in particular a cruise control and path following system [3]. It can be directly deployed using the modules previously developed and interfaced with the multi-body simulation software.

As an example, a road-tire interaction model is used with Adams/View to obtain a realistic movement of the car and be able to perform closed-loop testing. Once the algorithm was evaluated and tuned, thanks to the co-simulation, it can be tested on SujeeCar. The first step is an open-loop test in which steering wheel and wheel speed values are generated and loaded on the RCP unit. The scenario can then be run on the car. Later, the algorithm will be tested in closed-loop on the car. The results of the drive tests can as well be recorded and replayed with the simulation software, to verify for instance the accuracy of the dynamics model.

3.2 EOscc2 Test Platform (SujeeCar)

In addition to the several test benches for the wheel hub motors and brake system, a prototype of the car was built to be able to experiment with a real system at an early stage of the development (Fig.1). This was possible as the axles were early finished and fully functional thanks to their modular design. Thus, constructing the demonstrator platform mainly consisted of connecting these two axles with T-slot aluminum profiles. The length of the prototype corresponds to the case where EOssc2 is unfolded. Batteries are placed under the driver's seat and the electronic devices are gathered above the axles. To control the car, a laptop is used to interact with the RCP unit.

4 Performance Analysis Tests

4.1 Drive Mode Change

During drive mode changes or folding, it is important that the wheels roll at the same time to follow the movement. Since the axis of rotation of the steering movement is not perfectly in the middle plane of the wheel, it performs a circle with a so-called *scrub* radius (91 mm for EOscc2). Simulations were done to estimate the actuator forces while folding or switching between Ackermann and Turn-on-the-spot modes. It was tested with and without the wheels rolling at the same time. The results (Fig.4) were used to verify the correctness of the equations of movement.

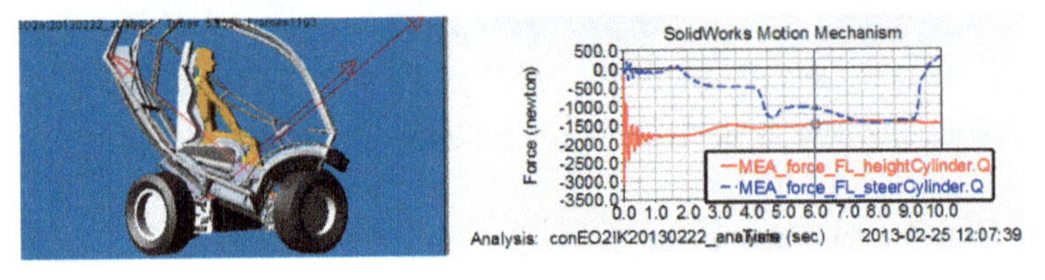

Fig. 4 Estimation of the forces applied to actuators while folding or switching between Ackermann and Turn-on-the-spot steering modes

4.2 Path Following Algorithm

A path following module was designed and implemented [3]. It consists of a proportional input-scaling feedback controller that uses the forward velocity and angular acceleration of the vehicle as control inputs. For this controller, a kinematic vehicle model is used to map from the path curvature specified by the given path to the vehicle's actual steering angle. This controller was interfaced to the rigid-body simulation software and tested in closed loop for different trajectories.

In this simulation, EOscc2 kinematic parameters are: 1.9 m for the wheelbase, the axle track equals 1.35 m, and the wheel radius is 0.325 m. The controller parameters were empirically tuned. The car is commanded to move with constant forward speed of 2.8 m/s. A desired path for the standard test track described in ISO3888-2 [5] is used. From the results (Fig.5), it can be seen that the system converges to the reference trajectory asymptotically. Once the vehicle converges, the vehicle follows the trajectory very closely. The convergence can also be seen in the error graph. These results show that the controller does not attain extremely large values, and is bounded which are essential properties for the real systems.

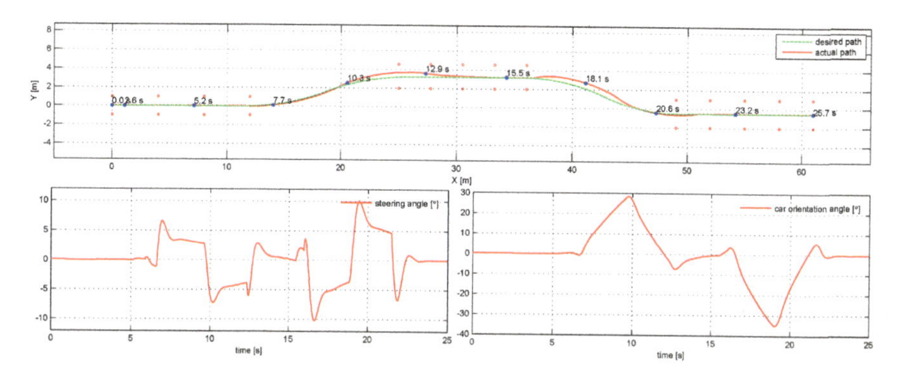

Fig. 5 Path following controller in Adams/View – MATLAB/Simulink cosimulation results for ISO3888-2 double lane change test track

4.3 Line Change Experiment

Because of lack of standard tests for autonomous vehicle driving skills, we borrowed tests from the conventional automotive industry as the Carnegie Mellon Red Team did for DARPA Grand Challenge [10]. We chose the same test method for EOscc2, in order to be able to compare the future results.

To measure maneuverability and stability of the vehicle with influence of different drivers, a lane change test is performed. A test track was constructed according to ISO 3888 Part 2 (Fig.6). For three different drivers, the test scenario was that each driver begins the run in the right lane, swerves into the left, and then immediately cuts back into the right. During the experiment, car and driver data were logged by the RCP unit (67 parameters in total) and the drives were recorded with four cameras (Fig.6). The data can be plotted or exported to the mechanical simulator for replay. Hence, the dynamic parameters of the car can be estimated and the accuracy of the multi-body simulation can be verified. A sample of the results is shown in the Fig.6 for trajectories of three drivers test run.

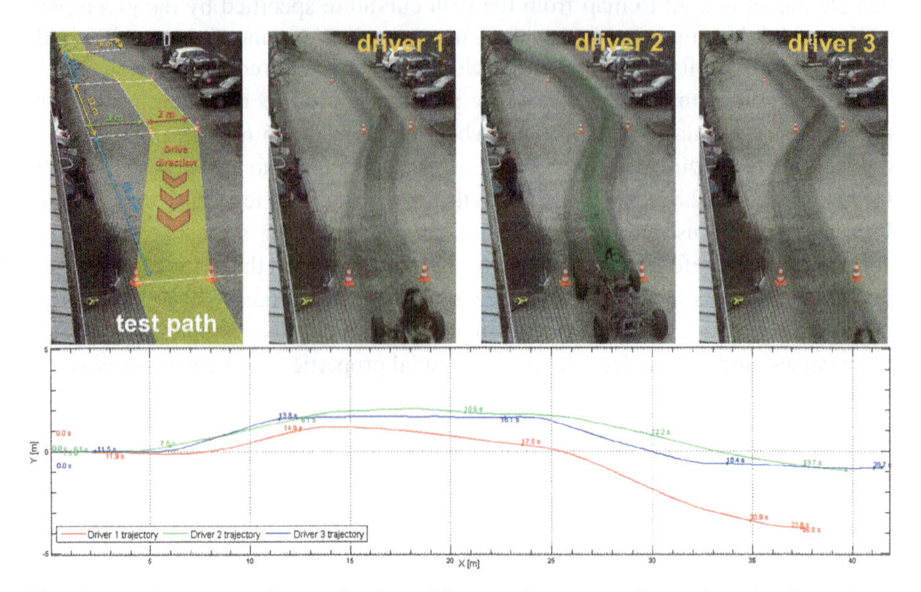

Fig. 6 Lane change experiments for three drivers and corresponding trajectories from simulation replay

5 Conclusion

In this paper, we present a practical framework and workflow for development and implementation and evaluation of vehicle software components. It covers all activities from unit testing of single components to field experiments with the real car. The described framework is based on RCP approach that is used for system development and enables modularity in the design of subsystems. The framework was successfully used for the development and implementation of actuators and motors low–level control in addition to middle–level algorithms. The performance of the designed modules is demonstrated through detailed simulations and experiments. In the conducted simulations, the car software modules are used in a SIL fashion and in the real experiments, the software modules are interfaced to the hardware, tested and verified.

From the presented performance analysis tests of the framework to validate hardware and software components and systems of the car, it is verified that this framework is an effective and adaptive solution as a development and test environment. It reduces effort, time, and costs significantly for implementing component prototypes of the target system. Because of the benefits of this framework and its capabilities as a realtime hardware interface, it is selected for the integration phase of the EOscc2 car as well as the development and optimization of most software components for the car control. The system will be used for future experiments on autonomous driving and for driver assistance modules especially addressing users (driver) modeling and adaption.

Acknowledgment. We thank all team members of the ITEM project [6] (in alphabetical order: Christian Oeckermann, David Grünwald, Haci Bayram Erdinc, Janosch Machowinski, Roman Szczuka, Sujeef Shanmugalingam, Sven Kroffke, Yong-Ho Yoo) who provided valuable comments, ideas, and assistance, which were essential to this study. This work is developed for the EO car which is evolved in the subproject "Innovative Technologies Electromobility (ITEM)" of main project "Model Region Electric Mobility (PMC)" – Module 2 "Intelligent Integration of Electric Mobility" and funded by the German Federal Ministry of Transport, Building and Urban Development (Grant Nr. 03ME0400G). Program coordination is carried out by the National Organization Hydrogen and Fuel Cell Technology (NOW GmbH).

References

[1] Jahn, M., Schröer, M., Yoo, Y.-H., Yüksel, M., Kirchner, F.: Concept of actuation and control for the EO smart connecting car (EO scc). In: Su, C.-Y., Rakheja, S., Liu, H. (eds.) ICIRA 2012, Part I. LNCS (LNAI), vol. 7506, pp. 87–98. Springer, Heidelberg (2012)

[2] Birnschein, T., Kirchner, F., Girault, B., Yüksel, M., Machowinski, J.: An innovative, comprehensive concept for energy efficient electric mobility - EO smart connecting car. In: ENERGYCON 2012. IEEE (2012)

[3] Ahmed, M., Yüksel, M.: Autonomous Path Tracking Steering Controller for EO Smart Connecting Car. In: Proceeding of the World Congress on Multimedia and Computer Science 2013 (ICIAR-13). IEEE (2013)

[4] Ahmed, M., Yoo, Y.-H., Kirchner, F.: A cosimulation framework for design, test and parameter optimization of robotic systems. In: Joint Conference of the 41st International Symposium on Robotics and the 6th German Conference on Robotics (ISR/ROBOTIK 2010). VDE Verlag (2010)

[5] Lundahl, K., Åslund, J., Nielsen, L.: Vehicle dynamics platform, experiments, and modeling aiming at critical maneuver handling. Technical report, Linköping University (2013)

[6] ITEM–Project Web Page, http://robotik.dfki-bremen.de/en/research/projects/item.html (accessed January 08, 2014)

[7] Broggi, A., et al.: Autonomous vehicles control in the VisLab Intercontinental Autonomous Challenge. Annual Reviews in Control (2012)

[8] Guizzo, E., How Google's Self-Driving Car Works. IEEE Spectrum, http://spectrum.ieee.org/automaton/robotics/artificial-intelligence/how-google-self-driving-car-works (accessed January 24, 2014)

[9] ROboMObil-System Architecture and Safety - DLR, http://www.dlr.de/rm/desktopdefault.aspx/tabid-8001/13698_read-34737 (accessed January 27, 2014)

[10] Urmson, C., Whittaker, W., Harbaugh, S., Clark, M., Koon, P.: Testing driver skill for high-speed autonomous vehicles. Computer 39(12) (2006)

[11] Wright, M., Jaques, B., Morea, S.: A Brief History of the Lunar Roving Vehicle. NASA: Marshall Space Flight Center (2002)

[12] Mitchell, W.J., Borroni-Bird, C., Burns, L.D.: Reinventing the Automobile: Personal Urban Mobility for the 21st Century. The MIT Press (2010)

[13] Hiriko, driving mobility, http://www.un.org/esa/dsd/susdevtop-ics/sdt_pdfs/meetings 2012/statements/espiau.pdf (accessed January 26, 2014)

[14] Ahmed, M., Oekermann, C., Kirchner, F.: Cosimulation Environment for Mechanical Design Optimization with Evolutionary Algorithms. In: International Conference on Artificial Intelligence (ICAI 2014). IEEE (2014)

Application of Li-Ion Cell Aging Models on Automotive Electrical Propulsion Cells

**Davide Tarsitano, Federico Perelli, Francesco Braghin,
Ferdinando Luigi Mapelli and Zhi Zhang**

Abstract. In this paper the capacity fade of a Li-Ion battery for electric and hybrid vehicles is studied. The battery lifetime is a crucial characteristic for the usage of this technology on Full Electrical Vehicle (EV) or Hybrid Electrical Vehicle (HEV). Thanks to low costs and easy electronic devices design for small Li-Ion battery tests, many studies have established life prediction models.

The aim of this paper is to verify whether these models can be used to predict capacity fade for a high capacity Li-Ion battery for full electric and hybrid vehicles, or not.

During this study a test bench has been developed to control charge and discharge cell current. At last a comparison between different models will be provided.

Keywords: Li-Ion battery, automation propulsion cell, aging model, battery life prediction.

1 Introduction

At present, hybrid and electric vehicles are increasingly being used both for public and private transport. Thanks to high efficiency electric motors and new innovative powertrain configurations, fuel consumption and pollution can be reduced [1] [2]. Also overhead costs are reduced.

Due to high battery costs, critical aspects for this vehicles design are energy storage sizing and durability. In fact after degradation of performances it is necessary to replace the batteries.

There are many types of technologies for making HEV cells, but up to now the most used one are Li-Ion cells based on Iron Phosphate technology [3]. As consequence, it is useful to investigate which aspects affect battery performances. Many

D. Tarsitano(✉) · F. Perelli · F. Braghin · F.L. Mapelli · Z. Zhang
Politecnico di Milano, Department of Mechanics, via La Masa 1, Milan, Italy
e-mail: {davide.tarsitano,federico.perelli,francesco.braghin,ferdinando.mapelli}@polimi.it,
zhangzhiroom@163.com

J. Fischer-Wolfarth and G. Meyer (eds.), *Advanced Microsystems for Automotive Applications 2014*, Lecture Notes in Mobility,
DOI: 10.1007/978-3-319-08087-1_23,

studies have indicated the total Ah-throughput, the operating temperature, Depth of Discharge (DOD) and C-Rate as major parameters for durability alteration.

In this paper two different models have been considered. The first one [4] suggests a square root relationship between number of cycles and battery capacitance. The second one [5] is more complex and relates the lost capacity to temperature with an Arrhenius law and to a Ah-throughput power

These aging models and many others in literature have been created thanks to tests made on small batteries, for example those which are used by common electrical devices.

In EV/HEV applications, the same technology can be used, but higher levels of capacity are needed because of the higher amount of total energy and power installed. Due to complex chemical reaction and non-linear scale effect, it is impossible to automatically assert that the same models created for small cells can estimate high capacity battery lifetime.

During this study a test bench has been established in order to measure a EV/HEV battery voltage, current and temperature during charge and discharge phases. Thanks to post processing it is possible to evaluate the degradation of battery capacitance. The test bench is composed by a power supply, an electronic load, the data acquisition system and a personal computer as controller. Thanks to the developed software it is possible to control the battery current during all steps of the test. The data acquisition system can store all battery information, such as voltage, current and temperature. A post-processing software has been created in order to calculate all needed parameters, which will be compared to literature models.

2 Test Bench

The test bench is composed by 5 components:

- Battery
- Micro Processor (real-time controller)
- Programmable Power Supply (used as Battery Charger)
- Electronic Load
- Data acquisition system
- PC

As shown in Fig. 1 and Fig. 2 the Micro Processor (MP) reads references from computer output.

During the discharge phase MP controls the electronic load in order to have the battery output current equal to the reference imposed by computer. During charge phase, electronic load does not work; whereas the power supply is switched on, and automatically controls the battery charging current. Thanks to Data acquisition system and to MP, this test bench permits to have a real time control of battery current and voltage. All security devices are included in MP and in power supply.

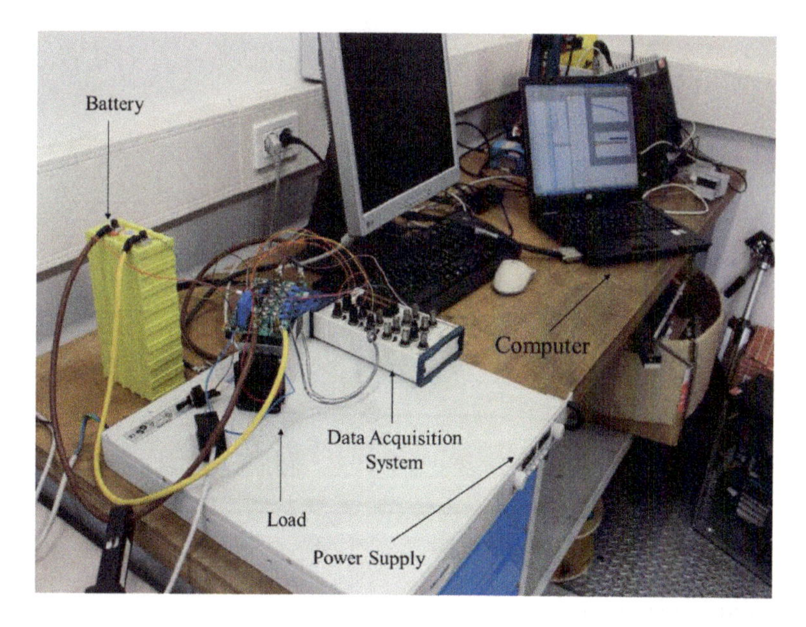

Fig. 1 Test Bench components

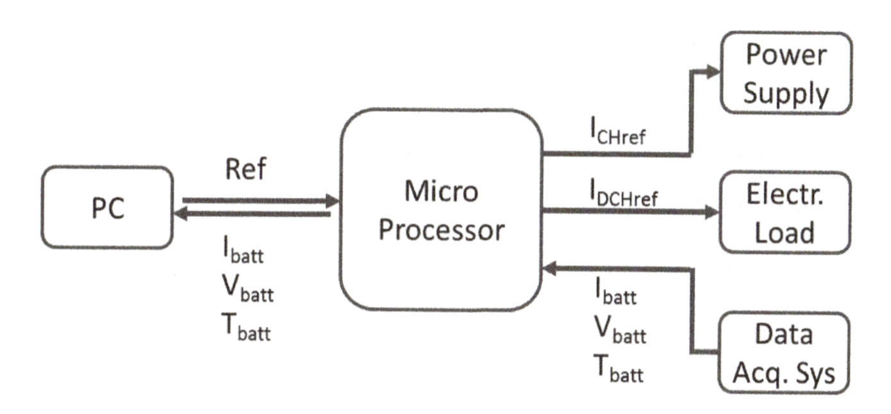

Fig. 2 Test Bench logical scheme

The electronic load is composed by three non-switching MOSFET. As shown in Fig. 3 it is possible to control this component's internal resistance by regulating gate voltage. This component works next to the linear zone, and never reaches the saturation. As consequence the electronic load can regulate the transient current.

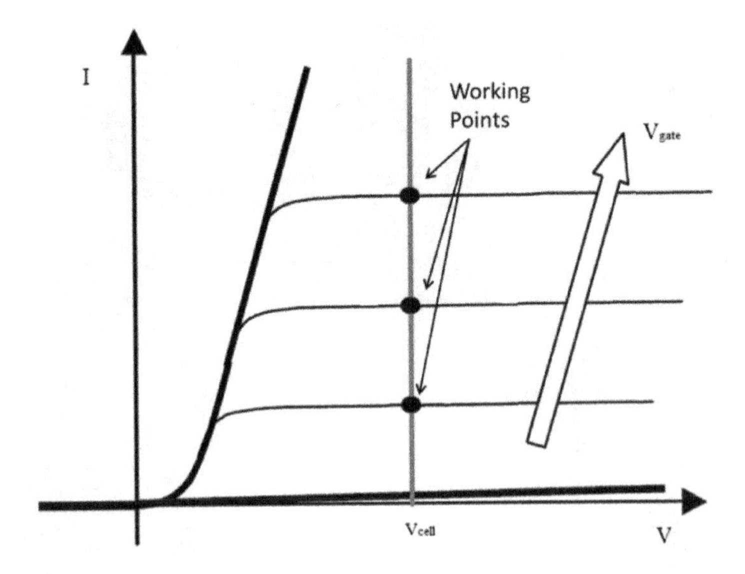

Fig. 3 MOSFET regulation

In Table 1 are shown most important parameters of the Test Bench.

Table 1 Test bench

	Values
Maximum Discharge Current	80A
Maximum Charge Current	120 A
Minimum Voltage	2.5 V
Maximum Voltage	4.2 V

3 Models

The extensive usage of batteries in electric vehicles requires establishing an accurate model [6] [7] [8] of battery aging and life. During the battery lifetime, its performance slowly deteriorates because of the degradation of its electrochemical constituents. Thus power and energy fade associated with impedance rise and capacity loss, respectively. Most of these processes cannot be studied independently and occur at similar timescales, complicating the investigation of aging mechanisms. Many aging models are available in literature [3] [4] [9] but all of them have been developed and validated on small capacity cells (of capacity up to 10Ah).

Two different models have been studied.

3.1 Capacitance Retention

This model [3] identifies the number of cycles as the main influent factor for battery capacitance retention. Many studies on small Li-Ion cells assert that the percentage loss of capacitance trend is linearly extrapolated with the square root of the number of cycles. In such a manner this relationship can be expressed as

$$C_t = C_{in} - d_t\sqrt{N}$$

where N is the number of cycles, d_t is the degradation rate constant, C_{in} and C_t are the percentage initial capacity and percentage current capacity related to nominal one as

$$C_{in} = \frac{c_{in}}{C_n} \qquad C_t = \frac{c_{act}}{C_n}$$

where c_{in} is the capacity calculated on first charge and discharge cycle acquired data, c_{act} is the capacity at a certain number of cycles (N), and C_n is the nominal capacity of the battery.

3.2 Capacity Fade

According to this model [4] the capacity fade can be expressed at each C-Rate as a function of battery temperature, time, Depth Of Discharge (DOD) and number of cycles.

$$Q_{loss} = f(T, t, DOD, CRate, N)$$

where Q_{loss} is the percentage loss of capacity related to nominal one, T is the battery temperature and N is the number of cycles.

Because of the fact that the total Ah-throughput at a fixed C-Rate is proportional to time as

$$A_h = N * DOD * C_n$$

it is possible to relate capacity fade only to temperature and total Ah-throughput.

The model proposed uses an Arrhenius and a power relationship to express respectively temperature and Ah-throughput dependence.

$$Q_{loss} = B * exp\left(\frac{E_a}{RT}\right) * (A_h)^z$$

where B is the pre-exponential factor, E_a is the activation energy in $Jmol^{-1}$, R is the gas constant, T is the absolute temperature and z is the power low factor.

As consequence the percentage capacity at a certain number of cycles can be expressed as

$$C_t = C_{in} - Q_{loss}$$

4 Results

The battery used for the test is a 60 Ah Li-Ion cell. In Table 2 all Data Sheet information are reported.

Table 2 Battery Data Sheet

	Values
Nominal Capacitance	60 Ah
Maximum Voltage	4.0 V
Minimum Voltage	2.8 V
Max Charge Current	3 C
Mac Discharge Current	3 C
Operating Temperature	-45 °C / 85 °C

As shown in Fig. 4 the battery cycle is composed by four parts. The discharge phase is characterized by constant 1C reference current. Once reached the minimum voltage starts the first rest phase. Thereafter the charge phase lasts up to maximum voltage achievement. Discharge current is fixed at 1C. At last the second rest phase finishes the cycle. Every rest phase lasts ten minutes, during which the battery current reference is set to zero.

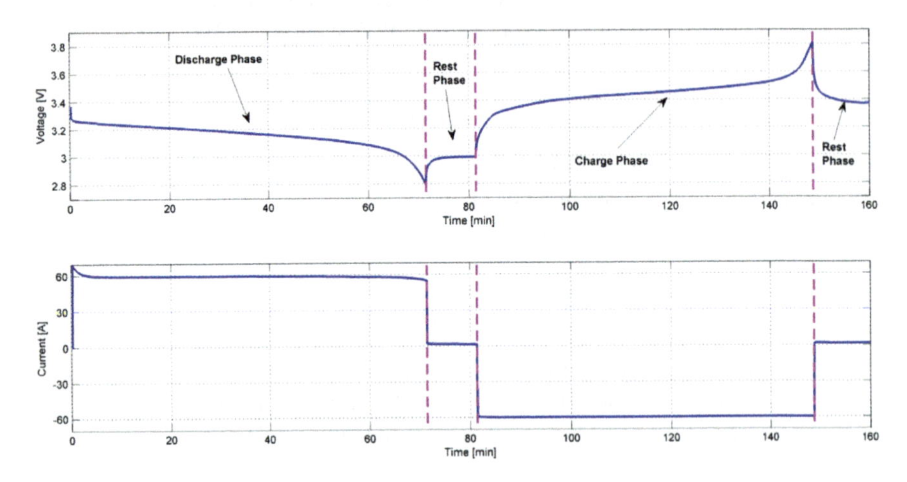

Fig. 4 Cycle Battery Voltage and Current

To investigate battery capacity fade phenomenon a threshold for end-of-life had to be chosen.

Many literature studies set the 80% of the initial capacity as threshold to identify battery lifetime.

At the beginning of this test the initial capacity has been calculated. It results higher than the nominal one (Data Sheet C_n), as it is known to be a normal characteristic for EV/HEV cells. For this type of usage, it is referred to data sheet values to establish whether the battery is still in good condition or not. As consequence it has been decided to set the achievement of the 80% of nominal capacity as threshold for battery end-of-life, even if it is in contrast with literature studies.

4.1 Capacitance Retention

For the Capacitance Retention model d_t and C_{in} parameters have been identified.

$$d_t = 0.7872 \qquad C_{in} = 119{,}4$$

This model's law results

$$C_t = 119{,}4 - 0.7872\sqrt{N}$$

4.2 Capacity Fade

For the Capacity Fade model B and z parameters have been identified,

$$B = 34017 \qquad z = 0{,}5074$$

which bring to the following relationship

$$Q_{loss} = 34017 * exp\left(\frac{E_a}{RT}\right) * (A_h)^{0{,}5074}$$

In Fig. 5 results of post-processing and model implementation are shown. The C_t parameter represents the percentage of battery capacity related to nominal one.

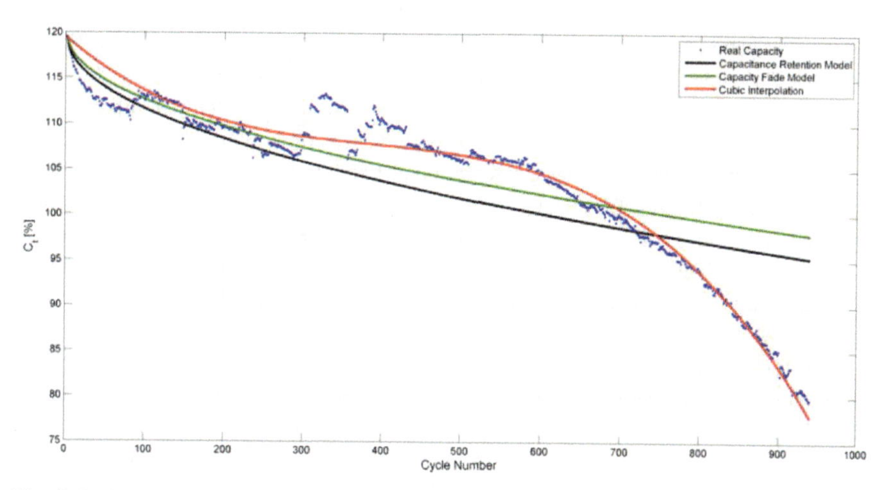

Fig. 5 Cycle Battery Voltage and Current

Both models do not successfully fit empirical data, especially at the end of battery life. In addition to this, C_t trend seems to be more similar to a cubic than to a root square function. Therefore in Fig. 5 is show a cubic interpolation, which clearly represent the real capacity fade trend.

The identified equation has been calculated as interpolation of experimental data with the function

$$C_t = C_{in} + c * N + b * N^2 + a * N^3$$

where a, b and c parameters result

$$a = -1{,}48 * 10^{-7} \quad b = 1{,}72 * 10^{-4} \quad c = -0.074$$

5 Conclusions

The pertinence of small Li-Ion battery aging models on a EV/HEV battery was discussed in this paper. Two models for performance degradation have been compared with results of a large-capacity cell test.

In conclusion it is possible to assert that literature models for low capacity Li-ion batteries cannot be used to predict EV/HEV Li-Ion cell lifetime. For sure complex chemical reaction and non-linear scale effect are the main causes for these different behaviors, and the fact that battery end-of-life threshold is set to the 80% of nominal capacity instead of initial one emphasizes these phenomenon.

A higher number of tests would surely help investigating capacity fade causes and creating new aging models.

References

[1] Mapelli, F.L., Tarsitano, D.: Energy control for plug-in hev with ultracapacitors lithium-ion batteries storage system for fia alternative energy cup race. In: 2010 IEEE Vehicle Power and Propulsion Conference (VPPC), pp. 1–6 (2010)

[2] Mapelli, F.L., Tarsitano, D., Annese, D., Sala, M., Bosia, G.: A study of urban electric bus with a fast charging energy storage system based on lithium battery and supercapacitors. In: 8th International Conference and Exhibition on Ecological Vehicles and Renewable Energies (EVER), pp. 1–9 (2013)

[3] Vetter, J., Novák, P., Wagner, M., Veit, C., Möller, K., Besenhard, J., Winter, M., Wohlfahrt-Mehrens, M., Vogler, C., Hammouche, A.: Aging mechanisms in lithium-ion batteries. Journal of Power Sources 147(1-2), 269–281 (2005)

[4] Uno, M., Tanaka, K.: Accelerated aging testing and cycle life prediction of supercapacitors for alternative battery applications. In: 2011 IEEE 33rd International Telecommunications Energy Conference (INTELEC), pp. 1–6 (2011)

[5] Wang, J., Liu, P., Hicks-Garner, J., Sherman, E., Soukiazian, S., Verbrugge, M., Tataria, H., Musser, J., Finamore, P.: Cycle-life model for graphite-LiFePO4 cells. Journal of Power Sources 196(8), 3942–3948 (2011)

[6] Pradai, E., Di Domenico, D., Creff, Y., Bernard, J., Sauvant-Moynot, V., Huet, F.: Physics-based modeling of LiFeP04-graphite Li-ion batteries for Power and Capacity fade predictions: Application to calendar aging of PHEV and EV. In: Vehicle Power and Propulsion Conference (VPPC), pp. 301–308 (2012)
[7] Sauer, D., Wenzl, H.: Comparison of different approaches for lifetime prediction of electrochemical systems — Using lead-acid batteries as example. Journal of Power Sources 176(2), 534–546 (2008)
[8] Marano, V., Onori, S., Guezennec, Y., Rizzoni, G., Madella, N.: Lithium-ion batteries life estimation for plug-in hybrid electric vehicles. In: IEEE Vehicle Power and Propulsion Conference, pp. 536–543 (2009)
[9] Majima, M., Ujiie, S., Yagasaki, E., Koyama, K., Inazawa, S.: Development of long life lithium ion battery for power storage. Journal of Power Sources 101, 53–59 (2001)

Part V
Components and Systems

Visualisation Functions in Advanced Camera-Based Surround View Systems

Markus Friebe and Johannes Petzold

Abstract. This paper presents an overview of visualisation functions in camera based surround view systems. Video capturing and data transmission from the cameras into image processing unit will be discussed. The article shows how to correct camera pose and lens distortions with special focus on wide angle lenses. The driver will expect well stitched images and a constant and correct brightness in the visualised surrounding area. Image post processing techniques in order to overcome these artefacts are presented and subjectively compared to different state of the art visualisation functions. The paper also presents a video watchdog for important functional safety requirements. Finally an outlook to future visualisation functions is given.

Keywords: 360° view, surround view, advanced driver assistance system, free view point selection, brightness correction, image stitching, projection model.

1 Introduction

Autonomous driving is one of the major aims in automotive applications. A key technology to reach that aim are camera based surround view systems. Besides long range and short range surround sensing sensors like radar or ultrasonic, camera based sensors can cover more detailed information in short distance. The system consists of four fisheye cameras. The camera mounting positions are shown in Figure 1. Using this configuration, the system provides the driver with a 360° field of view around the vehicle. One of the main applications of surround view systems is a virtual camera view from bird's eye perspective, of the top of the vehicle and the surrounding scene. This is very helpful for all kind of parking manoeuvres.

M. Friebe(✉) · J. Petzold
Continental AG, Business Unit ADAS, Segment Surround View,
Johann-Knoch-Gasse 9, 96317 Kronach, Germany
e-mail: {markus.friebe,johannes.petzold}@continental-corporation.com

J. Fischer-Wolfarth and G. Meyer (eds.), *Advanced Microsystems for Automotive Applications 2014*, Lecture Notes in Mobility,
DOI: 10.1007/978-3-319-08087-1_24,

In many state of the art surround view solutions texture information is missing, where image content of adjacent cameras is overlapping. This is visible in Figures 2 and 3.

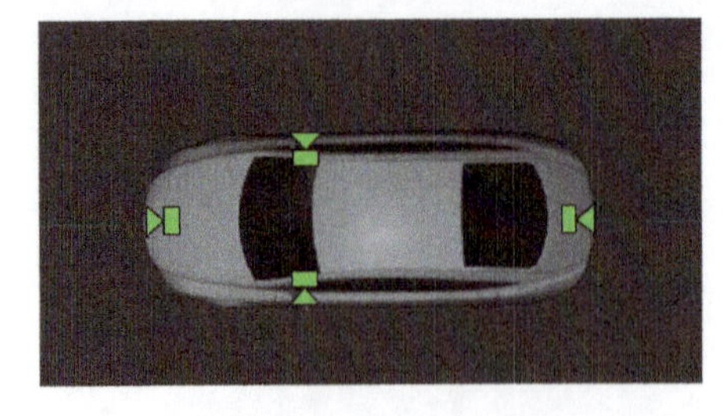

Fig. 1 Camera mounting positions around the car

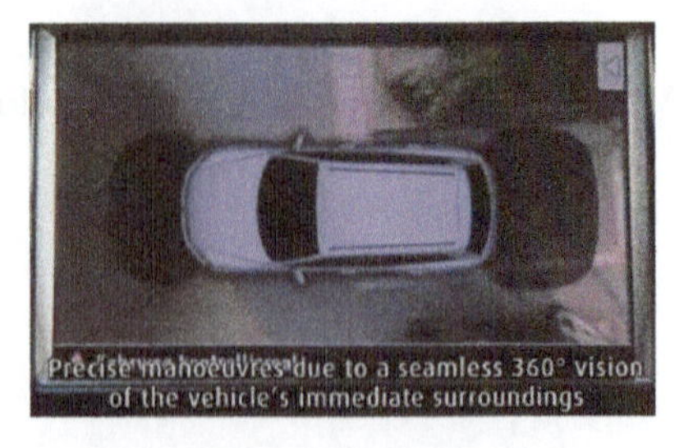

Fig. 2 Valeo 360° bird's eye view [1]

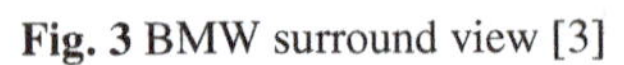

Fig. 3 BMW surround view [3]

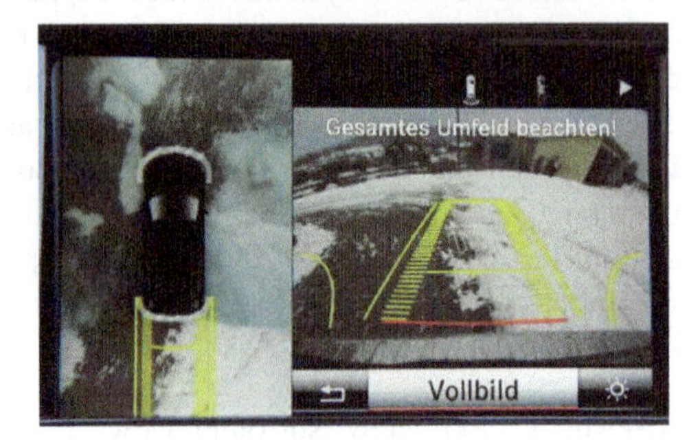

Fig. 4 Mercedes-Benz surround view [2]

Also brightness differs between neighbouring cameras in the final surround view image. Brightness differences are visually annoying in Figure 4.

In chapter 2 the proposed video data chain is shown. Image processing functions are discussed in chapter 3 and finally an outlook is presented in chapter 4.

2 Video Data Chain

The video data chain is composed of video signal sources, the surround view processing unit and the signal sink. The block diagram is visible in Figure 5. The video signals are captured from cameras with wide angle lenses with a horizontal field of view of 185°. The image sensor signals are compressed by a MJPEG encoder [7] and transmitted via Ethernet [4] to the surround view processing unit. The surround view processing unit consists of MJPEG decoder to decompress the video signals and provide the image information to the computer vision and image processing functions. To enable texture back projection, a 3D projection model is used by computer vision and image processing functions. The surround view output video signal is rendered by a graphic processing unit (GPU).

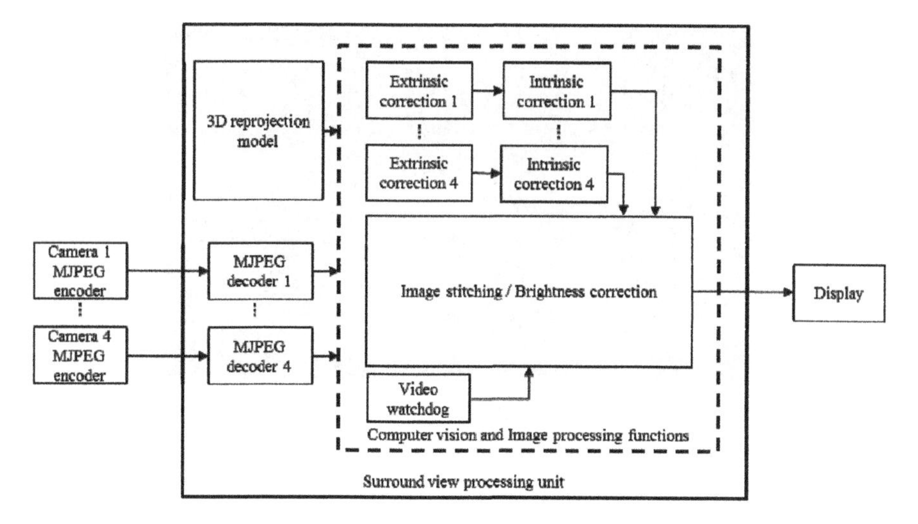

Fig. 5 Video data chain

Functional safety is one of the main requirements in surround view systems. Therefore a video watchdog is necessary. We modified the MJPEG stream in order to send additionally to image signals the frame number information. The decoder decodes the frame number information and the block video watchdog is verifying the frame numbers. In case of lost camera frames, image stitching will be interrupted.

3 Image Processing

This chapter describes the proposed image processing shown in the dashed box in Figure 5. Cameras pose and lens correction with respect to 3D projection model and vehicle coordinate system are presented in chapters 3.1 and 3.2 respectively. Image stitching and brightness correction are discussed in chapters 3.3 and 3.4.

3.1 Correction of Camera Pose and Orientation

In the block computer vision and image processing functions camera pose and orientation is corrected. This is also called extrinsic correction. The 3D projection model, where captured video signals are visualised is shown in Figure 6 and describes a bowl around a virtual vehicle. This model is based on a vehicle coordinate system illustrated in Figure 7. The origin is the middle point of the front axle of the vehicle at the ground plane. The positive x-axis appears from front to rear, the positive y-axis shows from left to right side and the positive z-axis from bottom to up direction of the vehicle.

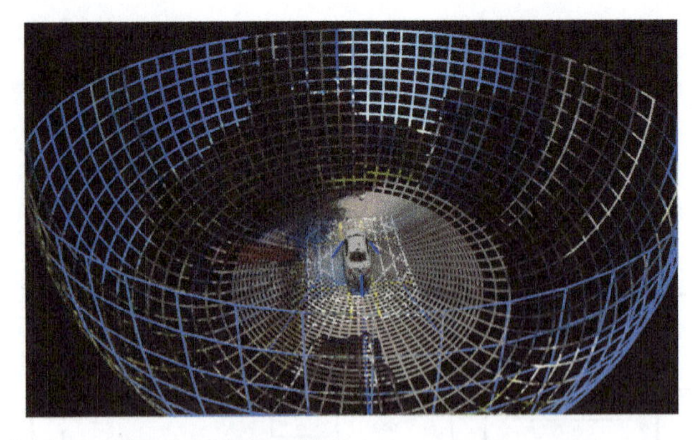

Fig. 6 3D projection model around the vehicle

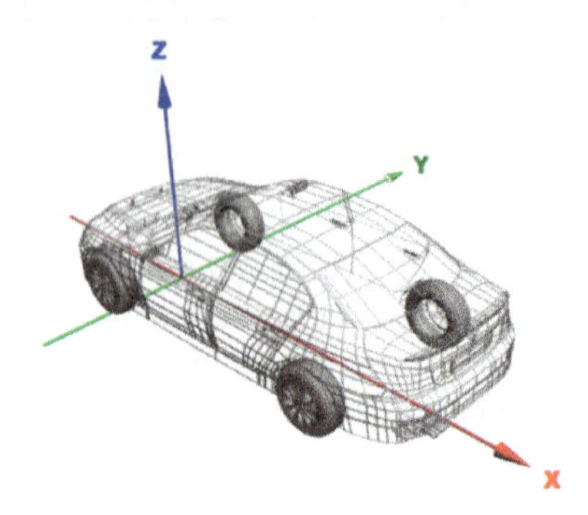

Fig. 7 Vehicle coordinate system

The transformation from 3D point on the 3D projection model in vehicle coordinate system

$$X_V = (X, Y, Z, 1)^T$$

to camera coordinate system

$$X_C = (X_C, Y_C, Z_C, 1)^T$$

can be described with the equation

$$X_C = MX_V \quad \text{with } M = \begin{pmatrix} R & t \\ 0 & 1 \end{pmatrix}. \tag{1}$$

Further explanations can be found in [5]. M represents the parameters from extrinsic calibration, the rotation matrix R and translation vector t. Transformed 3D points from vehicle coordinate system into camera coordinate system are used to correct the following mentioned lens distortions.

3.2 Correction of Lens Distortions

After transformation of vehicle coordinates to camera coordinates, lens distortions are corrected. This is also called intrinsic correction. Different to the proposal in [5], we correct only the radial-symmetric distortions. Between x_C and y_C, a normalised point in camera coordinate system out of equation (1) and radial-symmetric distorted image sample X_R^{radial} exist the following relationship:

$$X_R^{radial} = X_t + \begin{pmatrix} x_C \\ y_c \end{pmatrix} \left(\sum_{i=0}^{K} \lambda_i \cdot r^i \right) \text{ with } r = \sqrt{x_C^{\,2} + y_C^{\,2}}. \qquad (2)$$

λ_i are parameters estimated from intrinsic camera calibration and X_t the vector to the centre pixel of an image. Strong lens distortions are visible in Figure 8 on straight lines from calibration pattern. Using lens correction parameters λ_i and equation (2) a corrected image looks like Figure 9.

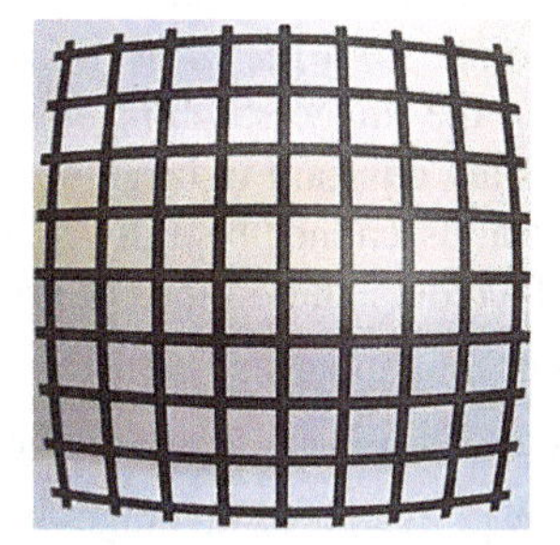

Fig. 8 Image with lens distortions

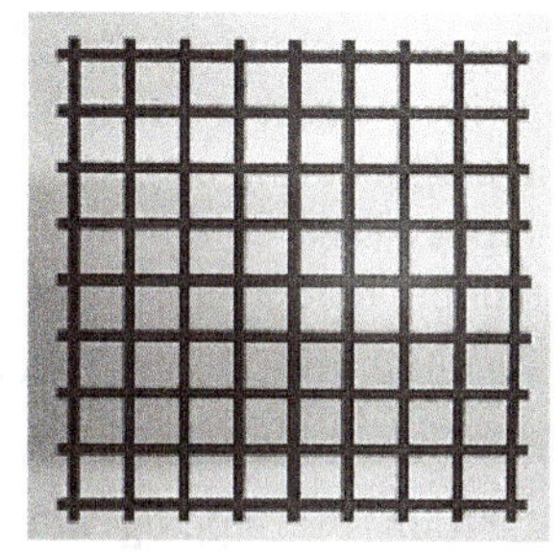

Fig. 9 Image with corrected lens distortions

Strong lens distortions are corrected and straight lines from calibration pattern look straight. For this example, we achieve an average projection error of 0.21 pixels, minimum error of 0.01 pixels and maximum error of 1.04 pixels.

Compared to our intrinsic calibration technique, which only use one input image, the method [6] uses more than one input images. A single image method is an advantage especially for high volume camera manufacturing like in automotive industry.

3.3 Image Stitching and Overlapping Areas

This chapter describes image stitching and the signal processing in overlapping areas of neighbouring cameras. Image stitching needs the relationship between 3D point in the vehicle coordinate system and the image sample for all four cameras. This relationship was previously discussed in chapters 3.1 and 3.2. Using equations (1) and (2), a 3D point X_V uses projected texture information from corresponding lens distorted image sample X_R^{radial}. This projected texture information is visualised including the virtual vehicle in Figure 10.

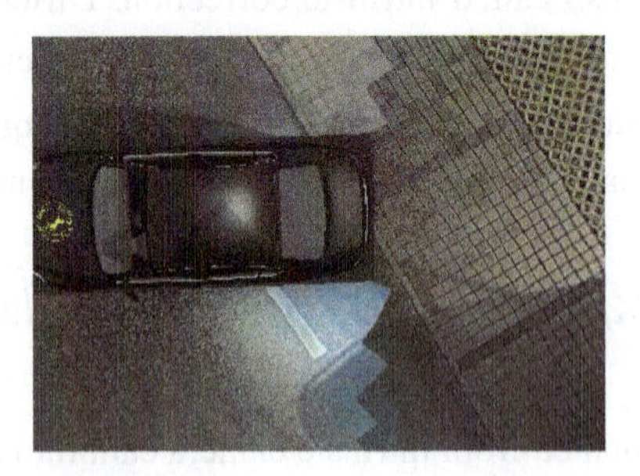

Fig. 10 No overlapping area

The area behind the vehicle is rendered with texture information from the rear camera. The area on left and right side of the vehicle is rendered with texture information from the left and right camera respectively. Image stitching quality depends on the following constraints: The 3D points from the projection model matches with the real world, image samples along the image stitching border are visible in neighbouring cameras and both cameras are constant in brightness, contrast, sharpness and colour. Because these constraints cannot be fully achieved, image stitching borders are visible in Figure 10. To reduce these kinds of artefacts, image stitching overlapping areas A_n between neighbouring camera images shown in Figure 11 are introduced. Overlapping areas transparency is marked in yellow. Within overlapping areas $(X,Y,Z) \in A_n$, weighted average between neighbouring projected camera images is computed using equation:

$$\begin{aligned} I_{overlap}(X,Y,Z) = {} & \alpha(X,Y) \cdot I_1(X,Y,Z) \\ & + (1-\alpha(X,Y)) \cdot I_2(X,Y,Z) \end{aligned} . \tag{3}$$

$\alpha(X,Y)$ represents a weighting coefficient dependent on spatial distance to the border of the overlapping area. $I_1(X,Y,Z)$ and $I_2(X,Y,Z)$ incorporates the projected textures of neighbouring cameras. Image content outside the overlapping area remains unchanged. Simulation result for defined overlapping area is illustrated in Figure 12.

Image stitching borders are less visible compared to non-overlapping areas in Figure 10. A parallel processing unit uses (3) for output video rendering. This speeds up video processing time.

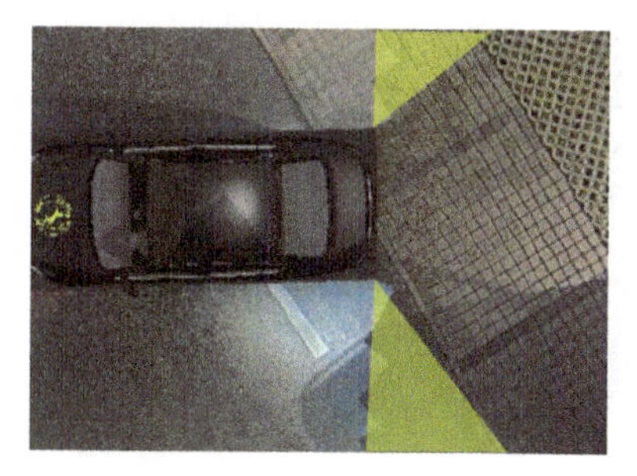

Fig. 11 Visualised overlapping areas

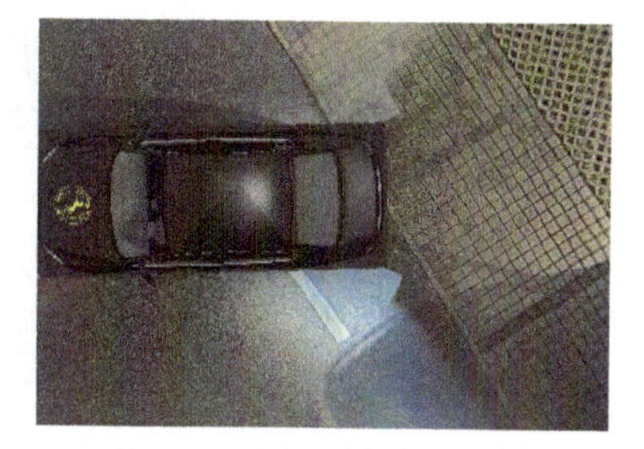

Fig. 12 Used overlapping area

3.4 Brightness Correction

In surround view system each camera has its own automatic white balance and gain control. Due to varying image content and illumination of each camera, different image properties in brightness, contrast and colour occur in stitched images. The result is a surround view image with many different brightness, contrast and colour levels around the vehicle as shown in Figure 13 and Figure 14. Cameras on the right and left hand side are mainly looking on green area and road respectively. White balance parameters from left and right camera are different to parameters from rear and front camera.

Fig. 13 Without brightness correction

Fig. 14 Without brightness correction

The brightness correction algorithm is composed of an analysis and correction component. In the analysis component, for each camera image an average brightness in predefined overlapping areas B_n is computed. One of the four corresponding areas n, where brightness in neighbouring cameras is analysed is shown in Figure 15. The arrows are visualising matched overlapping areas in vehicle coordinate system from left and rear camera. Based on average brightness levels in overlapping areas, correction functions along left and right side of the vehicle are computed. Images captured from front and rear camera remains unchanged.

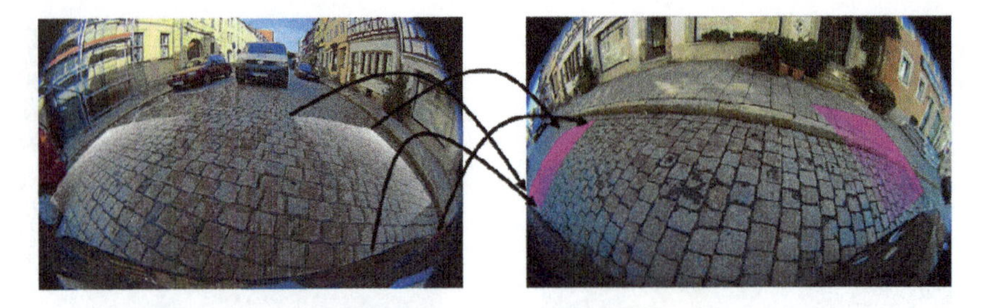

Fig. 15 Corresponding brightness analysis areas

Surround view images after brightness correction are shown in Figure 16 and 17. Here, only the brightness in left and right camera image are adjusted using an additive correction function.

Fig. 16 With brightness correction

Fig. 17 With brightness correction

Compared to Figures 13 and 14 less artefacts in neighbouring cameras are visible, especially in those areas where camera images are stitched. Comparing the results from approach [1] and [2] shown in Figures 2 and 4 respectively, image stitching borders are less visible. A parallel processing unit uses the correction functions for output video rendering. This further enhances video processing time.

3.5 Free Virtual View Point

One of the main advantages of surround view system using a GPU is free virtual view point selection. In addition to the static virtual view on the top of the vehicle shown in Figure 2-4, the driver is able to change the virtual view point for the surround view. The simulation results in Figure 18 and 19 visualise driving along a kerbstone. The driver can change the view point according the driving situation. In challenging crossing traffic situations like shown in Figure 20, a changed virtual view point provides the driver a better view to the surrounding scene than from driver's view perspective.

Fig. 18 Virtual view moved to the right side

Fig. 19 Virtual view moved to the left side

Currently used 3D projection models lead to object distortions of the oncoming vehicle. Changing the 3D projection model in Figure 21 avoids these object distortions.

Fig. 20 Virtual crossing traffic view

Fig. 21 Improved virtual crossing traffic view

Figure 22 shows a virtual view of a parking situation. Currently used 3D projection models lead to object distortions of the wheel. Changing the 3D projection model in Figure 23 avoids this object distortion.

Fig. 22 Virtual parking view

Fig. 23 Improved virtual parking view

As seen in the simulation results, modifiing the 3D projection model provides the driver a more realistic view of the surrounding area.

4 Conclusion and Outlook

This article provided an overview about the video data chain in camera based surround view systems with a special focus on visualisation functions. Simulation results show that using image stitching with overlapping areas and brightness correction improves image quality in surround view images.

Future work will concentrate on animated virtual vehicle, new augmented reality functions and modifications on 3D projection surface in order to enhance visualisation functions for driver assistance surround view systems. Also high efficiency video standards [8] will be investigated to enhance image quality.

References

[1] Valeo, 360° Bird's eye view (2011), http://valeovision.com/innovation
[2] Mercedes-Benz, Die 360-Grad-Kamera in der Praxis (2012), http://blog.mercedes-benz-passion.com/2012/12/die-360-grad-kamera-in-der-praxis/
[3] BMW, Kamerasysteme (2010), http://www.bmw-muenchen.de/sync/showroom_rebrush/de/de/newvehicles/7series/sedan/2010/showroom/safety/camera_system.html
[4] Institute of Electrical and Electronics Engineers, IEEE Standard for Layer 2 Transport Protocol for Time-Sensitive Applications in Bridged Local Area Networks, IEEE P1722 Draft D6 (September 2013)
[5] Winner, H., Hakuli, S., Wolf, G.: Handbuch Fahrerassistenzsysteme., Vieweg + Teubner ISBN 978-3-8348-0287-3, Darmstadt (2009)
[6] Scaramuzza, D., Martinelli, A., Siegwart, R.: A Toolbox for Easy Calibrating Omnidirectional Cameras. In: Proceedings to IEEE International Conference on Intelligent Robots and Systems (IROS 2006), Beijing China, October 7-15 (2006)
[7] ITU, ISO/IEC 10918-1:1993(E) CCIT Recommendation T.81 (1993), http://www.w3.org/Graphics/JPEG/itu-t81.pdf
[8] Sullivan, G.J., Ohm, J., Han, W.-J., Wiegand, T.: Overview of the High Efficiency Video Coding (HEVC) Standard. IEEE Transactions on Circuits and Systems for Video Technology 22(12), 1649–1668 (2012)

Evaluation of Angular Sensor Systems for Rotor Position Sensing of Automotive Electric Drives

Jens Gächter, Jürgen Fabian, Mario Hirz, Andreas Schmidhofer and Heinz Lanzenberger

Abstract. The objective of this paper is the generic approach for assessment and evaluation of different position sensor technologies used for automotive electric traction drive applications. The information of the position angle is essential for the robustness and the quality of the control respectively the reliability of the overall electrified powertrain. The rotor position sensor is a key component, which has to be considered with best application's usage. A systematic testing within automotive requirements can support the technology selection as well as the performance evaluation in a very early stage of the electric drive system development. First an overview about automotive relevant rotor position sensors is given. Based on this the testing parameter, the test bench capabilities and a test result of a selected end of shaft position sensor is described.

Keywords: rotor position sensor, angular position measurement, electric drive, magneto-resistive effect, resolver.

1 Introduction

1.1 Motivation

Most electric machine types for electrified powertrains need the exact angular position of the rotor. Fig. 1 depicts a schematic diagram of a typical electric drive with three phase electric machine architecture. The control of the machine

J. Gächter(✉) · J. Fabian · M. Hirz
Graz University of Technology, Institute of Automotive Engineering,
Inffeldgasse 11/2, 8010 Graz, Austria
e-mail: {jens.gaechter,juergen.fabian,mario.hirz}@tugraz.at

A. Schmidhofer · H. Lanzenberger
MAGNA Powertrain AG & Co KG, Project House Europe,
Frank Stronach-Straße 3, 8200 Albersdorf, Austria
e-mail: {andreas.schmidhofer,heinz.lanzenberger}@magnapowertrain.com

J. Fischer-Wolfarth and G. Meyer (eds.), *Advanced Microsystems for Automotive Applications 2014*, Lecture Notes in Mobility,
DOI: 10.1007/978-3-319-08087-1_25,

depends on the electric machine type. For common synchronous types it is the field-orientated control. Thanks to their high efficiency and high power densities, permanent magnet synchronous machines are actually the preferred choice of OEMs for electric drives in hybrids or full electric vehicles.

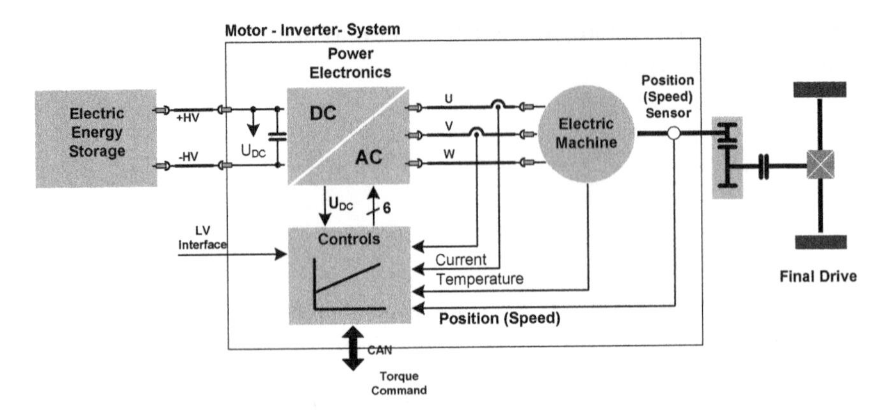

Fig. 1 Powertrain architecture of an electric vehicle

The control generates the optimum phase shift between the stator coil induced magnetic field and the rotor magnetic field. In addition to this mainstream permanent magnet machine several further electric machines are coming into consideration, like switched reluctance, current excited rotor or flux switching types. All these types need a robust, high-precision and high-resolution rotor position measured by the sensor for their controls. Sensor errors or delay will lead to torque ripple, reduced efficiency, reduced maximum torque capability, or in a worst case scenario to a stranded vehicle. Beside the technical features the sensor system must be cost effective. Preliminary studies and testing can follow up all these considerations in an early stage of development. A universal sensor test bench offers the best solution for assessment of relevant rotor position sensor types on reproducible conditions.

1.2 Rotor Position Sensor Types

Different angular sensor technologies are available for measuring the rotor position. The principle of operation can be divided into optical, inductive, magnetic, or capacitive methods. The harsh environmental conditions in the vehicle require a robustness of the sensor system against temperature, vibration, mechanical shock, mechanical tolerances and external magnetic fields. Therefore, optical based sensors are not applicable due to limited temperature of optoelectronics and its lower robustness against pollution. Generally, three main sensor technologies are suitable: magnetic field sensors on the basis of the magneto-resistive principle, sensors on the basis of rotatable alternating magnetic fields (Resolver RES) and eddy-current sensors.

1.2.1 Magneto-Resistive Sensors

A common characteristic of these sensors is the sensing of the amplitude and direction of a magnetic field. Fig. 2a shows a typical configuration. The sensor chip detects the magnetic field of the permanent magnet, which is preferable round in form and mounted coaxially on the rotating shaft. This configuration is only applicable fulfilling the geometrical preconditions at a "free" rotor side of the electric drive. That is why this arrangement is called end of shaft (EOS), and is preferably applied to the non-drive end as realized in [1].

The most common used effects are the anisotropic magneto-resistance (AMR) and the giant magneto-resistance (GMR) effect. One drawback of AMR angular position sensors is their natural limitation to an unambiguously detectable angular range of 0°-180°. In comparison, the GMR sensor has the advantage of the natural 360° range of unambiguity in angle sensing and the higher sensitivity to magnetic fields in speed of rotating sensing. For both technologies the sensor chip consists of two full bridges (see Fig. 2b), of which one supplies a cosine-wave signal and the other a sine signal depending on the external field direction. The angle of rotation is calculated using the arctan function from the ratio of the two sine wave and cosine wave sensor signals.

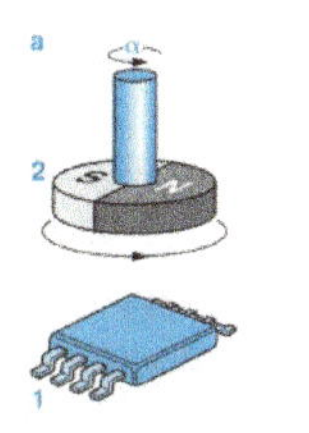

a. Sensor principle

1... sensor chip 2...permanent magnet

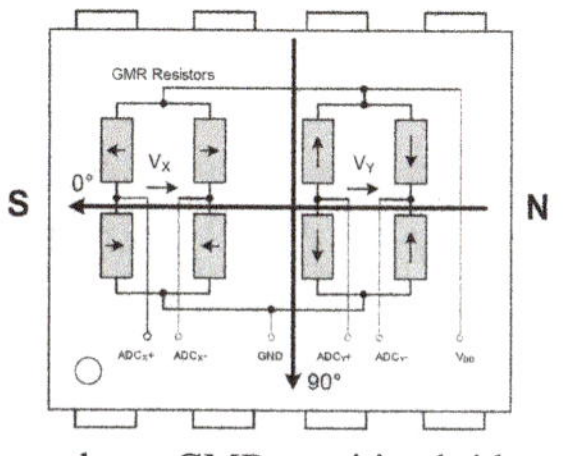

b. GMR sensitive bridges

Fig. 2 a. End of Shaft (EOS) with magneto-resistive sensor principle [2], b. GMR sensor chip with sensitive bridges [3]

1.2.2 Resolver

The resolver is a special realization of a differential transformer with one primary and two secondary windings, where the coupling between primary and secondary winding depends on the angular position between both windings. The construction is similar to an electric machine with the primary winding on the rotor and the secondary windings on the stator. By applying an alternating voltage on the rotating excitation winding, a sine and cosine voltage occurs in the two orthogonal stator windings. An external circuitry provides the excitation signal and processes the two output signals. For automotive an enhanced robust version combining also the variable reluctance principle is used. The excitation winding and the two output windings are placed in the stator. The rotor has a special shape effecting change of the air gap and multiple angle signals in the output windings. Fig. 3 shows the elliptical shape of the rotor that generates twice the signal period per revolution (2X type).

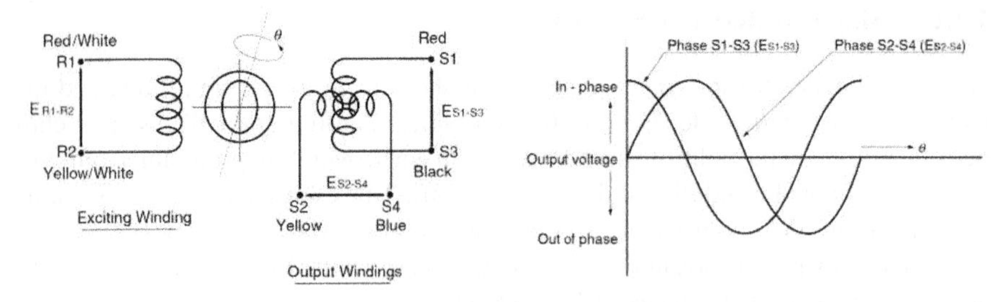

Fig. 3 Schematic resolver principle and corresponding output winding voltages [4]

The excitation winding is fed by high frequency voltages. Beside a buffer amplifier the enhanced hardware includes the Resolver to Digital Conversion (RDC) chip, where the excitation signal generation, the demodulation of the resolver output signals and the diagnosis functionality is implemented. The stator-rotor system is off the circuitry and very robust to harsh environment. The sensor system is well proven and represents therefore the most often applied technology for a coaxial high density integration.

1.2.3 Eddy Current Sensors

When an electrically conductive flat or curved disk (sensor target in aluminum or copper) approaches a coil to which high frequency alternating current has been applied, it has an effect upon the coil's equivalent resistance and its inductance. This variable impedance effect is evaluated by an ASIC sensor electronics, which is located close to the target. High operating frequency is required to keep current consumption low. The eddy current modulation depends on the shape of the target and the relative movement between the target and the sensor. For electric drives a very thin sine-cosine shaped layer is applied to the rotating shaft. The sensor system can be integrated for rotors with high diameter in radial and axial direction. The sensor electronics is fixed mounted and covers 1X of the target structure. Fig. 4 shows a 6X type with axial sensing principle. The sensor is very robust against electromagnetic interference.

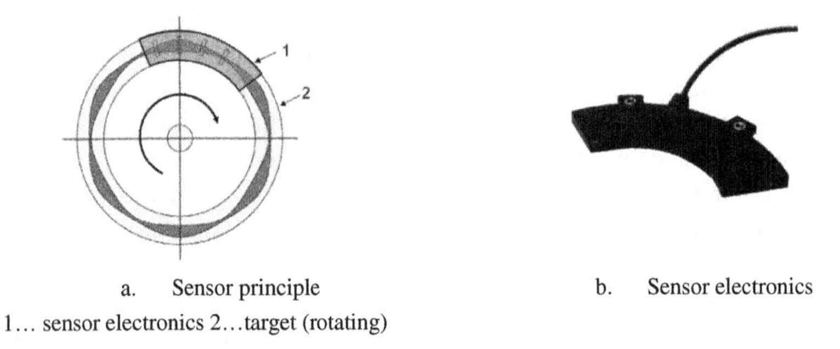

a. Sensor principle

1… sensor electronics 2…target (rotating)

b. Sensor electronics

Fig. 4 a. Eddy current sensor, b. Sensor electronics hardware [5]

2 Sensor Evaluation and Testing

2.1 Automotive Requirements

The speed range of electric drives is defined by the vehicle powertrain topology. Table 1 gives a rough overview about the maximum speed level for different applications. The sensor's requirements are strongly depending on the position of the rotor shaft and the packaging situation.

Table 1 Overview maximum speed for application

	Max speed [rpm]	*Application*	*Electric machine position*
Low Speed Range	2000	Direct Axle Drive	In-wheel motor without gearbox
Medium Speed Range	7500	Hybrid Drive	Crank shaft integrated ICE coupled via AMT, Gen-Sets
High Speed Range	15000	Electric Axle	Coupled via reduction gearbox
Very High Speed Range	24000	Hybrid Drive Electric Drive	Belt-driven integrated Coupled via high gear ratio reduction gearbox

Table 2 shows the sensor system key requirements covering all electric drive configurations. The ambient temperature is the main environmental requirement beside the noise vibration and harshness (NVH). Depending on packaging the sensing device is exposed to temporarily up to 180°C, the corresponding evaluation electronics can be located at lower temperature for instance in the inverter control electronics.

Table 2 Sensor overall specification

	min	*max*	*tolerance*	*unit*
Total speed range *)	-24000	+24000	< 1%	rpm
Position range	0	360	1.5	°el
Ambient temperature sensing device	-40	+160 (180 temp)	-	°C
Ambient temperature electronics	-40	+125	-	°C

*) usually there is a preferred direction for complete range, for instance ICE coupled machines

The mechanical tolerances for the sensor system integration are also fundamental with respect to the series production. That is why it is very important to determine the influence for accuracy due to geometrical mounting variations.

2.2 Test Bench Requirements

For a comprehensive sensor evaluation and assessment, the sensor testing parameters are defined to verify the functionality of the sensor device (=device under test, DUT), as shown in Table 3. The testing is emphasized to investigate the effect of displacement in three dimensional variation including tilt considering mechanical effects. An additional requirement is enabling the test of most common rotor position sensor types with compatible mounting fixtures.

Table 3 Test bench specification

	min	*max*	*accuracy*	*unit*
Total speed range	-24000	+24000	< 1%	rpm
Speed gradient within 0-6000 rpm	-10000	+10000	-	rpm/s
Position reference sensor range	0	360	< 0.1 @ 24000 rpm	° mech
Ambient temperature DUT (thermal chamber)	-40	+160 (180 temp)	< 1	°C
Temperature gradient DUT ambient	-5	+5	-	K/min
Axial displacement (z)	0	+15	0,01	mm
Lateral displacement (x, y from center position)	-15	+15	0,01	mm
Tilt	-45	+45	$< 10^{-5}$	° mech

2.3 Test Bench Setup

Fig. 5 depicts the block chart of the test bench. The DUT is mounted to a special fixture comprising minimized thermal expansion effects. The size of the thermal chamber for DUT is designed to test maximum 160x160 mm PCBs populated for instance by an EOS sensor chip. The maximum value of the reference sensor signal at maximum speed is 1.44 MS/s. For DUT testing, the data acquisition unit is capable of sampling with maximal 5 MS/s depending on speed and DUT resolution.

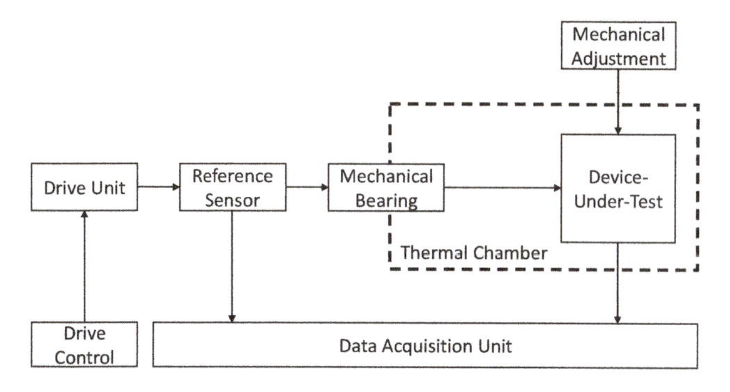

Fig. 5 Test bench overview

Fig. 6 shows the test bench set up for the EOS sensor measurement. The permanent magnet is carried by a non-magnetic material adapter. Due to the flexible mounting fixation sensor configurations with different magnets can be investigated very efficiently. The PCB including the sensor chip is mounted to a special holder coupled with the mechanical adjustment mechanism.

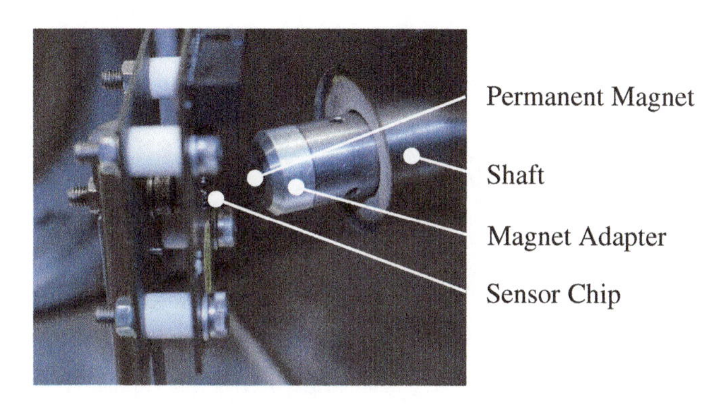

Fig. 6 End of shaft measurement arrangement

The test setup comprises also the resolver system measurement capabilities with a generic RDC and interface electronics similar to the EOS measurement for easy implementation and testing. In this case the resolver rotor is mounted on a special adapter on the axle and the resolver stator is fixed with a special holder coupled with the mechanical adjustment mechanism.

3 Measurement Method and Results

By comparing the results of the reference sensor and the angle calculation of the DUT, it is possible to obtain and evaluate the characteristics of the tested sensor principle. Due to the difficulty of evaluating the angle measurement error by obtaining the typical ramp signals, the error is calculated with the quality factor *E*. This factor represents the integral of the absolute value from the angle difference of the reference sensor φ_{Ref} and the DUT φ_{DUT} over one period *T*:

$$E = \int_{t}^{t+T} |\varphi_{Ref} - \varphi_{DUT}|\, dt \tag{1}$$

The advantage of this definition is that a single value can be used to determine the quality of the angular position sensor signal. For online calculation of the quality factor *E* in the data acquisition unit, it is necessary to perform an angle offset compensation between the reference sensor and the DUT. Exemplary, a measurement scenario of a very strong misaligned AMR-sensor can be seen in Fig. 7, where the sine and cosine sensor signals (at the top), and the calculated angle (at the bottom) are illustrated. Due to the strong misalignment through the adjustment device, the deviation of the ideal sine shape is very large. This results in a significant error in relation to the reference sensor signal.

It is also important to obtain the different signal offsets and the different amplitudes in the sine and cosine signals, because this causes an additional angular error and has to be compensated in the data acquisition unit.

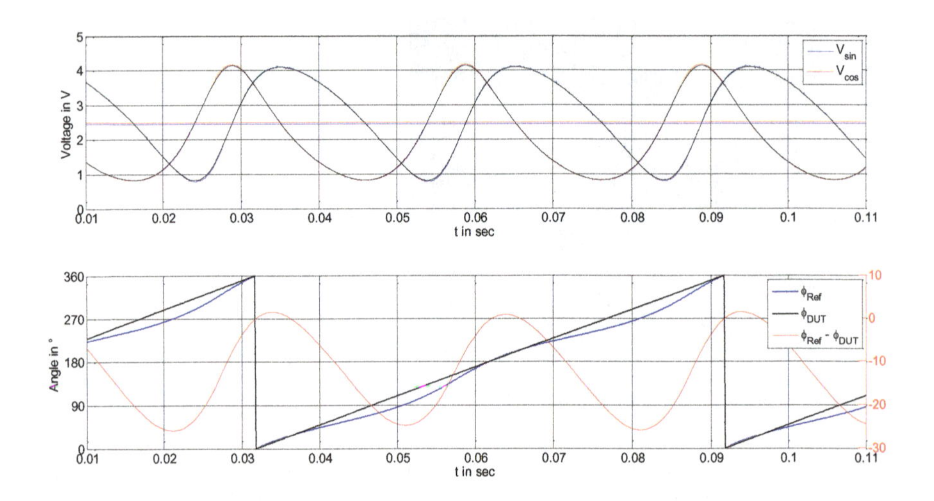

Fig. 7 Misaligned AMR Sensor and corresponding angle characteristics

By adjustments via the adjusting device the angular error can be reduced step-by-step using the relationship (2). The optimal alignment can be reached, the so called sweet spot of the sensor. As shown in Fig. 8, this leads to nearly perfect matched curves of the reference and the DUT angles and the angle measurement error in this position is smaller than a half degree.

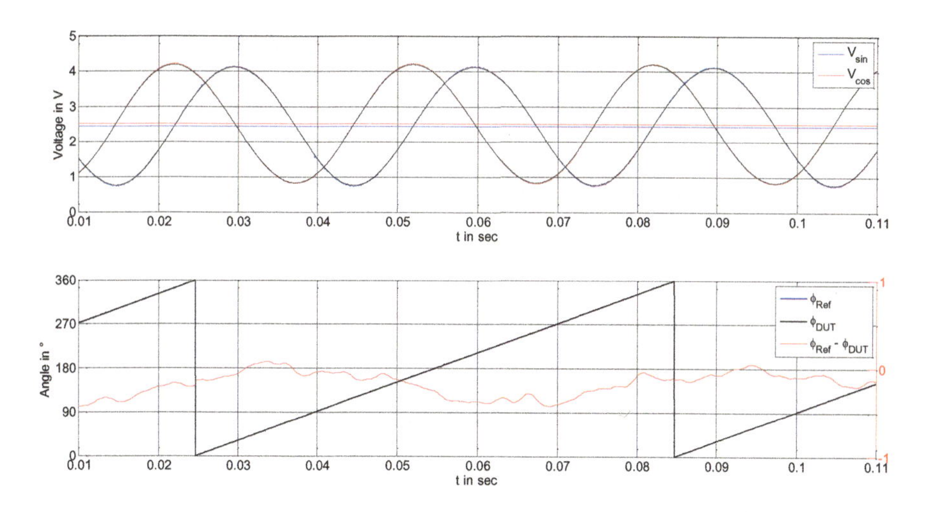

Fig. 8 Aligned AMR sensor and corresponding angle characteristics

Another possibility of the test bench in combination with the adjustment device is to examine the sensitivity of the sensor in terms of displacement in lateral x- and y-direction. In this scenario the error *E* is also calculated according to formula (1), as a function of the displacement in both directions, as depicted in Fig. 9.

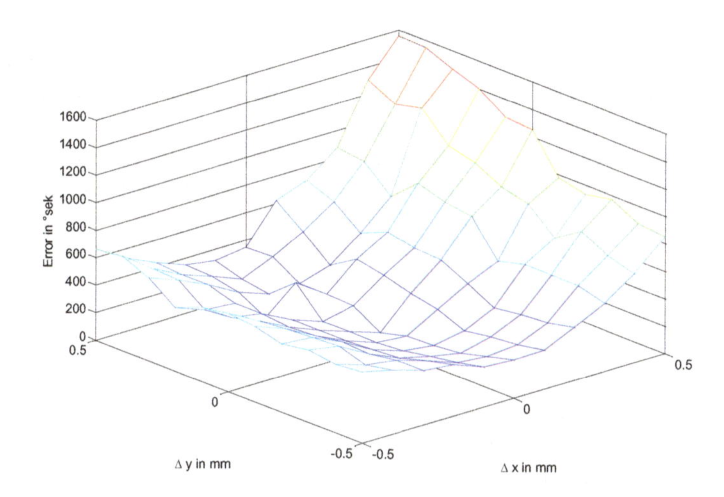

Fig. 9 Sensor characteristic as a function of displacement in x- and y-direction

The measurement scenario was started in the sweet spot where the quality factor *E* has its minimum. From this position, the displacement of the sensor was varied in 0.1 mm steps in both directions. This means that the value of the quality factor is lowest in the origin. Unexpectedly, the sensor – in this case the AMR sensor – is more sensitive along the x- rather than the y-direction, which means that the error is not symmetric in relation to displacement in x- and y-direction. These results are important for emulating mechanical tolerances.

4 Conclusion and Outlook

This paper describes an experimental method for performance evaluation of automotive angular sensors. Most common sensor principles and the automotive requirements are summarized. With respect to this, a test bench is established for selecting the most suitable sensor system under automotive conditions. The investigated parameters are ambient temperature, displacement in axial and lateral direction, tilt, and supply voltage. First results of an AMR based end-of-shaft sensor are shown. Displacements in x- and y-direction effect into high sensitive dependency for the angle error. Examining all further relevant parameter leads to an extensive and unmanageable amount of measurement scenarios. Further investigations are focused to methods for determination of dominating parameters and their limitations. More sensor types, especially resolvers will be evaluated on the test bench.

References

[1] Schmidhofer, A., Horvat, J., Gabriel, T., Lanzenberger, H., Prix, D., Bichler, M.: Highly Integrated Power Electronics for a 48 V Hybrid Drive Application. In: EPE 2013 ECCE Europe15th European Conference on Power Electronics and Applications, Lille, France, September 3-5 (2013)

[2] Bosch, R.: Bosch Automotive Electrics and Automotive Electronics: Systems and Components, Networking and Hybrid Drive, 5th edn. Springer Vieweg, Plochingen (2014) ISBN 978-3-658-01784-2

[3] Angle Sensor – GMR-Base Angle Sensor Infineon TLE5012, Datasheet, V 1.1, page 10 (2012), http://www.infineon.com

[4] TamagawaSinglesyn Datasheet, Catalogue No. 12 -1579N2, http://www.bomatec.ch

[5] Ebbesson, C.: Rotatory Position Sensors – Comparative study of different rotatory position sensors for electrical machines used in an hybrid electric vehicle application. Master thesis, Lund University, Department of Industrial Electrical Engineering and Automation, Sweden (2011)

Future Trends of Advanced Power Electronics and Control Systems for Electric Vehicles

Jürgen Fabian, Jens Gächter and Mario Hirz

Abstract. Nowadays, advanced power electronics have efficiencies of more than 95 % and give enhanced possibilities of power conversion. Nevertheless, the efficiency of electric propulsion systems can be raised in the future, but the potential for efficiency increase in electric systems is limited. Ongoing advances in power electronics can support an improvement of control systems within automotive drive units. New developed or designed components enable systems that are more efficient as well as effective control strategies. In this paper, advantages and disadvantages of different configurations are specified, evaluated and discussed. Practical examples of electric converter systems are given explicitly and conclusions for the future are made.

Keywords: power electronics, control systems, efficiency, converter technologies, electric vehicles.

1 Introduction

Discussions about the optimal technology of propulsion systems for future ground vehicles have been raising over the last few years. Those who are advocating conventional internal combustion engines are faced with the fact that fossil fuels are limited. Others favour hydrogen fuel as the solution for the future, either in combination with combustion engines or as an energy carrier for fuel cells. Finally, there are battery-electric or hybrid propulsion systems in use, gaining more and more popularity worldwide. However, the change to electric technologies does not only include a modification of drive train components, it requires a fundamental technology turnaround by implementation of a complex system with different characteristics [1], as shown in Figure 1 and Table 1.

J. Fabian(✉) · J. Gächter · M. Hirz
Graz University of Technology, Institute of Automotive Engineering,
Inffeldgasse 11/2, 8010 Graz, Austria
e-mail: {juergen.fabian,jens.gaechter,mario.hirz}@tugraz.at

J. Fischer-Wolfarth and G. Meyer (eds.), *Advanced Microsystems for Automotive Applications 2014*, Lecture Notes in Mobility,
DOI: 10.1007/978-3-319-08087-1_26, © Springer International Publishing Switzerland 2014

Fig. 1 Components of battery electric vehicles (Hybrid4AII i-BSG 48V) [2]

Table 1 Changes in the drive train of electric vehicles compared to vehicles with conventional internal combustion engines [3]

Components not needed	*Adapted components*	*Additional components*
Combustion engine	Transmission	Electric machine
Fuel-injection system	Wheel suspension	Power electronics
Tank	Drive line	Battery system
Clutch	Air conditioning	
Exhaust-gas system	Water pump	
Auxiliary equipment	Thermal insulation	

Although OEMs are currently able to offer and deliver green technologies, future market predictions still remain uncertain. As a long-term commitment for clean mobility beyond internal combustion engines, the hybrid electrical vehicle

Vehicles	Power Electronics	Onboard Charger (AC)	Battery: Module	Battery: Battery	Battery: GEN2	Battery: GEN3	E-Motors
E-Cell Sprinter, AMG SLS	80/100kW	22kW	6 Modules à 33 Z				TEM
PI-Hybrid S-Class		3,3 kW	6 Modules à 20 Z				GEM 80kW
Hybrid E-Class S-Class	35/55kW	Not necessary	4 Modules à20 Z				20kW
E-Cell Smart Electric drive		standard 3,3kW; optional 22kW	3 Modules a'33 Z				

Fig. 2 A modular system as precondition for economic efficiency [5]

shows the highest market share for alternative powertrains due to its wide driving range and high flexibility [4]. Different practical examples of electric and hybrid modular systems are shown in Figure 2.

2 Developments in Power Electronics

In general, power electronics refers to conversion and control of electric power. Active electronic circuits operate with solid-state semiconductor devices where these power electronic components, such as transistors, act as switches. Nowadays, power electronics have efficiencies of more than 95 % and give advanced possibilities of power conversion. The converter, which feeds the electrical machine, plays a key role in the propulsion system. The complex and dynamic control of electric machines fall to the converter, as well as the battery management, and the charging method [3], [6-8].

A further trend to improve power density can be seen in an integration of the electronic components on the electric machine thus avoiding connecting cables, connectors, and contacts to reduce weight, and losses. Mechanical vibrations coming from the electric machine are of disadvantage for the power electronics system in terms of life time expectancy.

Examples of energy consumption of battery-electric vehicles are given in Figure 3. To achieve a high efficiency of the machine, the switching frequency of the electronic components has to be high, as illustrated in Figure 4. It compares the old technology represented by a pulsed switch, whereas the future technology is based on power electronics with a special control (e.g. with pulse width modulation, PWM).

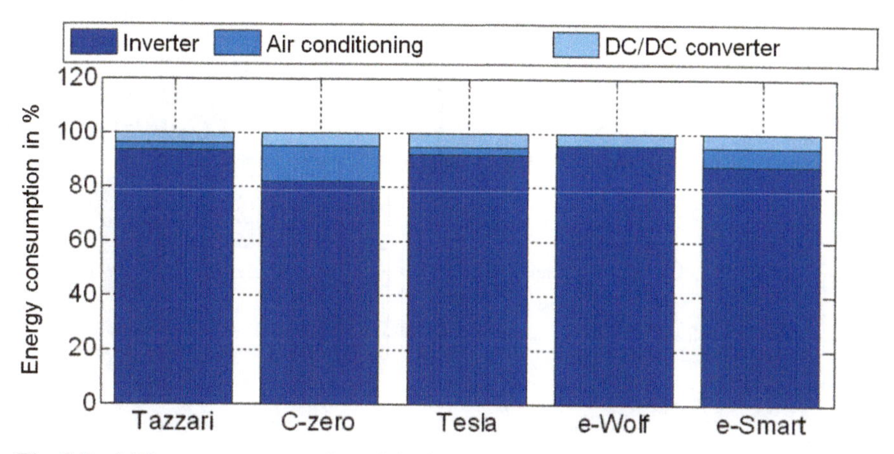

Fig. 3 Real-life energy consumption of the inverter, air conditioning, and DC/DC converter of different battery-electric vehicles, winter 2012, Ø 2.3°C (max=12.7°C, min=-12.9°C), according to [14]

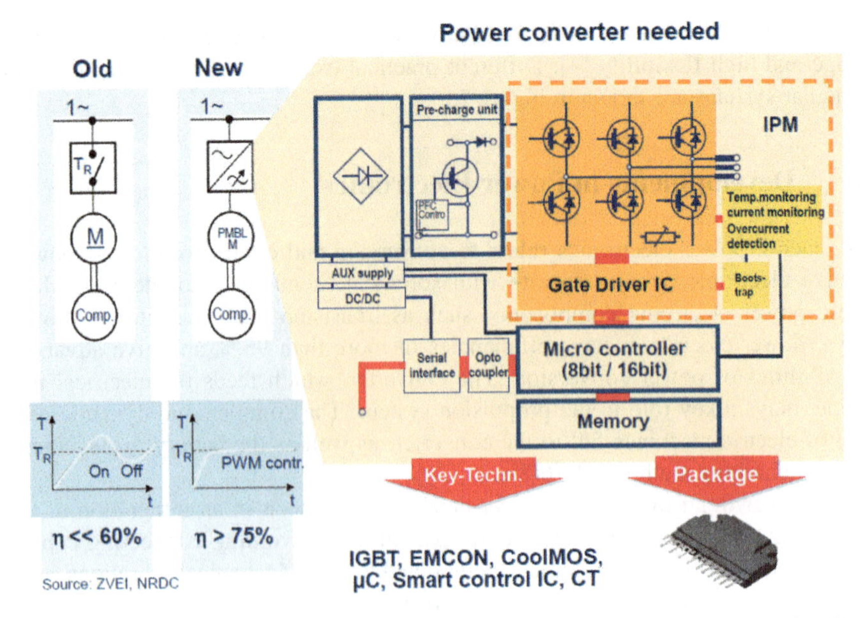

Fig. 4 Power savings through effective electric motors with speed control, e.g. for flow control [9]

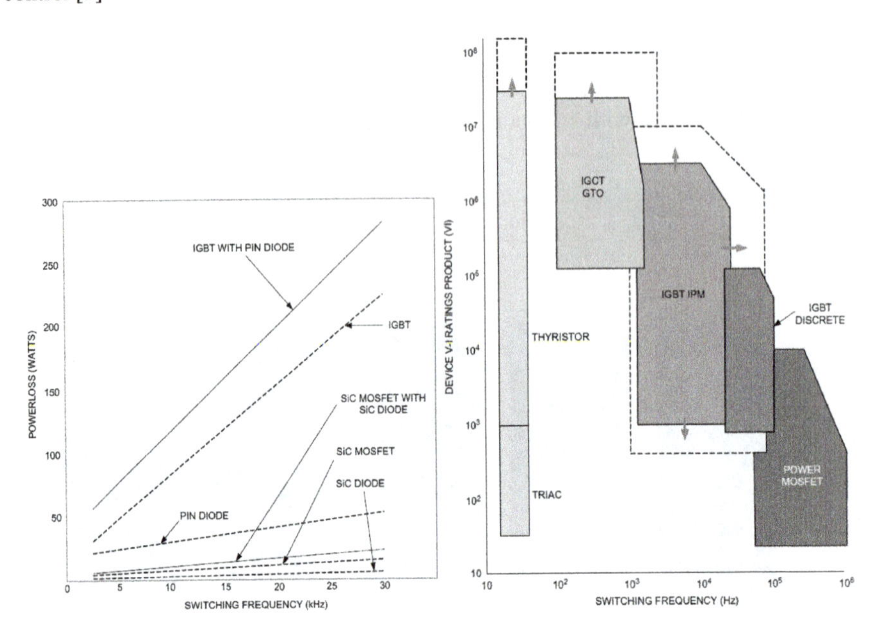

Fig. 5 Comparison of total power loss of different power electronics components for a half-bridge PWM inverter, V = 400 V, I_L = 15 A, (left), power frequency capability of current electronic devices and their future trends (in log-log scale) [7], [10] (right)

However, at higher switching frequencies the switching losses also increase, as depicted in Figure 5 (left). Therefore, power electronics has to be designed according to the following criteria: efficiency, constructional space, weight, and costs.

Figure 5 (right) shows the current state of the art and future trends of power frequency capability of different power electronics components.

Using wide band gap materials allows the production of power semiconductor devices with lower on state losses, higher switching speed, and higher permissible operating temperature. Provided that new packaging technology will allow to operate the semiconductors at elevated temperatures without sacrificing life time expectancy, new semiconductor devices based on high band gap materials will further increase the efficiency and power density of the inverters. However, qualifying these components for use in surface based vehicles, as well as reducing production cost, will be some of the challenges to be taken in future inverter development [1].

3 Control Systems and Strategies for Electric Vehicles

One major motivation for using variable speed drives at induction machines are the energy savings, which can be achieved with speed control either by means of variable voltage or variable frequency. Figure 6 depicts the power circuit for electric vehicle drives along with its block diagram, which usually is a torque-controlled system. Efficiency improvements are achieved through the machines control, via the PWM voltage-fed inverter, turning the battery-fed DC supply voltage into a variable AC voltage with variable frequency, supplying the AC machine. Within a closed-loop, the microcomputer/DSP-based inverter/motor controller receives the torque commands as well as feedback signals (phase current and rotor angle), and generates PWM signals for the inverter.

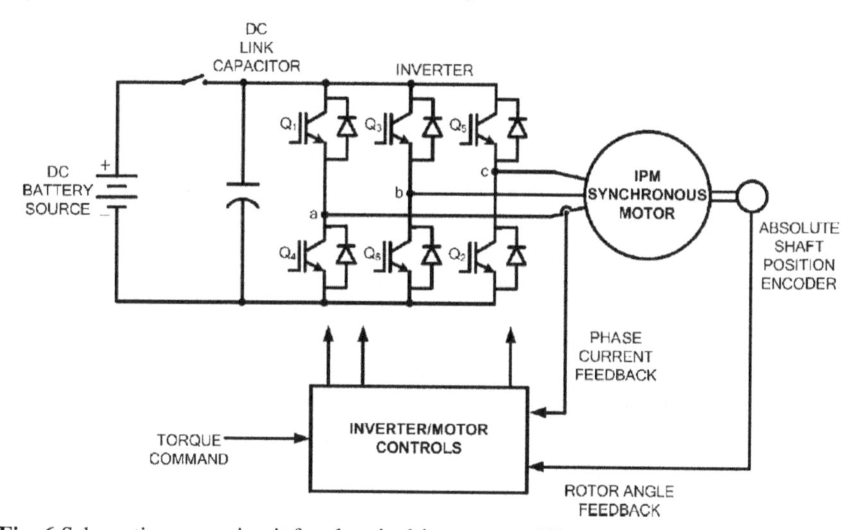

Fig. 6 Schematic power circuit for electric drive systems [7]

Figure 7a illustrates the torque-speed relation of an IPM machine (motor mode), with its phase voltage (V_s), and stator flux (ψ_s). Two operation modes are possible: PWM mode within the constant torque area, and square wave mode within the constant power field weakening mode.

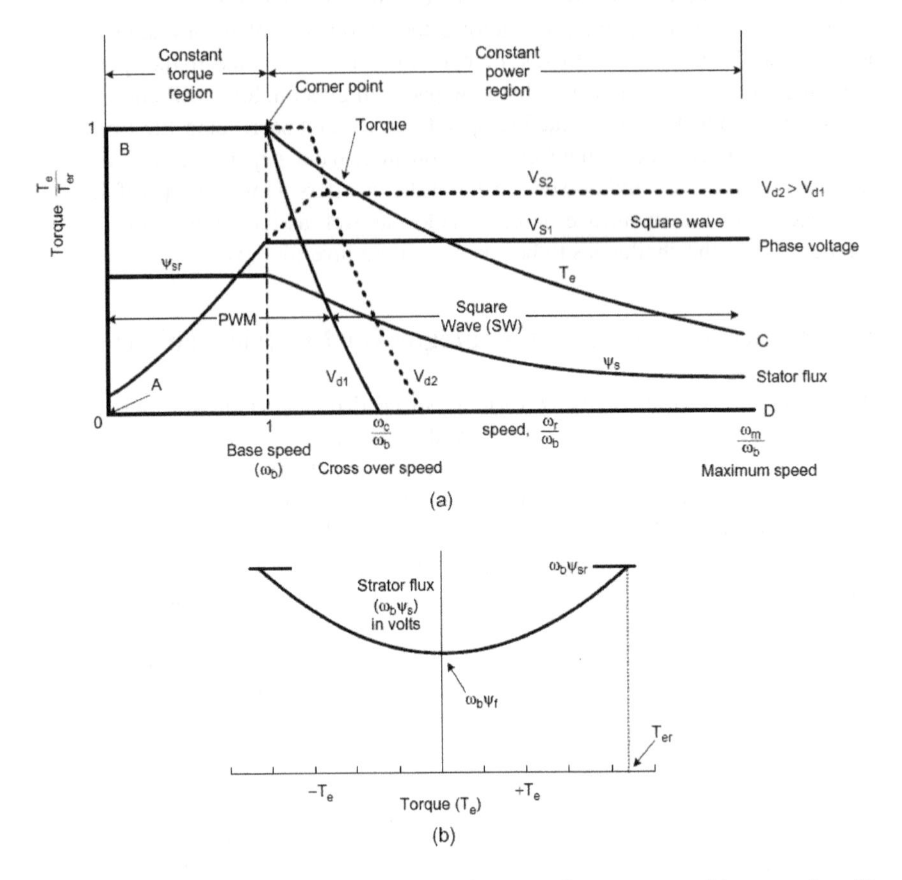

Fig. 7 (a) Torque-speed curve (motor mode), (b) stator flux program with torque for efficiency improvement [7]

The boundary area between both modes varies with the battery voltage. With higher battery voltage, the phase voltage level increases, and the maximum power output increases correspondingly. Hence, for efficiency optimization the machine operates with programmable stator flux with torque (Figure 7b). Therefore, the boundary area is slanting, because of the flux-speed relation $V_s = \omega_r\psi_s = \omega_b\psi_{sr} = \omega_c\psi_f$, where ψ_{sr} equals the stator flux at rated torque, ψ_f symbolizes the magnet flux, ω_b equals the base speed, and ω_c represents the crossover speed [1], [7].

4 Future Trends of Connected Powertrain Concepts for Electric Vehicles

Technology-related efficiency optimization of electric driven cars includes far-reaching measures which concern different areas, beside others, e.g. the reduction of air drag, rolling resistance, recuperation, and weight, as well as the optimization of electric components and mechanical systems. Besides these improvements, the operational strategy of electric propulsion systems and those of auxiliary equipment plays an important role in view of efficiency optimization. Modern electric cars are equipped with enhanced control systems, which enable a load dependent control of auxiliary units to decrease the energy consumption as a function of the driving situation and environmental boundary conditions, as illustrated in Figure 8 and Figure 9.

Fig. 8 Examples for real-life energy consumption of battery-electric vehicles [14]

Energy recovering during deceleration phases gives the possibility to recuperate a certain share of kinetic energy. The relatively high energy density of short-duration braking sequences challenges the development of technologies, which are able to convert the deceleration energy into electric energy. Due to their loading characteristics, state of the art battery systems, as they are applied in automotive applications, are not able to gather that high energy densities within short durations. Alternative technologies for short-term energy recuperation include capacitors and mechanic-electrical high-speed flywheel systems.

An additional technology to reduce the energy consumption of electric vehicles treats an optimization of the operation strategy under consideration of the road topology. In combination with future-oriented navigation systems, the course of the road along an expected route including traffic density, curve progression, and inclination/declination can be considered for the computation and optimization of energy demand.

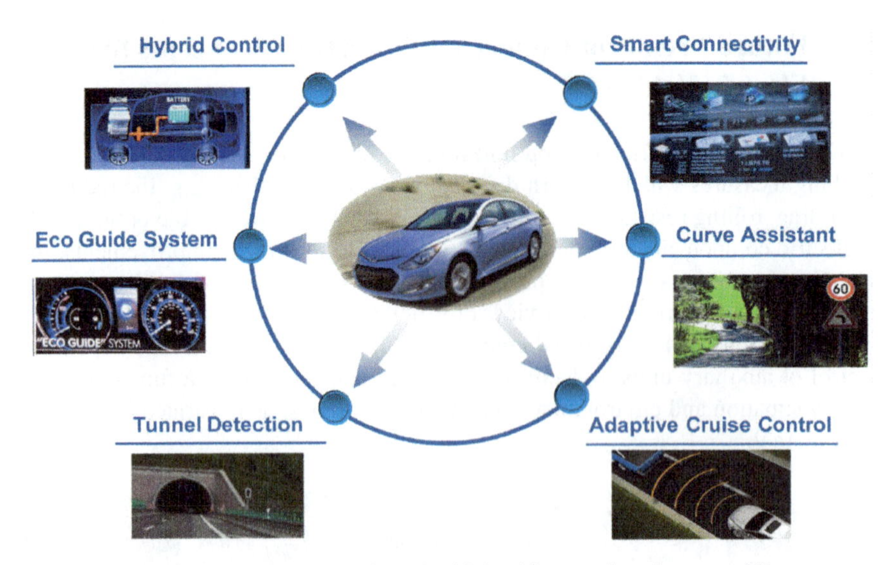

Fig. 9 Connected powertrain: interaction of vehicle, driver, and environment [4]

Figure 10 gives an overview of the above mentioned innovative concepts and strategies. The next generation of electrical vehicles (HMI systems) will predict the driver's next action with the help of advanced driver assistance systems and data from navigation systems to continuously improve fuel efficiency (or battery capacity respectively) by means of eco-operational mode [1], [4-5], [11-12].

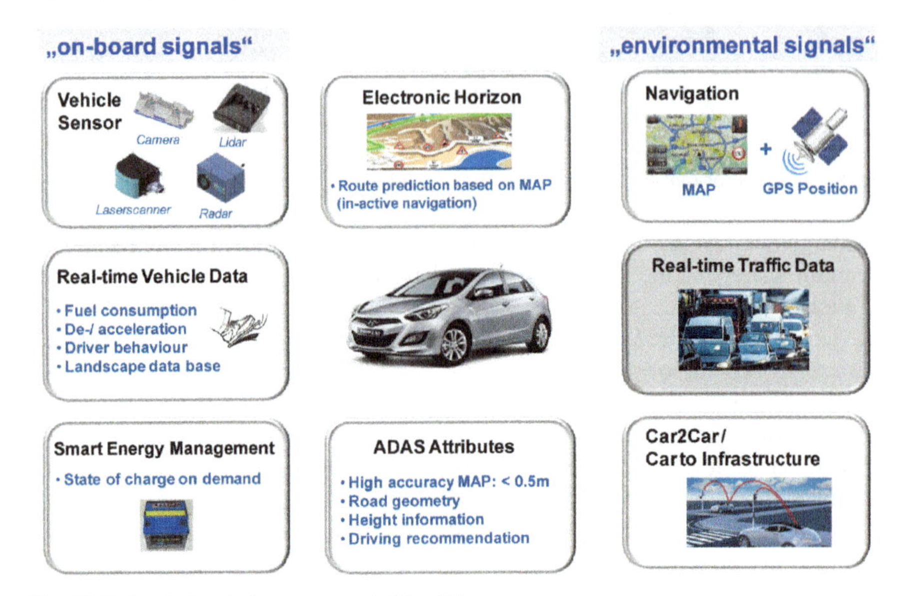

Fig. 10 Related signals for connected drive [4]

In conclusion, the proposed operations strategy ensures efficient use of the available system energy with the help of an intelligent control algorithm. This can result in a reduced size of the battery pack, which can further lower the costs, and the total weight of battery-electric vehicles. Beyond that, adaption to the consumer's individual way of driving guarantees improved customer acceptance, as well as improved fuel consumption, and driving pleasure. The potential strongly depends on the implemented operation strategy and the used information about the vehicle's environment, in general, up to 12 % efficiency increase is possible [13-15].

5 Summary

Cars driven by electric or hybrid technologies have the advantage that a high torque potential is provided from the start, hence the vehicle acceleration can be higher in comparison with conventional propulsion concepts. The speed-torque characteristic of electric motors is nearly ideal for the use in automotive applications, which leads to an advantageous drivability. The efficiency of electric propulsion systems can be raised in the future, but due to the high level in general, the potential for further efficiency increase is limited. The type of applied electric motors, the operational characteristics as well as the energy and torque density have a main influence on the weight and consequently these parameters have to be considered regarding the optimization of efficiency behaviour. An important area of research includes the battery systems. Besides material-related investigations to achieve high power and energy density, the weight, the charging and discharging characteristics, as well as lifetime-related aspects are in the focus of numerous R&D projects. Nevertheless, the limited energy density of battery systems remains an unsolved problem, which restricts the driving range of electric cars nowadays. An important influencing factor on the efficiency and thus on the driving range of electric vehicles is included in the operation strategy, which shows a considerable potential for further optimization. The effective implementation of recuperation, demand-related control of supplementary devices, and forward-looking driving strategies will support a further optimization of the energy consumption during the next years. In this way, far reaching research activities in the area of electric drives, battery systems, and control strategies for automotive applications will provide a fundamental basis for the further extension of electrically driven cars on worldwide markets [1].

References

[1] Fabian, J., Hirz, M., Krischan, K.: State of the Art and Future Trends of Electric Drives and Power Electronics for Automotive Engineering. In: SAE World Congress, Detroit (April 2014)

[2] Durrieu, D., Criddle, M., Webster, M., Menegazzi, P., Wu, Y., Decoster, S., Benchetrite, D.: Intelligent Electrification with combination of Electric Super-charger and Extender Stop-Start System 12V-48V. In: 25th International AVL Conference "Engine & Environment", Graz, Austria (September 2013)
[3] Kampker, A., Vallée, D., Schnettler, A.: Elektromobilität – Grundlagen einer Zukunftstechnologie. Springer (March 2013) ISBN 978-3-642-31985-3
[4] Yang, S.: Hyundai-Kia Motor's Powertrain strategy for Connection to the Environment. In: 25th International AVL Conference "Engine & Environment", Graz, Austria (September 2013)
[5] Kohler, H.: The Connected Powertrain: Further Challenges and Potential. In: 25th International AVL Conference "Engine & Environment, Graz, Austria (September 2013)
[6] Hüttl, R.F., Pischetsrieder, B., Spath, D.: Elektromobilität: Potenziale und wissenschaftlich-technische Herausforderungen, acatech – Deutsche Akademie der Technikwissenschaften (October 2010) ISBN 978-3-642-16253-4
[7] Bose, B.K.: Power Electronics and Motor Drives: Advances and Trends. Elsevier Inc. (2006) ISBN 978-0-12-088405-6
[8] Rashid, M.H.: Power Electronics Handbook. Academic Press (2001) ISBN 0-12-581650-2
[9] Pairitsch, H.: Leistungselektronik für die Energiewende. In: OGE Workshop, Graz, Austria (October 2013)
[10] Satoh, K., Yamamoto, M.: The present state of the art in high power semiconductor devices. Proc. IEEE 89, 813–821 (2001), doi:10.1109/5.931470.
[11] Kampmann, S., Pöchmüller, W.: Energy Management in the Connected Powertrain. In: 25th International AVL Conference "Engine & Environment", Graz, Austria (September 2013)
[12] Fitzgerald, A.E., Kingsley, C., Umans, S.: Electric Machinery, 6th edn., p. 203. McGrawHill ISBN 0-07-112193-5
[13] Kraus, H., Ackerl, M., Karoshi, P., Fabian, J., Ringdorfer, M.: A new Approach to an Adaptive and Predictive Operation Strategy for PHEVs. In: WKM-Symposium, Bochum, Germany (June 2014)
[14] Rumbolz, P.: Untersuchung der Fahrereinflüsse auf den Energieverbrauch und die Potentiale von verbrauchsreduzierenden Assistenzfunktionen. In: ÖVK work-shop, Graz, Austria (January 2014)
[15] Bär, T., Aidel, J., Zöllner, J.M.: Szenenbasierte Fahrstilerkennung durch probilistische Auswertung von Fahrzeugdaten. In: 5th Conference on Driving Assistance. TÜV-Süd Akademie GmbH, Munich (May 2012)

Author Index

Subject Index

CPSIA information can be obtained at www.ICGtesting.com
Printed in the USA
LVOW02*1425100714

393760LV00001B/14/P

9 783319 080864